Studies in Computational Intelligence

Volume 994

Series Editor

Janusz Kacprzyk, Polish Academy of Sciences, Warsaw, Poland

The series "Studies in Computational Intelligence" (SCI) publishes new developments and advances in the various areas of computational intelligence—quickly and with a high quality. The intent is to cover the theory, applications, and design methods of computational intelligence, as embedded in the fields of engineering, computer science, physics and life sciences, as well as the methodologies behind them. The series contains monographs, lecture notes and edited volumes in computational intelligence spanning the areas of neural networks, connectionist systems, genetic algorithms, evolutionary computation, artificial intelligence, cellular automata, self-organizing systems, soft computing, fuzzy systems, and hybrid intelligent systems. Of particular value to both the contributors and the readership are the short publication timeframe and the world-wide distribution, which enable both wide and rapid dissemination of research output.

Indexed by SCOPUS, DBLP, WTI Frankfurt eG, zbMATH, SCImago.

All books published in the series are submitted for consideration in Web of Science.

More information about this series at https://link.springer.com/bookseries/7092

Youssef Baddi · Youssef Gahi · Yassine Maleh ·
Mamoun Alazab · Loai Tawalbeh
Editors

Big Data Intelligence for Smart Applications

 Springer

Editors
Youssef Baddi ⓘ
Chouaib Doukkali University
El Jadida, Morocco

Yassine Maleh ⓘ
Sultan Moulay Slimane University
Beni Mellal, Morocco

Loai Tawalbeh ⓘ
Texas A&M University
San Antonio, TX, USA

Youssef Gahi ⓘ
National School of Applied Sciences
Ibn Tofail University
Kenitra, Morocco

Mamoun Alazab ⓘ
Charles Darwin University
Casuarina NT, Australia

ISSN 1860-949X ISSN 1860-9503 (electronic)
Studies in Computational Intelligence
ISBN 978-3-030-87956-3 ISBN 978-3-030-87954-9 (eBook)
https://doi.org/10.1007/978-3-030-87954-9

This Springer imprint is published by the registered company Springer Nature Switzerland AG
The registered company address is: Gewerbestrasse 11, 6330 Cham, Switzerland

Preface

For a large majority of companies, the value of data is no longer in doubt. Calculating an application's usage rates, measuring the time spent on social networks, or even analyzing its usage are now some of the best practices accepted and implemented by the most significant number. But while companies understand the importance of collecting and analyzing their data, few know how to process and use it. However, knowing how to use this data is a prerequisite for integrating Artificial Intelligence into its products and services.

It is worth noting that technologies such as Artificial intelligence and Big Data are evolving even faster. They are rapidly growing by holding great promise for many sectors. The real revolutionary potential of these technologies mainly relies on their convergence. Big Data and Artificial Intelligence are two technologies that are inextricably linked, to the point that we can think about Big Data Intelligence. AI has become ubiquitous in many industries where intelligent programs relying on big data transform decision-making. Increased agility, smarter business processes, and better productivity are the most likely benefits of this convergence.

Big data, which is still poorly exploited, is nevertheless the black gold of AI. Artificial Intelligence is the logical continuation of our data analysis methods and techniques, an extension of Business Intelligence, followed by Big Data and Advanced Analytics. The so-called intelligent machine needs a massive amount of data analyzed and cross-referenced to draw innovative, even creative, capabilities close to the human brain's functioning.

Big Data Intelligence for Smart Applications comprises many state-of-the-art contributions from scientists and practitioners working in big data intelligence and smart applications. It aspires to provide a relevant reference for students, researchers, engineers, and professionals working in this area or those interested in grasping its diverse facets and exploring the latest advances in machine intelligence and data analytics for sustainable future smart cities applications. More specifically, the book contains 15 chapters. Accepted chapters describe and analyze various applications of machine intelligence and big data for smart applications, such as smart cities, Internet of Things, cybersercurity, vision systems, and healthcare, to mitigate COVID-19.

We want to take this opportunity and express our thanks to the contributors of this volume and the editorial board for their tremendous efforts by reviewing and providing interesting feedbacks. The editors would like to thank Dr. Thomas Ditsinger (Springer, Editorial Director, Interdisciplinary Applied Sciences) and Prof. Janusz Kacprzyk (Series Editor-in-Chief), and Ms. Swetha Divakar (Springer Project Coordinator) for the editorial assistance and support to produce this important scientific work. Without this collective effort, this book would not have been possible to be completed.

El Jadida, Morocco   Prof. Youssef Baddi
Kenitra, Morocco   Prof. Youssef Gahi
Khouribga, Morocco   Prof. Yassine Maleh
Darwin, Australia   Prof. Mamoun Alazab
San Jose, USA   Prof. Loai Tawalbeh

Contents

Data Quality in the Era of Big Data: A Global Review 1
Widad Elouataoui, Imane El Alaoui, and Youssef Gahi

Adversarial Machine Learning, Research Trends and Applications 27
Izzat Alsmadi

Multi-agent Systems for Distributed Data Mining Techniques:
An Overview 57
Mais Haj Qasem, Amjad Hudaib, Nadim Obeid,
Mohammed Amin Almaiah, Omar Almomani,
and Ahmad Al-Khasawneh

Time Series Data Analysis Using Deep Learning Methods
for Smart Cities Monitoring 93
Giuseppe Ciaburro

A Low-Cost IMU-Based Wearable System for Precise Identification
of Walk Activity Using Deep Convolutional Neural Network 117
Amartya Chakraborty and Nandini Mukherjee

Facial Recognition Application with Hyperparameter Optimisation 141
Hannah M. Claus, Cornelia Grab, Piotr Woroszyllo,
and Patryk Rybarczyk

Internet-Assisted Data Intelligence for Pandemic Prediction:
An Intelligent Framework 173
H. M. K. K. M. B. Herath

NHS Big Data Intelligence on Blockchain Applications 191
Xiaohua Feng, Marc Conrad, and Khalid Hussein

Depression Detection from Social Media Using Twitter's Tweet 209
Rifat Jahan Lia, Abu Bakkar Siddikk, Fahim Muntasir,
Sheikh Shah Mohammad Motiur Rahman, and Nusrat Jahan

A Conceptual Analysis of IoT in Healthcare 227
Muhammad Azmi Umer, Muhammad Taha Jilani, Asif Rafiq,
Sulaman Ahmad Naz, and Khurum Nazir Junejo

**Securing Big Data-Based Smart Applications Using Blockchain
Technology** 241
Rihab Benaich, Imane El Alaoui, and Youssef Gahi

**Overview of Blockchain-Based Privacy Preserving Machine
Learning for IoMT** 265
Rakib Ul Haque and A. S. M. Touhidul Hasan

**Big Data Based Smart Blockchain for Information Retrieval
in Privacy-Preserving Healthcare System** 279
Aitizaz Ali, Muhammad Fermi Pasha, Ong Huey Fang, Rahim Khan,
Mohammed Amin Almaiah, and Ahmad K. Al Hwaitat

**Classification of Malicious and Benign Binaries Using Visualization
Technique and Machine Learning Algorithms** 297
Ikram Ben Abdel Ouahab, Lotfi Elaachak, and Mohammed Bouhorma

**FakeTouch: Machine Learning Based Framework for Detecting
Fake News** 317
Abu Bakkar Siddikk, Rifat Jahan Lia, Md. Fahim Muntasir,
Sheikh Shah Mohammad Motiur Rahman, Md. Shohel Arman,
and Mahmuda Rawnak Jahan

About the Editors

Youssef Baddi is a full-time Assistant Professor at Chouaïb Doukkali University UCD EL Jadida, Morocco. Ph.D. Thesis degree in computer science from ENSIAS School, University Mohammed V. Souissi of Rabat, Morocco, since 2016. He also holds a Research Master's degree in networking obtained in 2010 from the High National School for Computer Science and Systems Analysis—ENSIAS-Morocco-Rabat. He is a member of the Laboratory of Information and Communication Sciences and Technologies STIC Lab, since 2017. He is a guest member of the Information Security Research Team (ISeRT), and Innovation on Digital and Enterprise Architectures Team, ENSIAS, Rabat, Morocco. Dr. Baddi was awarded as the best Ph.D. student in University Mohammed V. Souissi of Rabat in 2013. Dr. Baddi has made contributions in the fields of Group Communications and protocols, information security and privacy, Software-defined network, the Internet of Things, Mobile and Wireless Networks Security, Mobile IPv6. His research interests include Information Security and Privacy, the Internet of Things, Networks Security, Software-defined Network, Software-defined Security, IPv6, and Mobile IP. He has served and continues to serve on executive and technical program committees and as a reviewer of numerous international conferences and journals such as Elsevier Pervasive and Mobile Computing PMC and International Journal of Electronics and Communications AEUE, and Journal of King Saud University—Computer and Information Sciences. He was the General Chair of IWENC 2019 Workshop and the Secretary member of the ICACIN 2020 Conference.

Youssef Gahi is an Associate Professor in computer science at the National School of Applied Sciences, University ibn Tofail, Kenitra, Morocco. He received the M.Sc. and Ph.D. degrees in computer science from the University Mohammed V in collaboration with the University of Ontario Institute of Technology in 2008 and 2013. Before starting his academic career in 2017, Dr. Gahi worked for many international consulting firms, from 2008 to 2017, as a software engineer, solution architect, and IT senior consultant in Mobile, J2EE, and Big Data technologies. As a researcher, he authored more than 40 scientific contributions in peer-reviewed journals and conferences. His research topics are focused on Big Data management, Big Data quality,

Big Data security, Recommendation Systems, and Cloud computing models for data privacy.

Yassine Maleh (http://orcid.org/0000-0003-4704-5364) is a cybersecurity professor and practitioner with industry and academic experience. He is a Ph.D. degree in Computer Sciences. Since 2019, He is working as a professor of cybersecurity at Sultan Moulay Slimane University, Morocco. He worked for the National Port agency (ANP) in Morocco as a Senior Security Analyst from 2012 to 2019. He is a senior member of IEEE, member of the International Association of Engineers, and the Machine Intelligence Research Labs. Dr. Maleh has made contributions in the fields of information security and privacy, Internet of Things security, and wireless and constrained networks security. His research interests include information security and privacy, Internet of Things, networks security, information system, and IT governance. He has published over 80 papers (book chapters, international journals, conferences/workshops), 16 edited books, and 3 authored books. He is the editor in chief of the International Journal of Smart Security Technologies. He serves as an associate editor for IEEE Access (2019 Impact Factor 4.098), the International Journal of Digital Crime and Forensics, and the International Journal of Information Security and Privacy. He was also a guest editor of a special issue on 'Recent Advances on Cyber Security and Privacy for Cloud-of-Things' of the International Journal of Digital Crime and Forensics, Volume 10, Issue 3, July–September 2019. He has served and continues to serve on executive and technical program committees and as a reviewer of numerous international conferences and journals such as Elsevier Ad Hoc Networks, IEEE Network Magazine, IEEE Sensor Journal, ICT Express, and Springer Cluster Computing. He was the publicity chair of BCCA 2019 and the general chair of the MLBDACP 19 symposium.

Mamoun Alazab is an Associate Professor in the College of Engineering, IT and Environment at Charles Darwin University, Australia. He received his Ph.D. degree is in Computer Science from the Federation University of Australia, School of Science, Information Technology and Engineering. He is a cybersecurity researcher and practitioner with industry and academic experience. Dr. Alazab's research is multidisciplinary that focuses on cybersecurity and digital forensics of computer systems, including current and emerging issues in the cyber environment like cyber-physical systems and the internet of things, by taking into consideration the unique challenges present in these environments, with a focus on cybercrime detection and prevention. He looks into the intersection of machine learning as an essential tool for cybersecurity, for example, for detecting attacks, analyzing malicious code, or uncovering vulnerabilities in software. He has more than 100 research papers. He is the recipient of a short fellowship from the Japan Society for the Promotion of Science (JSPS) based on his nomination from the Australian Academy of Science. He delivered many invited and keynote speeches, 27 events in 2019 alone. He convened and chaired more than 50 conferences and workshops. He is the founding chair of the IEEE Northern Territory Subsection: (Feb 2019—current). He is a Senior Member of the IEEE, Cybersecurity Academic Ambassador for Oman's Information Technology

Authority (ITA), Member of the IEEE Computer Society's Technical Committee on Security and Privacy (TCSP). He has worked closely with government and industry on many projects, including IBM, Trend Micro, the Australian Federal Police (AFP), the Australian Communications and Media Authority (ACMA), Westpac, UNODC, and the Attorney General's Department.

Loai Tawalbeh completed his Ph.D. degree in Electrical & Computer Engineering from Oregon State University in 2004, and M.Sc. in 2002 from the same university with GPA 4/4. Dr. Tawalbeh is currently an Associate professor at the department of Computing and Cyber Security at Texas A&M University-San Antonio. Before that he was a visiting researcher at University of California-Santa Barbra. Since 2005 he taught/developed more than 25 courses in different disciplines of computer engineering and science with focus on cyber security for the undergraduate/graduate programs at: NewYork Institute of Technology (NYIT), DePaul's University, and Jordan University of Science and Technology. Dr. Tawalbeh won many research grants and awards with over than 2 Million USD. He has over 80 research publications in refereed international Journals and conferences.

Data Quality in the Era of Big Data: A Global Review

Widad Elouataoui, Imane El Alaoui, and Youssef Gahi

Abstract Currently, Big Data is gaining wide adoption in the digital world as a new technology able to manage and support the explosive growth of data. Indeed, data is growing at a higher rate due to the variety of the data-generating adopted devices. In addition to the volume aspect, the generated data are usually unstructured, inaccurate, and incomplete, making its processing even more difficult. However, analyzing such data can provide significant benefits to businesses if the quality of data is improved. Facing the fact that value could only be extracted from high data quality, companies using data in their business management focus more on the quality aspect of the gathered data. Therefore, Big data quality has received a lot of interest from the literature. Indeed, many researchers have attempted to address Big data quality issues by suggesting novel approaches to assess and improve Big data quality. All these researches inspire us to review the most relevant findings and outcomes reported in this regard. Assuming that some review papers were already published for the same purpose, we believe that researchers always need an update. It is worth noting that all the published review papers are focused on a specific area of Big data quality. Therefore, this paper aims to review all the big data quality aspects discussed in the literature, including Big data characteristics, big data value chain, and big data quality dimensions and metrics. Moreover, we will discuss how the quality aspect could be employed in the different applications domains of Big data. Thus, this review paper provides a global view of the current state of the art of the various aspects of Big data quality and could be used to support future research.

W. Elouataoui (✉) · Y. Gahi
Engineering Sciences Laboratory, National School of Applied Sciences, Kenitra, Morocco
e-mail: widad.elouataoui@uit.ac.ma

Y. Gahi
e-mail: gahi.youssef@uit.ac.ma

I. E. Alaoui
Telecommunications Systems and Decision Engineering Laboratory, Ibn Tofail University, Kenitra, Morocco
e-mail: imane.el.alaoui@uit.ac.ma

© The Author(s), under exclusive license to Springer Nature Switzerland AG 2022
Y. Baddi et al. (eds.), *Big Data Intelligence for Smart Applications*,
Studies in Computational Intelligence 994,
https://doi.org/10.1007/978-3-030-87954-9_1

Keywords Big Data Quality · Big Data Quality Approach · Big Data Value
Chain · Big Data Quality Dimensions · Big Data Characteristics

1 Introduction

With the digital transformation of companies, data has become the foundation on
which many organizations rely to differentiate themselves and develop a sustainable
competitive advantage. However, this data-driven strategy requires a solid infrastructure to collect, store, and large-scale process data. Furthermore, as the number
of devices generating data increases at a high rate, the volume of that induced data
increases. According to a study conducted by the research firm IDC, the world's data
will grow from 23 zettabytes in 2017 to 175 zettabytes (175 billion terabytes) by 2025
(Reinsel et al. 2018). Thus, Big data has emerged as a new data ecosystem able to
support, process, and manage this explosive growth of data.

Big data may be defined as "the way we gather, store, manipulate, analyze, and
get insight from a fast-increasing heterogeneous data" (Taleb et al. 2018). Indeed,
big data is multi-source data. It is a mix of structured and unstructured data that
combines various data types, usually noisy and inaccurate. Hence, in big data, the
collected data is often poor-quality data that cannot be directly used, cleaned, and
transformed into useful information. For this reason, data quality has always been
a critical concern of companies believing that data is only helpful if it is of high
quality.

Indeed, data quality is a broad and multifaceted concept that can impact several
aspects of a company's business. According to a survey conducted by TDWI
(Abdullah et al. 2015; TDWI 2021), the three most common problems caused by
low data quality are: the time lost to validate and fix data errors, the extra cost, and
the unreliability of the extracted data, so managers get less confident in analytics
systems and cannot trust the generated reports. Furthermore, Big data is highly used
by companies to analyze customer's behavior and understand their insights. A survey
about the benefits of high-quality data conducted by TDWI has shown that the most
common service derived from high-quality data is customer satisfaction, a priority
for marketers (Abdullah et al. 2015). Indeed, understanding customer insights and
needs allows companies to focus their efforts on the right customer segment and gain
a competitive advantage. However, not ensuring a high quality of data in such a case
will negatively impact the company's strategic planning. It may lead managers to
make the wrong decisions instead of the right ones.

Data quality was introduced long before big data emerged and was applied first
to traditional data stored in relational databases. However, big data have emerged
new quality issues that traditional data quality techniques cannot cope with. In Batini
et al. (2015), have shown that the evolution from data quality to big data quality is
related to the fact that big data involves various new data types, data sources, and
application domains. Thus, the concept of big data quality has been addressed by
many studies that have attempted to suggest new approaches to assess and improve

big data quality. Therefore, this paper aims to review the most recent research focused on Big data quality to explore the current state of the art and highlight some possible future work directions. As we have mentioned earlier, the purpose of this paper is to cover all Big data quality aspects addressed in the existing literature. Thus, each section of this paper is intended for a specific Big data quality aspect.

The rest of this paper is organized as follows: Section 2 defines the Big data characteristics introduced in the literature. Section 3 highlights big data quality dimensions and metrics. Section 4 describes the different phases of the big data value chain and presents the quality dimensions that should be considered in each step. In Sect. 5, we present the adopted research methodology. Section 6 summarizes and provides a classification of the most recent Big data quality improvement approaches followed by the key findings that emerged out of the review. Section 7 describes big data application domains where data quality is of critical concern. Finally, we discuss some remaining Big data quality issues and highlight some possible future work directions.

2 Big Data Characteristics

While it is true that the term big data refers to a massive amount of data that are continuously increasing, this does not necessarily mean that big data is only about data volume. Indeed, implementing a big data strategy involves storing, analyzing, processing, and monitoring data. Therefore, numerous challenges are raised, such as handling and visualizing data, integrating data from various data sources, securing data, and so on. In Espinosa et al. (2019), have reviewed all Big data challenges addressed in the existing literature and have shown that the high growth rate, the lack of tools, and data management challenges are the most common Big data issues. This variety in Big data issues is related to big data's nature and its multiple characteristics, a trending research topic.

The first Big data characteristics were introduced in 2001 by Doug Laney that have defined big data in terms of the 3 V's: Volume, Velocity, and Variety. Afterward, SAS (Statistical Analysis System) has introduced two new characteristics, i.e., Variability and complexity. Furthermore, Oracle has defined big data in 4 V's, i.e., Volume, Velocity, Variety, and Value. In 2014, Data Science Central, Kirk Born described big data in 10 V's, i.e., Volume, Variety, Velocity, Veracity, Validity, Value, Variability, Venue, Vocabulary, Vagueness (Failed 2016a). Later, more V's were suggested in the literature, such as Vulnerability, Virality, Verbosity, etc. In 2018, the number of proposed characteristics already exceeded 50. According to a study presented in Dhamodharavadhani et al. (2018), this number will increase to reach 100 V's by 2021.

It is worth noting that not all these characteristics are used at once. Instead, big data characteristics were explicitly introduced to meet some application requirements and are therefore only used in particular big data application domains.

In Pros (2013), the authors have highlighted the emerging Big data characteristics. In addition, they have suggested the big data V's that should be considered in some big data application domains for better data analytics, such as social media, IoT, and weather prediction. In Table 1, we explain the most common characteristics of Big Data and highlight their impact on Big Data quality.

Ensuring the quality of data with such characteristics is a significant challenge. The quality issues faced when dealing with big data are highly related to big data V's. Thus, data quality aspects should be reviewed and adapted to deal with the new challenges raised by Big data characteristics. In the next section, we highlight how data quality could be ensured in a Big data context and the quality dimensions that should be considered.

3 Big Data Quality

Given the particular characteristics of big data, new challenges are introduced and need to be explored. Indeed, the emergence of big data has raised a new challenge related to understanding Big data quality requirements. Thus, before implementing any quality approach, a deep understanding of the vital quality requirements must be considered when managing big data solutions. In Failed (2016b), the authors have suggested a novel approach to define Big data systems' quality requirements. The proposed method consists of associating each data characteristic with all quality attributes, which leads to a global assessment of data quality without missing any quality aspect. To better understand Big data quality requirements, we should talk about its properties called Data Quality Dimensions (DQD's).

Data quality dimensions (DQD's) could be defined as "a set of data quality attributes that represent a single aspect or construct of data quality" (Ramasamy and Chowdhury 2020). The most common data quality dimensions (DQD's) are Accuracy, Completeness, Consistency, Freshness, Uniqueness, and Validity. However, more data quality dimensions were suggested in the literature to cover other data quality aspects, such as security, traceability, navigation, and data decay.

Any data quality dimension can be quantified using metrics. Indeed, data quality metrics allow measuring how data meet the quality requirements and provide a concrete meaning to quality aspects. In Arolfo and Vaisman (2018), have presented the mathematical definition of each quality dimension. They have defined data quality metrics for dimension D as "a function that maps an entity to a value, such that this value, typically between 0 and 1, indicates the quality of a piece of data regarding the dimension D."

It is worth noting that there is a high dependency between data quality dimensions strongly related to each other. Thus, focusing only on one size without considering its correlation with the different measurements may not be a practical approach for supporting data quality. In Sidi et al. (2013), the authors have thrown light on the existing relation between DQD's. They have suggested a framework that allows

Table 1 Big data characteristics

Characteristic	Elucidation	Meaning	Impacts on data quality
Volume	What is the size of the data?	A noticeable characteristic of big data. It refers to the amount of the generated data in Exabytes, Zettabytes, or even Yottabytes	Data Volume raises challenges related to data quality processing and management tools that should handle a massive amount of noisy and inconsistent data within a reliable time
Variety	How heterogeneous are the data?	Big data incorporates multiple data sources (social media, IoT, mobile devices, …) that are usually of different data types such as geolocation data, sensor data, social media data, …	Data Variety leads to heterogeneous data with inconsistent data formats and incompatible data structures and semantics, adding more complexity to manage data
Velocity	How fast is data generated?	It refers to the rate of data generation: The speed at which data is produced, analyzed and processed	High-Velocity data requires processing tools with high performance and could raise timeliness issues, especially for time-sensitive processes
Veracity	How much could data be trusted?	It is the truthfulness and the accuracy of the generated results. This characteristic is of high importance, especially for business decisions making	Data could contain anomalies or inaccurate values, which impact the reliability and the truthfulness of data
Value	Is the generated data valuable?	Refers to the business value that could be derived from data. It consists of specifying the benefits that processing data could provide	Extracting valuable insights from data is not always obvious and may be challenging for organizations and data managers
Variability	What is the rate of change of data meaning?	Sheds light on the dynamic aspect of data. It refers to the frequency at which the meaning of data changes	Data with high Variability raise challenges related to data freshness and timeliness

(continued)

Table 1 (continued)

Characteristic	Elucidation	Meaning	Impacts on data quality
Visualization	How is it challenging to develop a meaningful visualization of data?	It refers to the process of presenting the extracted information in a readable form, more comfortable to understand	Visualizing a big scale of data in a static or dynamic form may be challenging. In Agrawal et al. (2015), Agrawal et al. have presented the challenges and opportunities of big data visualization
Validity	Is data suitable and accurate for the intended use?	Refers to the correctness of the input data and its suitability for its intended purpose	Data Validity should be inspected before data processing as invalid data are useless and cannot support decision making
Volatility	How long could data be kept and considered as applicable?	Refers to the duration of usefulness and relevancy of data	Outdated data are irrelevant and may bias data analysis
Viability	How relevant is the data?	This process consists of "uncovering the latent, hidden relationships among these variables." to "confirm a particular variable's relevance before investing in the creation of a fully-featured model. IRJET (2021)	"Big Data should have the capability to be live and active forever, and able for developing, and to produce more data when need. Khan et al. (2018)
Vocabulary	What terminology is used to describe data?	Refers to data terminology that describes data's content and structure	Data readability and availability are highly related to the vocabulary and the wording used
Vagueness	How much data is available?	Confusion over the meaning of big data. It is about its nature, content, tools availability, tools selection, etc. (Dhamodharavadhani et al. 2018)	Unclear and non-understandable data could not be used effectively
Venue	What kind of platforms and systems are used to manage data?	Various types of data arrived from different sources via different platforms like personnel systems and private & public cloud (IRJET-2021)	Big data sources are not always credible and may provide inaccurate information (e.g., social media)

(continued)

Table 1 (continued)

Characteristic	Elucidation	Meaning	Impacts on data quality
Viscosity	How difficult is it to work with data?	Refers to the difficulty to use or integrate the data (Failed 2018)	Big Data Complexity may have a direct impact on data usability and effectiveness

detecting dependencies between quality dimensions using statistical methods and data mining.

Data quality dimensions could be classified based on multiple criteria. One of the most common classifications of DQD's was suggested by Wand et al. in Wand and Wang (1996), which have defined four categories of DQD's:

- Intrinsic: implies that data have quality in their own right.
- Contextual: denotes that data is driven by context.
- Representational: related to the format and the meaning of data.
- Accessibility refers to the extent to which data is available.

Many other classifications of DQD's were proposed in the existing literature (The Challenges of Data Quality and Data Quality Assessment in the (Big Data Era 2021; Zhang et al. 2017; Alaoui et al. 2019a). As illustrated in Fig. 1, we present the most common DQD's that we group into four aspects:

- Reliability: The extent to which data is dependable and can be trusted.
- Availability: Implies that the appropriate level of access to data is provided.

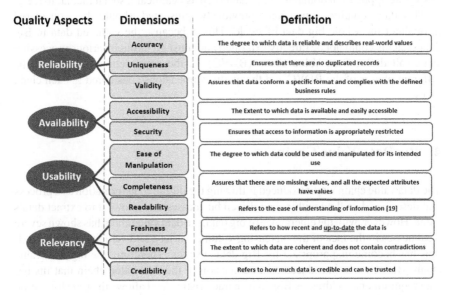

Fig. 1 Data quality dimensions

- Usability: Refers to the extent to which data could be effectively used.
- Relevancy: Refers to the extent to which data is pertinent and suitable for its intended use.

One of the most critical questions in the literature is whether standard quality dimensions and metrics used to evaluate data quality could always be practical in big data contexts. Most of the studies on big data quality use the conventional quality dimensions in their quality approaches. In Juddoo and George (2018), the authors have attempted to identify the most critical quality dimensions in big data by evaluating the similarity of 38 data quality dimensions in 34 articles that have dealt with big data quality. The obtained results have shown that Accuracy, Usefulness, and Confidence (conventional quality dimensions), are the most cited dimensions in the studied articles. However, this does not remove the need to redefine and adapt the standard quality dimensions to fit Big data quality requirements. In Merino et al. (2016), Caballero et al. have shown the importance of extending data quality aspects to address the new challenges introduced by big data. Moreover, they have defined unique data quality aspects that should be considered in Big data context (Contextual adequacy, Temporal adequacy, and operational adequacy). Also, they have highlighted how DQD's could be employed to address the 3 V's (Volume, Velocity, Variety) of big data using a novel assessment model.

Despite the many studies that tackle big data, few studies address Big data quality issues focusing on data quality dimensions. Indeed, the earliest study that has tackled data quality dimensions in a Big data context was published in late 2013 (Ramasamy and Chowdhury 2020). Therefore, more research focus is required in this area. In addition, more quality dimensions related to big data need to be explored, especially in specific application domains of big data such as healthcare, social media, IoT, etc.

The data quality dimensions previously presented should be considered throughout the whole big data lifecycle. This is because the gathered data in Big data systems could not be directly exploited and must go through various processing steps called Big Data Value Chain (BDVC). In the next section, we present the different phases of the BDVC and the quality dimensions considered in each chain phase.

4 Big Data Value Chain

The data value chain was introduced following the digital transformation as a process that describes data flow within data-driven businesses and allows it to extract data's value. However, with the emergence of big data, the data value chain has shown severe weaknesses in the face of the increasing volume, variety, and veracity of big data. It thus has become inappropriate for Big data systems. Therefore, the big data value chain (BDVC) was introduced as an extension of the data value chain that fits big data requirements, addresses Big data characteristics, and allows the transformation of raw data into valuable insights (Faroukhi et al. 2020).

Fig. 2 Big data value chain model proposed by Faroukhi et al. (2020)

Many big data value chain models were suggested in the literature (Alaoui et al. 2019b; Failed 2013; Curry 2015). In the following, we describe the different phases of the most exhaustive and most recent BDVC in the literature proposed by Faroukhi et al. (2020). The BDVC can be divided into seven steps, as shown in Fig. 2. Finally, we present the quality issues that should be addressed and the quality dimensions to consider for each phase.

4.1 Data Generation

This first step refers to the process of producing data. The generated data may be structured, semi-structured, or unstructured depending on the source type (human-generated, approach mediated, machine-generated) (Arolfo and Vaisman 2018). Then, data are gathered and recorded through an acquisition process.

4.2 Data Acquisition

This step consists of collecting data from various sources and storing it before bringing any changes to the original data. As the gathered data could be of multiple sources and types, it is essential to identify the integrity and variety of data sources since each data type is processed differently.

Data sources' accuracy must be considered in this phase, and the security restrictions must be identified (Failed 2014). In addition, the freshness of the collected data must be considered, especially if the data source is interactive and dynamic, such as social media data (Alaoui et al. 2019a).

4.3 Data Pre-processing

The preprocessing phase is a crucial step in BDVC. It allows cleaning the data gathered in the previous stage from inconsistencies and inaccurate values, thus transforming the data into a valuable and practical format. One of the most common issues that may be faced in this phase is noisy and incomplete data. This could be resolved using clustering algorithms and some data filtering techniques (Zhang et al.

2017). In this phase, data quality is highly improved due to the multiple subphases that this phase involves, such as data transformation, data integration, data cleaning, and data reduction (Failed 2017):

- Data transformation: Refers to the process of converting the structure of data into a more appropriate format to make it more valuable.
- Data integration: Consists of combining various types of data from multiple data sources into a unified dataset, easy to access.
- Data cleaning: The process of ensuring that data is correct and free from anomalies by removing the corrupted and duplicated values.
- Data reduction: This process identifies a model representing the original data in the simplest possible form to reduce the required storage capacity. This process includes multiple techniques, such as data noise reduction, data compression, and other dimensionality reduction techniques (Failed 2017).

Many quality dimensions must be considered in this phase, such as completeness, consistency, uniqueness, and validity (Alaoui et al. 2019b; Failed 2014).

4.4 Data Storage

After the preprocessing phase, data is stored to keep a clean copy of data for further use. However, it is worth noting that the storage system could also affect data quality. Indeed, if a Big data system requires real-time analysis, data timeliness could have a high impact on the freshness and the relevancy of the stored data. In such a case, quality aspects such as storage penalty, storage currency, and availability should be considered in this phase (Alaoui et al. 2019b). Finally, after data storage, data can be processed using the appropriate tools.

4.5 Data Analysis

Data processing is the most critical stage of the whole process. It consists of analyzing and manipulating data previously cleaned to identify the unknown correlations and patterns and transform data into proper knowledge. Different big data analysis techniques could be used in this phase, such as machine learning, deep learning, and data mining.

One of the most common challenges faced in this stage is to ensure the reliability of the obtained results and improve the prediction capabilities' accuracy. This issue could be addressed using more accurate algorithms and models such as Bayesian classifiers, association rule mining, and decision trees (Zhang et al. 2017).

In this phase, it is difficult to identify the quality aspects that should be considered since the data processing is positively related to the context of the analysis and the intended use. However, some standard quality dimensions were suggested by many researchers to ensure a high quality of data during this phase, such as accuracy, validity, and usability (Alaoui et al. 2019b; Failed 2014; Serhani et al. 2016). In Failed (2014), the authors have suggested three general principles that should be considered when assessing data quality:

- The system that performs the analysis should not influence the outcome.
- A version control system must be put in place.
- The quality assessment results must be checked against predefined quality criteria.

4.6 Data Visualization

The analysis results are visualized in a readable form in this phase, easier to understand, using visual elements such as graphs, maps, and dashboards. This would help managers to explore data more efficiently and support them in their decision-making.

This phase has received less attention in the literature than the other phases of BDVC. This stage's quality aspects are more holistic and related to user satisfaction and quality of data representation (Serhani et al. 2016). In Failed (2014), the authors have introduced new dimensions to assess the quality of this last process, such as complexity, clarity, and selectivity.

4.7 Data Exposition

This last step consists of sharing and exposing insights and value generated throughout each phase of the BDVC. For example, the extracted insights could be used to sell specific services, shared with the public as open data, or shared internally to improve organizational performance (Faroukhi et al. 2020).

Each phase of the big data value chain applies some changes to the data. Therefore, it is essential to perform a continuous data quality assessment and monitor BDVC to ensure that these changes are not affected. Thus, in the following figure, we present the different data quality dimensions that should be considered in each phase of the BDVC (Fig. 3).

Many approaches were suggested in the literature to improve the different DQD's throughout the BDVC. In the next section, we report the most recent quality approaches introduced and classify these approaches based on multiple criteria.

5 Research Methodology

A systematic literature review of research on big data quality was conducted following the guidelines stated in Tranfield et al. (2003), where Tranfield et al. have suggested a review methodology that consists of 3 main stages:

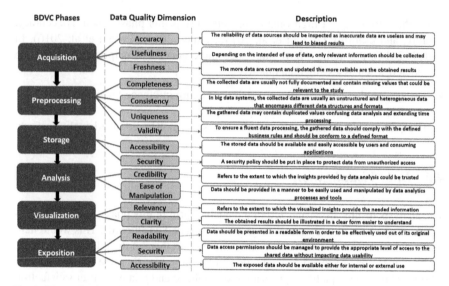

Fig. 3 The projection of the most common DQD's on the different phases of the BDVC

- Planning the review: identifying the need for a review and preparing a review proposal.
- Conducting the review: searching and selecting articles.
- Reporting the review: reporting the main findings and making conclusions.

To conduct the review, primary research was performed using first, broader keywords such as "Big Data Quality," "Big Data Quality Approach," "Big Data Quality Framework," and then more detailed keywords such as "Big data Accuracy," "Big Data Consistency" and "Big data completeness." The research was performed on the following online libraries:

- IEEE Xplore (http://ieeexplore.org/)
- ACM digital library (http://dl.acm.org/)
- Science Direct (http://www.sciencedirect.com/)
- Springer (http://www.springer.com/gp/)
- Research Gate (https://www.researchgate.net/).

The papers were firstly filtered out based on their titles. Thus, papers with titles that do not contain the selected keywords were excluded leaving 105 studies. Then, we excluded irrelevant papers based on their abstracts, and 40 articles were kept for our research. Finally, we included articles that were:

- Proposing approaches to assess or improve Big data quality.
- Proposing methods applied to one or different phases of the BDVC.
- In English.
- Available.
- Published in digital libraries.

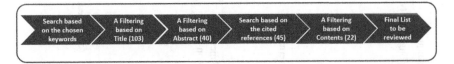

Fig. 4 The methodology followed for the literature search

Then, secondary research was conducted by reviewing the references of the selected studies. Five more studies were thus added to the selected articles. Finally, the chosen studies were fully read, and 22 studies were selected for our review. Figure 4 represents the methodology followed for the literature search. A summary of the selected studies will be presented in the next section, followed by the key findings of this review paper.

6 Big Data Quality Approaches

This section goes over the most recent and available researches that have suggested novel approaches to assess and improve Big data quality throughout the different phases of the BDVC. A summary of these studies is described in Table 2, followed by the key findings that emerged out of the review. The selected papers were categorized based on multiple criteria, namely:

The addressed BDVC phases: The most addressed phases of the BDVC are the preprocessing and the processing phases. Indeed, few studies address the whole BDVC. This is linked to the fact that most data processing is carried out during these two phases, especially in big data systems.

The considered DQD's: The most considered DQD's are Accuracy, Completeness, Consistency, and Timeliness. These quality dimensions are standard for any data quality assessment and are of significant importance whether in a big data context or not.

The addressed application domain: Out of the 22 studies reviewed, only seven focus on a specific big data application domain. The most addressed big data domains are social media and healthcare.

The used techniques: In a big data context, assessing the quality of the whole data is costly in time, money, and effort. Thus, most of the selected studies use data sampling for data quality assessment, which allows them only to analyze a representative sample of data and generalize the obtained results to the entirety of data. In addition, machine learning is yet another widely used technique to measure data quality and extract valuable insights from data.

Based on the available studies, the reviewed approaches presented above lead us to raise some challenging issues that need more research focus and which we summarize in the following points:

Table 2 A summary of Big data quality approaches

References	Year	Phases	DQD's	Domain	Main idea and findings	Used Techniques
Soni and Singh (2021)	2021	Preprocessing	Accuracy, Reliability, Timeliness	–	An approach that aims to reduce the computation time and space could be used when applying quality measures to unstructured data. This approach relies on data sampling and allows measurement: a "believability factor," time is taken to execute the sampled dataset, and the Mean Absolute Error of the sampled dataset	Data sampling, Quality measures
Failed (2020)	2020	The whole BDVC	Credibility and Accuracy	–	It aims to improve data quality and especially data credibility using a new data governance model based on data traceability that considers the different ownership of data	A data governance model
Faroukhi et al. (2020)	2020	The whole BDVC	Accuracy, Credibility, Completeness, Readability, Freshness, Consistency, Storage Penalty, Uniqueness, Conformity, Referential Integrity, Normalization	–	A unified approach that that extends BDVC with security and quality aspects. According to three processes, the suggested framework was perceived: data quality, data security, and orchestration. Thus, in each phase of the BDVC, the quality aspects, the security functions, and the orchestration processes to consider were presented	Security Functions, Orchestration process

(continued)

Table 2 (continued)

References	Year	Phases	DQD's	Domain	Main idea and findings	Used Techniques
Faroukhi et al. (2021)	2020	The whole BDVC	Shareability	–	The authors have suggested a generic model that includes data monetization with the Big Data Value Chain. The proposed model aims to monetize and share insights and data as a service at different levels of data maturity	Cloud computing and Analytical capabilities
I. E. Alaoui et Y. Gahi (2019)	2019	The whole BDVC	Accuracy, Freshness, Consistency, Storage Penalty, Uniqueness, Conformity, Integrity	Social media	It aims to show the impact of the quality metrics on sentiment analysis approaches. The method used consists of grouping tweets into multiple classes using hashtags	Using dynamics and contextual dictionaries based on hashtags, choosing the right keywords for data collect
A. Juneja et. and Das (2019)	2019	Preprocessing	Accuracy, Completeness, Timeliness	Traceability Monitoring	It consists of defining the data quality rules and performing data profiling. Then, a data quality assessment is performed to validate the data quality profile using data sampling	Auto-discovery quality rules Data profiling Data sampling
Mylavarapu et al. (2019)	2019	Preprocessing	Accuracy	–	It aims to assess the accuracy of data using machine learning. The suggested model addresses the contextual and the intrinsic categories of data accuracy	Machine learning

(continued)

Table 2 (continued)

References	Year	Phases	DQD's	Domain	Main idea and findings	Used Techniques
Xu et al. (2020)	2019	Preprocessing	Accuracy	–	It aims to clean data from unreliable data using an incorrect data detection method based on an extended application of local outlier factor (LOF)	The local outlier factor algorithm
Khaleel et al. and Hamad (2019)	2019	Acquisition and Preprocessing	Completeness and Consistency	–	It aims to improve data quality by converting unstructured data to structured data using a metadata approach, distributed data file system (Fragmentation algorithm), and quality assurance algorithms	Machine learning, Fragmentation algorithms
Taleb et al. (2019)	2019	Preprocessing	Accuracy, Completeness, Timeliness	–	A novel data quality profiling model that inspects the different data quality dimensions and relies on a quality profile repository that contains all information related to data quality such as data quality profile, data quality dimensions, …	Big data sampling and data quality profiling algorithms
Arolfo and Vaisman (2018)	2018	The whole BDVC	Readability, Completeness, Usefulness	Social Media	· Data quality is improved when data is filtered ·Data quality is affected by the number of data properties considered in the study	Apache Kafka Zookeper

(continued)

Table 2 (continued)

References	Year	Phases	DQD's	Domain	Main idea and findings	Used Techniques
Taleb et al. (2018)	2018	Preprocessing	Accuracy, Completeness	–	It consists of assessing the quality of unstructured data using a data quality repository where each data type is linked to the appropriate DQD's and features extraction methods	Data sampling and profiling Data mining
Cappiello et al. 2018)	2018	Preprocessing	Completeness, Consistency, Distinctness, Timeliness	Transportation	It provides a context-aware evaluation of data quality based on data profile information and user requirements	Data profiling
Zan and Zhang (2018)	2018	Preprocessing	Accuracy, Integrity, Normative, Consistency, Timeliness	Healthcare	It consists of performing a credibility test by removing unreliable data before embarking on the data quality assessment	Analytic hierarchy process (AHP)
Ardagna et al. (2018)	2018	Preprocessing	Accuracy, Completeness, Consistency, Distinctness, Precision	–	An adaptable approach that allows improving Big data quality based on the business requirements and the context of the use of data. The suggested model consists of 4 main components: a source analyzer, a data quality profiling module, a data quality assessment module, and a data quality adapter that allows selecting the best configuration depending on the main goal (budget minimization, time minimization, confidence maximization)	Data Sampling Machine learning approaches

(continued)

Table 2 (continued)

References	Year	Phases	DQD's	Domain	Main idea and findings	Used Techniques
Wahyudi et al. (2018)	2018	The whole BDVC	A generic approach	–	A generic process pattern model that could be used as a reference for improving Big data quality. The proposed model contains four main steps (discover, access, exploit, analyze). In addition, four variants of the proposed pattern addressing different quality issues were defined to tackle data with variable quality	Data access, preprocessing, and analysis techniques
Failed (2017)	2017	Preprocessing	Relevancy, Uniqueness	–	It aims to increase data quality and reduce data degradation in the preprocessing phase using a new algorithm that combines two feature selection strategies: RELIEF algorithm and mRMR algorithm	Machine learning
Taleb and Serhani (2017)	2017	Preprocessing	Accuracy, Completeness, Consistency	–	It consists of defining quality rules by associating the quality scores that do not meet the quality requirements with the appropriate preprocessing activities. Then, the quality rules are evaluated and optimized	Big data quality rules discovery algorithm
Serhani et al. (2016)	2016	The whole BDVC	Accuracy, Correctness, Completeness	Healthcare	It consists of performing a data-driven quality assessment before and after the preprocessing phase	Data mining Machine learning

(continued)

Table 2 (continued)

References	Year	Phases	DQD's	Domain	Main idea and findings	Used Techniques
Taleb et al. (2016)	2016	Preprocessing	A generic approach	–	A data quality assessment approach consists of 4 main steps: data sampling, data profiling, quality evaluation using a quality assessment algorithm, and data quality analysis. The suggested method uses representative data samples to reduce the processing time	Data sampling
Taleb et al. (2015)	2015	Preprocessing	A generic approach	–	A quality framework to support data quality profile generation for the preprocessing phase. The suggested framework consists of the following main components: a data quality profile selection, a data quality profile adapter, quality control, and a monitoring module	Data profiling, A preprocessing algorithm
Al-Hajjar et al. (2015)	2015	Preprocessing	Timeliness, usefulness, availability, consistency, accuracy	Social Media	A framework identifies the issues of quality analysis of big data on social media. It provides a mapping between big data analysis techniques and the satisfied quality control factors for several social media websites	Social media data analytics techniques

- As we have mentioned earlier, the preprocessing and the processing phases have received more attention in the literature than the other phases of the BDVC. This is because these phases are the most critical phases of the BDVC. However, this does not mean that the different phases could not impact the quality of the obtained results. Therefore, the existing approaches should be extended to englobe all BDVC phases to ensure that data quality is not degraded at any big data life cycle stage.
- One of the most critical challenges to be addressed by researchers is exploring and defining new quality dimensions related to big data rather than focusing only on the standard quality dimensions such as accuracy, completeness, and consistency. Thus, more DQD's such as security, traceability, and clarity should be addressed in the literature. Also, the existing dependencies between DQD's should be highly considered, as focusing on a specific data quality aspect may impact the other DQD's.
- Adopting a more business-oriented approach while assessing data quality is yet another challenging issue to be addressed. Indeed, there is a Big lack of more focused research that addresses Big data quality issues depending on the context in which data is used. Thus, the proposed approaches should be adapted to meet the quality requirements of the use of big data. This adaptation should be considered while building the data quality assessment framework and should not be limited to the experimentation phase.
- Addressing more big data V's is another exciting research direction to explore. Unfortunately, most of the reviewed approaches address only the most common big data characteristics such as Volume and Variety.
- The existing quality assessment frameworks are usually implemented using data sampling, data profiling, and machine learning techniques. However, the current assessment algorithms address only the quality rules defined before the assessment and could be extended to predict potential anomalies accurately.

As we have mentioned earlier, the BDVC must be redefined to address the particular characteristics of each Big data application domain. Therefore, the following section presents some big data application domains addressed in the literature in a Big data quality improvement context.

7 Data Quality and Big Data Applications Domains

Today, many sectors like healthcare, manufacturing, and transportation use big data analytics for data management. However, extracting value from big data remains a challenge for most industries that use big data. As each application domain has its characteristics, Big data quality issues are also related to the requirements of big data use. Therefore, the standard quality approaches could only be used as a roadmap and adapted to fit each sector's characteristics. Many studies that have tackled big data quality have focused on a specific application domain of big data.

One of the most common big data application domains addressed in the literature is social media. Indeed, many consumers use social media to share their thoughts and emotions. Therefore, social media data has become the primary source of data analysis techniques to understand customer insights, such as sentiment analysis and opinion mining. However, the extracted knowledge from such kind of data is not entirely reliable. Thus, these analysis techniques must be accompanied by a continuous data quality assessment for better decision-making. In Faroukhi et al. (2021), El Alaoui et al. have highlighted the essential Big data quality metrics that should be considered in each phase of the BDVC. Also, they have shown that the quality aspect of data directly impacts the reliability and performance of a sentiment analysis (SA) approach.

Moreover, the authors have measured the impact of considering each data quality metric on the accuracy of the SA model. The authors have used a quality-oriented approach that uses dynamics and contextual dictionaries based on hashtags. The suggested method was tested for the US 2016 election forecasting using Twitter data. In the same context, Pääkkönen et al. (2017) have proposed a new reference architecture for assessing social media data quality. The suggested approach extends previous research. The main contribution of this study is the extension and evaluation of the metadata management architecture. The proposed model consists of 3 components: quality management by defining quality rules, metadata management consisting of creating metadata, and quality evaluation using quality metrics defined by the quality rules.

Healthcare is yet another area where big data is widely used. Indeed, adopting big data analysis in healthcare allows avoiding preventable diseases, supporting clinical decision-making, and improving patient care. However, medical data is generally multi-source, unstructured, and heterogeneous. Thus, such kinds of data cannot be directly analyzed and processed. In Zan and Zhang (2018), the authors have suggested a 7-step methodology for assessing medical data quality. This study's main idea consists of performing a credibility test by removing unreliable data before embarking on the data quality assessment. Ensuring the high credibility of data will significantly affect the quality assessment process afterward, not only by improving the quality of the studied data but also by reducing data volume, a highly valuated factor in the Big data domain.

Moreover, the authors have presented an experimental simulation to check the credibility of medical data. In Serhani et al. (2016), have suggested a novel process to improve Big data quality and have assessed the proposed approach using a Sleep Heart Health Study dataset. This evaluation consists of two parts: In the first part, the authors have computed each attribute's completeness and consistency values. Then, they have visualized the improvement percentage of these metrics after the preprocessing phase using graphical charts. A second part was intended to evaluate the processing phase's quality based on some metrics such as accuracy and processing time. Data processing was performed using many classification methods; graphical charts were drawn to compare each classification model's accuracy and timely processing. Also, in Juddoo and George (2018), the authors have attempted to define the most relevant quality dimensions in the healthcare domain using Latent Semantic Analysis (LSA) to extract similarities between documents.

Banking is also one of the business domains where big data is making significant changes. However, compromising between data quality and data security remains a significant challenge in this sector. In Talha et al. (2019), have shed light on the existing tradeoff between data quality and safety. The authors have analyzed some points of convergence and divergence between data quality and data security in this study. Indeed, data quality and data security share some common properties, such as accuracy and integrity. However, there are some cases where data quality and data security cannot coexist. For instance, data security tends to restrict access to data, while improving data quality requires full access to information.

While multiple Big data quality approaches were suggested in the literature, this paper's big data application domains are not entirely explored and need more research. In addition, other big data application domains such as education, manufacturing, and government have never been addressed in the literature in a Big data quality improvement context.

8 Conclusion

Nowadays, big data is highly used for trade analytics by companies that use data to make decisions about their business. However, big data analytics could not be of great value if data is of poor quality. As the extracted data are usually unstructured and inaccurate, a data quality assessment is required when processing big data. This paper has reviewed the quality issues raised by big data V's and how these issues could be addressed using appropriate data quality dimensions throughout the whole BDVC. The existing quality approaches suggested in the literature, and their correlation with Big data application domain requirements are also discussed in this paper. This review led us to identify the following open research issues and challenges that can be further pursued in this area:

- The suggested quality approaches should be extended to address more big data V's.
- The standard quality dimensions should be redefined to fit Big data quality requirements
- The quality assessment should not be limited to the preprocessing and the processing stages and should englobe all BDVC phases.
- There is a Big lack of more focused research that address the characteristics of the specific context of the use of big data
- Other big data application domains such as education, manufacturing, and government should be addressed in the literature.
- New algorithms should be developed to discover the hidden anomalies and troubleshoot data mismatches.

References

N. Abdullah, S. A. Ismail, S. Yuhaniz, S. Mohd sam. *Data Quality in Big Data: a Review*, vol. 7, pp. 16–27, Jan. 2015

R. Agrawal, A. Kadadi, X. Dai, F. Andres, Challenges and Opportunities with Big Data Visualization. (2015). https://doi.org/10.1145/2857218.2857256

F. Arolfo, A. Vaisman, Data Quality in a Big Data Context: 22nd European Conference, ADBIS 2018, Budapest, Hungary, September 2–5, 2018, Proceedings, pp. 159–172 (2018). https://doi.org/10.1007/978-3-319-98398-1_11

I.E. Alaoui, et Y. Gahi, The impact of big data quality on sentiment analysis approaches. Procedia Comput. Sci. **160**, 803–810, janv. 2019. https://doi.org/10.1016/j.procs.2019.11.007

D. Al-Hajjar, N. Jaafar, M. Al-Jadaan et, R. Alnutaifi, Framework for social media big data quality analysis, in *New Trends in Database and Information Systems II*, ed. by N. Bassiliades, M. Ivanovic, M. Kon-Popovska, Y. Manolopoulos, T. Palpanas, G. Trajcevski, et A. Vakali (Springer, International Publishing, Cham, 2015), pp. 301–314. https://doi.org/10.1007/978-3-319-10518-5_23

D. Ardagna, C. Cappiello, W. Samá, et M. Vitali, Context-aware data quality assessment for big data. Future Gen. Comput. Syst. **89**, 548–562, déc. 2018. https://doi.org/10.1016/j.future.2018.07.014

C. Batini, A. Rula, M. Scannapieco, G. Viscusi, From data quality to big data quality. J. Database Manag. **26**(1), 60–82 (Jan. 2015). https://doi.org/10.4018/JDM.2015010103

E. Curry, The big data value chain: definitions, concepts, and theoretical approaches, in *New Horizons for a Data-Driven Economy: A Roadmap for Usage and Exploitation of Big Data in Europe* (2015). https://doi.org/10.1007/978-3-319-21569-3_3

C. Cappiello, W. Samá, M. Vitali, Quality awareness for a successful big data exploitation, in *Proceedings of the 22nd International Database Engineering & Applications Symposium on—IDEAS 2018* (2018). https://doi.org/10.1145/3216122.3216124

S. Dhamodharavadhani, G. Rajasekaran, R. Ramalingam, *Unlock Different V's of Big Data for Analytics* (2018)

J. Espinosa, S. Kaisler, F. Armour, W. Money, Big Data Redux: New Issues and Challenges Moving Forward. (2019). https://doi.org/10.24251/HICSS.2019.131

I. El Alaoui, Y. Gahi, R. Messoussi, Big data quality metrics for sentiment analysis approaches, in *Proceedings of the 2019 International Conference on Big Data Engineering* (2019). https://doi.org/10.1145/3341620.3341629

I. El Alaoui, Y. Gahi, R. Messoussi, *Big Data Quality Metrics for Sentiment Analysis Approaches*, p. 43 (2019). https://doi.org/10.1145/3341620.3341629

A. Faroukhi, I. El Alaoui, Y. Gahi, A. Amine, Big data monetization throughout Big Data Value Chain: a comprehensive review. J. Big Data **7**, 3 (2020). https://doi.org/10.1186/s40537-019-0281-5

A. Faroukhi, I. El Alaoui, Y. Gahi, et A. Amine, *Big Data Value Chain: A Unified Approach for Integrated Data Quality and Security*, p. 8 (2020). https://doi.org/10.1109/ICECOCS50124.2020.9314391

A.Z. Faroukhi, I. El Alaoui, Y. Gahi, et A. Amine, A novel approach for big data monetization as a service, in *Advances on Smart and Soft Computing* (Singapore, 2021), pp. 153–165. https://doi.org/10.1007/978-981-15-6048-4_14

IRJET-V4I957.pdf. Accessed 05 Apr. 2021. https://www.irjet.net/archives/V4/i9/IRJET-V4I957.pdf

A. Juneja, et N.N. Das, Big data quality framework: pre-processing data in weather monitoring application, in *2019 International Conference on Machine Learning, Big Data, Cloud and Parallel Computing (COMITCon)*, févr., pp. 559–563 (2019). https://doi.org/10.1109/COMITCon.2019.8862267.

S. Juddoo, C. George, Discovering the Most Important Data Quality Dimensions in Health Big Data Using Latent Semantic Analysis. (2018). https://doi.org/10.1109/ICABCD.2018.8465129

G. Kapil, A. Agrawal, R.A. Khan, A study of big data characteristics," in *2016 International Conference on Communication and Electronics Systems (ICCES)*, Oct. 2016, pp. 1–4. https://doi.org/10.1109/CESYS.2016.7889917

N. Khan, M. Alsaqer, H. Shah, G. Badsha, A.A. Abbasi, S. Salehian, The 10 Vs, Issues and challenges of big data, in *Proceedings of the 2018 International Conference on Big Data and Education*, New York, NY, USA, Mar. 2018, pp. 52–56. https://doi.org/10.1145/3206157.320 6166

M. Knight, What Is Big Data? *DATAVERSITY*, 05 Feb. 2018. https://www.dataversity.net/what-is-big-data/. Accessed 05 Apr. 2021

M.Y. Khaleel, et M.M. Hamad, Data quality management for big data applications, in *2019 12th International Conference on Developments in eSystems Engineering (DeSE)*, Oct. 2019, pp. 357–362. https://doi.org/10.1109/DeSE.2019.00072

S.S.B.T. Lincy, N.S. Kumar, An enhanced preprocessing model for big data processing: a quality framework, in *2017 International Conference on Innovations in Green Energy and Healthcare Technologies (IGEHT)*, Mar. 2017, pp. 1–7. https://doi.org/10.1109/IGEHT.2017.8094109

J. Merino, I. Caballero, B. Rivas, M. Serrano, M. Piattini, A Data Quality in Use model for Big Data. Future Gener. Comput. Syst. **63**, 123–130 (Oct. 2016). https://doi.org/10.1016/j.future.2015.11.024

G. Mylavarapu, J. P. Thomas, et K. A. Viswanathan, An Automated Big Data Accuracy Assessment Tool, in *2019 IEEE 4th International Conference on Big Data Analytics (ICBDA)*, mars 2019, p. 193-197. doi: https://doi.org/10.1109/ICBDA.2019.8713218.

I. Noorwali, D. Arruda, N.H. Madhavji, Understanding quality requirements in the context of big data systems, in *2016 IEEE/ACM 2nd International Workshop on Big Data Software Engineering (BIGDSE)*, May 2016, pp. 76–79. https://doi.org/10.1109/BIGDSE.2016.021

N. B. PROS, "The Missing Vs in Big Data: Viability and Value," *Wired*, May 06, 2013. Accessed: Apr. 05, 2021. [Online]. Available: https://www.wired.com/insights/2013/05/the-missing-vs-in-big-data-viability-and-value/

P. Pääkkönen et, J. Jokitulppo, Quality management architecture for social media data. J. Big Data **4** (2017). https://doi.org/10.1186/s40537-017-0066-7

A. Ramasamy, S. Chowdhury, *Big Data Quality Dimensions: A Systematic Literature Review*, May 2020. https://doi.org/10.4301/S1807-1775202017003

D. Reinsel, J. Gantz, J. Rydning, The Digitization of the World from Edge to Core, p. 28 (2018)

F. Sidi, P. Hassany Shariat Panahy, L. Affendey, M.A. Jabar, H. Ibrahim, A. Mustapha, Data quality: a survey of data quality dimensions, Aug. 2013. https://doi.org/10.1109/InfRKM.2012.6204995

R. Schmidt, M. Möhring, Strategic alignment of cloud-based architectures for big data, in *2013 17th IEEE International Enterprise Distributed Object Computing Conference Workshops*, Sep. 2013, pp. 136–143. https://doi.org/10.1109/EDOCW.2013.22

M. Serhani, H. El Kassabi, I. Taleb, R. Nujum, An Hybrid Approach to Quality Evaluation across Big Data Value Chain (2016). https://doi.org/10.1109/BigDataCongress.2016.65

S. Soni, A. Singh, *Improving Data Quality using Big Data Framework: A Proposed Approach* (2021)

The Challenges of Data Quality and Data Quality Assessment in the Big Data Era. https://datascience.codata.org/articles/https://doi.org/10.5334/dsj-2015-002/. Accessed 05 Apr. 2021

The UNECE Big Data Quality Task Team, A Suggested Framework for the Quality of Big Data, Dec. 2014

I. Taleb, M. Serhani, R. Dssouli, Big Data Quality Assessment Model for Unstructured Data (2018). https://doi.org/10.1109/INNOVATIONS.2018.8605945

TDWI Best Practices Report | Big Data Analytics, *Transforming Data with Intelligence* (2021). https://tdwi.org/research/2011/09/best-practices-report-q4-big-data-analytics.aspx Accessed 05 Apr. 2021

D. Tranfield, D. Denyer, P. Smart, Towards a methodology for developing evidence-informed management knowledge by means of systematic review. Br. J. Manag. **14**(3), 207–222 (Sep. 2003). https://doi.org/10.1111/1467-8551.00375

I. Taleb, M. Serhani et, R. Dssouli, *Big Data Quality: A Data Quality Profiling Model*, pp. 61–77 (2019). https://doi.org/10.1007/978-3-030-23381-5_5.

I. Taleb, M.A. Serhani, Big data pre-processing: closing the data quality enforcement loop, in *2017 IEEE International Congress on Big Data (BigData Congress)* (2017).https://doi.org/10.1109/bigdatacongress.2017.73

I. Taleb, H.T.E. Kassabi, M.A. Serhani, R. Dssouli, et C. Bouhaddioui, Big data quality: a quality dimensions evaluation, in *2016 Intl IEEE Conferences on Ubiquitous Intelligence & Computing, Advanced and Trusted Computing, Scalable Computing and Communications, Cloud and Big Data Computing, Internet of People, and Smart World Congress (UIC/ATC/ScalCom/CBDCom/IoP/SmartWorld)*, Toulouse, juill. 2016, pp. 759–765. https://doi.org/10.1109/UIC-ATC-ScalCom-CBDCom-IoP-SmartWorld.2016.0122

I. Taleb, R. Dssouli, et M.A. Serhani, Big data pre-processing: a quality framework, in *2015 IEEE International Congress on Big Data*, New York City, NY, USA, Juin 2015, pp. 191–198. https://doi.org/10.1109/BigDataCongress.2015.35

M. Talha, A.A. El kalam et, N. Elmarzouqi, Big data: tradeoff between data quality and data security. Procedia Comput. Sci. **151**, 916–922, Janv 2019. https://doi.org/10.1016/j.procs.2019.04.127

Y. Wand, R.Y. Wang, Anchoring data quality dimensions in ontological foundations. Commun. ACM **39**(11), 86–95 (Nov. 1996). https://doi.org/10.1145/240455.240479

A. Wahyudi, G. Kuk, et M. Janssen, A Process Pattern Model for Tackling and Improving Big Data Quality. Inf. Syst. Front **20**(3), 457–469, juin 2018. https://doi.org/10.1007/s10796-017-9822-7

X. Xu, Y. Lei, et Z. Li, An Incorrect Data Detection Method for Big Data Cleaning of Machinery Condition Monitoring , *IEEE Transactions on Industrial Electronics*, vol. 67, no. 3, pp. 2326–2336, Mar. 2020, https://doi.org/10.1109/TIE.2019.2903774.

S. Zan, X. Zhang, Medical data quality assessment model based on credibility analysis, in *2018 IEEE 4th Information Technology and Mechatronics Engineering Conference (ITOEC)* (2018). https://doi.org/10.1109/itoec.2018.8740576

G. Zhang, A data traceability method to improve data quality in a big data environment, in *2020 IEEE Fifth International Conference on Data Science in Cyberspace (DSC)*, July 2020, pp. 290–294. https://doi.org/10.1109/DSC50466.2020.00051

P. Zhang, F. Xiong, J. Gao, J. Wang, *Data Quality in Big Data Processing: Issues, Solutions and Open Problems*, p. 7 (2017). https://doi.org/10.1109/UIC-ATC.2017.8397554.

Adversarial Machine Learning, Research Trends and Applications

Izzat Alsmadi [ORCID]

Abstract The intelligence extracted through machine learning algorithms (MLAs) plays an important role in most of the smart applications and systems around us. Those MLAs make intelligence decisions on behalf of humans based on knowledge extracted from historical and current data. With such growth of MLA roles in human lives, the rise of adversarial attempts to manipulate those MLAs and influence their choices is not a surprise. The main goal of this paper is to present recent approaches, models and progresses in AMLs. Additionally, our goal is to focus on AML research trends and challenges.

Keywords Adversarial machine learning · Machine learning · General artificial intelligence · Cyber analytics

1 Introduction

There is strong evidence that machine learning models are susceptible to attacks that involve modification of input data intending to result in mis-classification. One more vulnerable model than others is neural networks, but there are many other categories of machine learning models that could fall prey to this kind of attack. To demonstrate the attack, we can think of a machine learning system that takes some input, which results in an accurate classification of that data input. On the other hand, it is possible for some malicious actor to artificially construct an input that will result in incorrectly classified input data when ingested by the machine learning example. Generating these artificial input data or adversarial examples can be done in several ways such as:

Supported by Texas A&M, San Antonio.

I. Alsmadi (✉)
Texas A&M, San Antonio, TX 78224, USA
e-mail: ialsmadi@tamusa.edu

- Fast gradient sign method
- One step target class method
- Basic iterative method
- Iterative least likely class method.

Each method has its advantages and disadvantages when used in an actual attack. For example, a fast gradient sign method is a simple and easy way to create adversarial examples, but this method's success rate is much lower than the others. After generating the adversarial examples, they can be used in two main forms of attacks that target these machine learning systems: Black-box versus white-box attacks. Black box attacks happen in a scenario where the malicious actor only has a small amount of information about the specifics of the model and how it works. However, in white box attacks, the attacker has full knowledge of the model and its parameters. There is one particularly tricky property of an adversarial example, and that is its transferability. Most adversarial examples can be transferred from one model to another, meaning that a malicious actor can develop one adversarial example designed to target their own test model. When it has been tested, use that example to target another model in a black box scenario.

Adversarial examples are inputs for machine learning models that allow an attacker to cause the target model to make a mistake. These examples are intentionally designed to make the model output a wrong reading, and in the world of machine learning at a huge scale, one exploitative example or sample can have a cascading effect on the model and whatever entity uses them.

To guard against adversarial input attacks, what is known as adversarial training is performed. Adversarial training is adding adversarial examples into the data set at the time of training the model. By injecting this adversarial example into the training set, we train the model to be resilient against this type of attack since it will know how to handle this kind of input when it is introduced. There have been experiments to show the effectiveness of ensemble training in improving the model's resilience against both black-box attacks and white-box attacks. Ensemble adversarial training has been shown to improve the resilience of the model from black-box attacks drastically. While, in the case of white-box attacks, adversarial ensemble training did not do much to improve the resilience of the model. The idea of transferred attacks seems particularly difficult to handle. Still, with the experiments done with ensemble training it was proven that with black box attacks it is possible to increase the models resilience to transferred attacks. However, new attack types have been suggested to increase the success rate of transferred attacks in a black box scenario. On the other hand, with white box attack, ensemble training, although it slightly increases resilience against transferred attacks, negatively affects overall resilience to white-box attacks. In conclusion, although this type of attack is serious against machine learning systems, with adequate training and especially ensemble training, we can dramatically reduce the effectiveness of such attacks. Sadly, its a continuous game between defenders that improve their algorithms, but attackers who will do the same. For example, there have been some suggested attacks that are theoretically robust against such training methods.

Evaluating the security of MLAs can involve the following processes:

- Identify attack classes to the system
- Measure the robustness of the system against these attack classes.
- Design and study defenses against potential attacks.

There are two main types of defense strategies: reactive versus proactive. One huge problem with securing machine learning systems is that adversarial examples are very difficult to detect. In particular, Neural Networks are susceptible to adversarial examples (Zhu et al. 2018a; Yu et al. 2018). Based on the experiments performed, they can trick a single classifier, which means that they can trick both the classifier and the detector. Moreover, it was shown no easy or concrete method to differentiate between an adversarial example and a legitimate input. Based on this, it can be concluded that the current security measures we possess are not adequate in order to defend against adversarial examples. Before we can even think about using machine learning in security applications and be confident in their performance, we need to design effective defenses against adversarial examples.

A secure machine learning algorithm is the one that can perform under adversarial conditions. These conditions deal with the possibility that an adversary can design training data to change the system to treat hostile data as legitimate (Barreno et al. 2010). In Barreno et al. (2010), the authors use these classes to develop a framework for security analysis. In security analysis, there are two metrics, security goals, and threat model. Security goals deal with two factors, integrity, how well a system can prevent attackers from reaching system assets and availability, how well a system can prevent attackers from interfering with regular operation. Threat models evaluate two factors attacker goals/incentives and capabilities. The taxonomy of these factors against learning systems are categorized into three additional categories:

- Influence (causative and exploratory), which describe the capacity of the attacker,
- Security violation (integrity and availability), the type of breach the attacker causes, and
- Specificity (target and indiscriminate), the attacker's intention.

2 Literature Review

Adversarial examples are evolving in machine learning security because the problem is ever-evolving and while can be addressed in the short term but cannot be solved in a simple manner. It requires the defender to continue winning the ever-progressing game with the attacker in order to get ahead of current adversarial examples. One of the more obvious places where adversarial machine learning could be used with disastrous results would be in applying self-driving cars (Kurakin et al. 2016). Having a two-ton missile controlled by an algorithm under the influence of what is basically an optical illusion isn't good for the future of self-driving cars. A stop sign is interpreted as a yield sign is a great adversarial model, and the self-driving software should absolutely be secured against this. The one-step fast gradient sign method (FSGM) is a way of simply creating an input data 'offset' closely related to the first gradient of a data point. This is a very simplistic methodology of creating adversarial input, which typically results in less success overall in a machine learning attack vector, but the nature of one-step adversarial generation permits the attack vector to be transitive to almost any model with equally successful attack rates via black-box methods. This approach increases the loss function of each step in the model. Alternative one-step methods are available, which increases the probability of a false classification versus the loss of information. Iterative methods of perturbation where data point gradients are adjusted multiple times prior to being presented to the model for classification are consistently more successful as an attack vector on the model being targeted as they are tuned to the intricacies of that model. This effect subsequently makes their capability to be transitive much less successful to other models due to their highly tuned nature. Obviously, there are other iterative perturbation methods available. The authors claimed that including one-step adversarial examples batched with the clean training data improved the robustness of the final model being able to identify one-step perturbations when presented in real life. The sacrifice was less than 1% loss of accuracy for clean data classification but did not improve robustness against iteratively created adversarial data. Along these same lines, increasing the model's depth further improved the robustness while reducing the accuracy discrepancy for clean data. As this ratio approaches a value of 1, all adversarial data is being identified and classified correctly.

Although it is easy to assume that adversarial examples would come from malicious actors, there is mounting evidence that adversarial examples may occur naturally, especially when inputs are from cameras or other sensors. Kurakin et al. (2016) demonstrate this in their study by printing out both clean and slightly adversarial images, taking their photo, cropping, and feeding them into a trained ImageNet Inception classifier. This method of manually taking photos of printed images, without control of lighting, camera angle, distance-to-page, is intentional and meant to introduce the "nuisance" variability that blurs the difference between the base and adversarial images. Overall, they found that even with trivial or slight perturbations, about 2/3 of the images would be top-1 mis-classified and 1/3 would be top-5 misclassified.

Adversarial training is the concept of training a machine learning system using examples of malicious data in order to make the system more resistant to such attacks and more capable of functioning in a hostile environment. In Adversarial Machine Learning at Scale, (Kurakin et al. 2016) describe a methodology to take adversarial training beyond small data sets and scale to large models. They also observe that "single-step" attacks are better for black-box attacks on systems and that adversarial training improves a system's performance against such attacks. In the base case, adversarial training is the inclusion of attack data in a model's training set. The paper states that it has been shown that machine learning models are vulnerable to attack during testing through the use of inputs that have been subtly manipulated to influence the model. However, it has also been shown that if such adversarial inputs are included in the training set, the model becomes more resistant to this type of influence and performs more reliably at test time. Prior to this paper, research in this space was restricted to relatively small data sets. This paper's primary contribution is to suggest methods for using adversarial training on much larger data sets. They use a large data set from ImageNet as an example for this study. The paper explains in detail the types of attack methods used in the experiment to generate adversarial data. The methods included various ways to manipulate the image data to it. They include adversarial examples in each training set they run, generating new examples to include with each iteration. Ultimately, the researchers found that including adversarial data in the training set brought the model's accuracy in identifying attacks in tests up to the level of tests on clean data. However, they found that the accuracy of the model on clean data decreased slightly when adversarial data was included in the training.

Kurakin et al. (2016) observed how adversarial examples might be used to train more robust models. Their idea of adversarial training, for the latter, is to inject adversarial examples in the same batches as their normal counterparts at every step of training, to allow models to train on the differences between the two simultaneously. They found that such a method significantly increased top-1 accuracy to 74% and top-5 accuracy to 92%, making said accuracy on par with accuracy on clean images. They found this method to be a more reliable safeguard for dealing with adversarial examples than structural considerations like choosing a more robust model architecture. The most influential factor for having a robust enough model is how many trainable parameters the architecture has. However, this consideration falls apart if a model is too small or too large, and it may not be worth the effort to look for a model within the "sweet" spot. Kurakin et al. (2016) focused on adversarial inputs to a model and how to train a model to protect itself from adversarial inputs. Specifically, the examples in this paper use deep learning neural nets for image recognition. The adversarial inputs can be created through slight alterations in the pixels in the image. Four different methods for generating these adversarial inputs are presented: fast gradient sign method (FSGM), one-step target class methods, basic iterative method, and iterative least-likely class method. FGSM is the most computationally efficient of the four but is least likely to be successful. One-step target class relies on the probability of a specific target class. The "step 1.1" variation is based on the target class that is least likely to be predicted by the neural net, and "step rnd" chooses a randomly selected target class. The remaining two methods apply small changes using FSGM or step 1.1 multiple times to create a successively better adversarial image on each iteration. To protect the model against adversarial inputs, the supposition is that we can introduce these inputs into the training data to teach the system to reject these inputs. The author chooses mini-batches of length 32, where 16 inputs are replaced with adversarial versions of the replaced inputs. It was not expected, but when the authors ran their experiments, they saw a small degradation in the accuracy of the model against the original dataset. The degradation was less than 1%. The biggest difference was in accuracy in the presence of adversarial inputs in the test data. In that case, the model trained with adversarial inputs performed much better than the model trained with clean input only. There is also a discussion of "label leaking". They had some ods of generating adversarial inputs that would result in accuracy improving as more adversarial inputs were added to the test data. This can be countered by choosing the one-step methods instead of the FSGM method. Another discovery reported is that deeper models performed better with the adversarial training data. Finally, they examined how well the training transfers when confronted with adversarial input generated with a different method. The outcome was that FSGM trained models did best at rejecting adversarial input.

Huang et al. (2015) paper also examines training deep neural networks with adversarial inputs to improve the accuracy of the network. This paper describes finding adversarial inputs using a linear approximation of a softmax layer g(x) when working with the formula that Szegedy et al. (2013) proposed. The paper describes in detail the algorithms needed to train the network using optimized adversarial inputs. It also discusses several reductions of formulas required to make them solvable. The paper goes on to demonstrate how the methods presented will perturb an image. Examples are shown where the original image is subtracted out of the perturbed image to show the noise image that was generated by three different methods. One interesting finding of the research is that more parameters in a model mean that the model is more resistant to adversarial examples. It is not quite known why this is, given the black box-like nature of models. However, a significant fraction of adversarial examples that fool one model is able to fool a different model. This property is called "transferability" and is used to craft adversarial examples in the black box scenario (Yan et al. 2018). This paper's results show that some of the transferability of the adversarial examples allow for the same exploit to be used in multiple models and that changing hyperparameters does not make much of a difference on the defensive side of the model. In order to get around some of the adversarial examples, some results were found that were quite interesting. One such finding was that deeper models are more benefited by adversarial training and that models with many layers actually change many of the aspects of the model's architecture. The number of filters and layers in the algorithms, like the number of iterations, had a diminishing marginal effect the higher it went, but an optimal level of what the factor Rho, (multiplier for the number of filters from standard) is showed that robustness of the model proportionally increases with the larger model size. This property of size increases theoretically until the model reaches an accuracy ratio of 1, but no models in this experiment were large enough to reach that precision. The main takeaways from the adversarial research are that model capacity used in conjunction with adversarial training is able to increase resilience to adversarial examples in machine learning.

Tramèr et al. (2017), paper indicated that models trained with adversarial data generated from the clean training data teach the automated one-step generator to be weak versus strong. This ultimately means the model behaves well when using 'white-box' attacks (those that know the features and model) but misbehave when encountering black-box attacks. Subsequently, black-box one-step adversarial data generation then showed an increased ability to be transitive between models directly countering Kurakin et al. (2016) findings. They subsequently suggest that perturbation creation be decoupled from the clean training data and leverage Ensemble Adversarial Training methods where adversarial data is passed through multiple perturbation stages to minimize correlation to the clean input training data effectively. This paper (Tramèr et al. 2017), presents an ensemble technique used for adversarial training that is more robust against black-box attacks on machine learning models.

This appears to be an ever-increasing event and is still relevant to machine learning systems today. The current philosophy on adversarial training is generally to use the model being trained in the creation of the adversarial dataset. Regardless of the method used for adversarial data creation, using the model to create the adversarial data always seems less effective in generalizing black-box types of attacks. Almost as though the models are over-fitting with regards to the adversarial data. The paper proposes using several static based single-step models in an ensemble to generate adversarial training data. This provides a more generic dataset making the model more robust against single-step attacks; however, it did not seem help against more costly multi-step attacks. It was cited though that other research has shown these multi-step attacks do not transfer well between models, so is less of an issue when considering black-box attacks. This form of adversarial training tended to make the models less accurate as compared to standard adversarial training and more susceptible to white-box single-step attacks. The ensemble technique's goal was to add robustness against single-step black-box attacks, which was shown to be true. The criticism here is levied at the vulnerabilities which still exist even after a model has been successfully trained using the best performing approach, the single-step (or one-step from the first paper). The researchers here suggest a black-box or white-box attack would be sufficient to exploit these problems. Again the goal is to champion the notion of robustness and not focus on too narrow an attack type. A black-box attack is one where perturbations are computed on one model (source) and transferred to others (targets). The paper finds that adversarial training increases robustness to white-box attacks and creates more errors for a black-box attack. Thus the models create false robustness that is only apparent in an isolated setting.

The paper (Xie et al. 2017) offered approach that ultimately minimizes the effects of perturbation with relation to the clean data—randomization of the input data. The easiest explanation is to visualize a small image as an input image. Their approach was to randomize the image's overall size by no more than 10% (random resizing) and then to randomly center the image within the bigger image size using RGB values of 0 on the border (random padding). In combination with typical adversarial training methods, this approach improves the robustness of the model as it effectively decouples the training data from the perturbations applied to the adversarial training samples. This method is unobtrusive to the overall model performance as simply two additional layers are added to the input of the model with minimal computational impact.

Wong et al. (2018) analyzed non-linear approaches to maximizing deep learning models' robustness and subsequently creating provable robustness. While a lot of work needs to occur in the libraries of existing models, they effectively demonstrated that some non-linear internal model functions will improve the robustness of the overall model. In summary, there is general agreement that including adversarial data into the training data at the time of model creation is a good approach to improving the security robustness of that model at the time of real use. The model's positive security results far outweigh any loss in the model's accuracy due to this adversarial training approach. Generation of the adversarial data needs to be decoupled from the clean training data to ensure the model doesn't identify with local minimums when stabilizing weight assignments.

There is the constant threat of malicious actors identifying and acting on weak points that may exploit a system or to render inoperable for all computer systems. This is becoming especially concerning for machine learning models and systems, which are often touted as achieving better, smarter results than existing "manual" heuristics and able to handle previously unseen information. Recent studies, however, indicate that the advantage of learning from unseen information may be used as a weak point to exploit and manipulate machine learning systems. One such attack approach is known as adversarial example training, where legitimate inputs are altered by adding small, imperceptible, perturbations that lead a trained classifier to mis-classify the altered input, while the input is still correctly classified by a human observer.

Papernot et al. (2016a) provide an example, where both would be classified as stop signs by any individual. However, they forced their neural network to classify the image on the right as a yield sign by applying a precise alteration on it. In their paper, they go on to propose a generalized "black-box" adversary attack that departs from contemporary methodologies. Instead of constructing adversarial examples using prior knowledge of a neural network and parameters or using some collected training set to train another attack model, they propose a model that can learn to perform a generalized attack from just using the original model's output. The adversary model collects a very small set of inputs representative of the entire domain, then selects an architecture to be trained that is applicable to the domain, and then trains said architecture by applying perturbations at certain components of the images to determine which perturbations affect saliency of the most, while being observably undistinguished. Using this methodology, they were able to generate adversarial examples that were misclassified by black-box neural networks and caused MetaMind, Amazon and Google classifiers to achieve misclassification rates of 84.24%, 96.19% and 84.24%, respectively.

Authors in Papernot et al. (2016a) went on to find a method to defensively reduce the effectiveness of adversarial samples from 95 to 0.5% through distillation. Distillation is a training method where knowledge from one neural network is transferred to another neural network. In this paper, however, they propose to use knowledge extracted from a neural network to improve its own resilience to adversarial samples. Essentially, by training a model on natural images and then transfer learning the same model on a dataset with adversarial examples, the model can be trained to be more resilient and robust against future adversarial examples. When applied on DNN for the CIFAR10 dataset, the model reduced the success of adversarial samples to 0.5% while maintaining an accuracy comparable to previously applied models.

In addition to CNNs and DNNs, reinforcement learning models such as Deep Q Networks (DQNs) can also be manipulated by adversarial examples. Common algorithms like DQN, TRPO, and A3C are open to being exploited by adversarial examples (Behzadan and Munir 2017). Almost all adversarial examples are indistinguishable to a human but have a huge effect on a ML model. These can lead to degraded performance even in the presence of perturbations too subtle to be perceived by a human, causing an agent to perform actions it would otherwise never do and cause the model to underperform. The two takeaways here are that it is possible to train a model on adversarial data to prepare it for malicious attacks, and the more sophisticated/deeper a model is, the more it will benefit from this approach. Given the task's nuanced nature, this makes intuitive sense, but seeing the successful results is very encouraging. A second paper, Towards Deep Learning Models Resistant to Adversarial Attacks, (Madry et al. 2017) expands upon this topic by conducting a similar study. Here the researchers focus more on deep neural networks, but also through the lens of image recognition. In this paper, the authors attempt, similarly to the original, to create models that are resistant to adversarial input attacks. The paper (Madry et al. 2017) investigates the apparent lack of robustness deep learning models have towards attacks. This is apparently a hot topic of research these days and is certainly a relevant one.

Boosting Adversarial Attacks with Momentum levy similar criticisms to the training methods of previous approaches (Dong et al. 2018). Namely that of poor "transferability" (i.e., poor generalization). The solution here is similar to the third paper with a focus on an ensemble approach. Their group's models are momentum-based iterative approaches, which they claim can fool white-box attacks and black-box ones. Dong et al. (2018) used similar attack vectors in attacking models they trained in their research. This research aims to analyze the defensive hardening created by adversarial training and develop strengthened black-box attack methods that can defeat these models. Again focused on the image recognition space, the authors explain the vulnerability neural networks and other models have to adversarial images, as well as the transferability of attacks across multiple models. They state that different machine learning algorithms tend to come up with the same decisions over time, which means that the same adversarial image that fools one model is likely to fool others. However, they find that adversarially trained models perform much better against these attacks, as found in Kurakin et al. (2016).

In Akhtar and Mian (2018), the authors chose to focus more on adversarial attacks against images. Basically, image NN models can be attacked by introducing a small amount of adversarial input into an image. The change in the image is nearly imperceptible to the human eye, but it significantly reduces the accuracy of the models. Several different types of adversarial attacks are discussed. Examples include one pixel attack, where one pixel is changed in the image to trick a classifier. Additionally, he discusses different methods for defense, such as the adversarial brute force adversarial training that is explored in the main paper discussed in this exam question.

It seems that a common technique for defending against adversarial example attempting to use iterative, or gradient descent, based attacks is to "mask" the gradient (Guo et al. 2018). This "masking" is sometimes intentional as part of the defense or a byproduct of the adversarial training method. Tsipras et al. (2018) acknowledge all of the research and efforts to attack and defend models using adversarial examples in training. They discuss the difficulties in crafting truly robust models and the fact that standard models are vulnerable to attack without adversarial training. The authors go on to discuss the cost of defending models from adversarial attacks. The most obvious cost they point out is the dramatic increase in training time required. However, beyond that, they also find that as more adversarial images are introduced into the training data set, the model's overall accuracy declines. They go on to discuss the opposing goals of high general accuracy and adversarial robustness. They describe a trade-off between security and correctness without making any definitive conclusions.

2.1   The Tolerance for False Positives Versus False Negatives

Cybersecurity is relying more and more on Machine Learning to identify and combat attacks on computer systems. Machine Learning systems enable defenders to respond quickly to newly discovered vulnerabilities and react to new attack vectors. However, attackers are also learning how to exploit Machine Learning systems. Barreno et al. (2010) presents a model for the evaluation of secure learning systems, algorithms that perform well in securing computer applications under adversarial conditions. Their goal is to define a taxonomy—a language that can be used to describe learning systems that can be used for cybersecurity purposes in adversarial conditions. In their research, (Biggio et al. 2012) leverage the paper (Barreno et al. 2010) taxonomy to describe an attack model in terms of goal, knowledge, and capability. In this research, the authors focus on the use of pattern recognition in machine learning and how pattern recognition can be manipulated in an adversarial environment. Huang et al. (2015) paper discussed how to increase dataset robustness and to prevent unwanted perturbations. The technique that they used was to introduce adversarial examples during an intermediate step intentionally. The study is trying to find a technique to handle strong adversarial examples from causing additional model mis-classifications, especially associated with deep-learning image classifiers. In their approach, they produced a min-max training procedure to increase robustness. A common problem with previous models that used logistic regression without regularization could not solve for linearly separable data, but this can be overcome by using an adversary during training. They designed a new method for finding adversarial examples based on linear approximation using the Jacobian matrix. Ultimately, the authors claim that their algorithm for min-max formulation of adversarial perturbation is better than previous works. They can produce better robustness and maintain solid accuracy.

We investigated AML literature based on specific subjects. Our goal is to describe current and future research trends in this area. We can see the following taxonomy of AML attacks.

3   Black Versus White-Box Attacks

From the perspective of attacker knowledge of target machine learning models, there are three categories of attacks on Machine Learning models, White Box, Black Box and Grey Box attacks. White Box attacks are adversarial attacks in which the attacker has intimate knowledge as to the inner workings of the model. This could range from knowing the features and target values to understanding something more complex such as the gradients weights in a deep learning model and understanding how to manipulate those values through data injection just enough to change a prediction. Black Box attackers are such that the adversary does not have knowledge of the inner workings of the model. These are much more difficult to conduct but there are theories, methodologies and some examples of attacks being orchestrated against a white box of a different model and then those attacks being carried out against a black box of a presumably different model. Overall, most adversarial attacks against Natural Language Processing (NLP) models are being used for message or news postings in order to spread misleading or hateful information as well as scenario's where spam defeating is desired or click baiting detection is needed.

There are several types of white box attacks that can be highly effective at manipulating the output of a machine learning model. Data poisoning attacks, particularly on sentiment analysis, are easy to accomplish and yield reliable results. This Adversarial attack is done by accessing the training data within a model and changing the target label from positive to negative or vice versa. The changes are very subtle only being done on a limited number of data rows for a word or sentence that is very similar, but not the same, as the values in which the attacker wants the model to be manipulated (Wallace et al. 2021). For example, if a product review about a new computer is negative containing words that would normally be classified as negative in a sentiment analysis model are labeled as positive and review for a TV that has similar wording and is truly negative will be labeled as positive. There are a few ways to defend against this type of attack. The first and simplest method is to stop training early before the weighting of the malicious changes to training data is significant. This technique works but cost in the form of accuracy. Perplexity analysis, a technique that analyzes and ranks sentence's structure can also be used to determine the validity of the sentences that are being used in the data and can help identify poising data injected into the training set. Another example of a white-box attack is the Hot Flip attack (Ebrahimi et al. 2017). This attack uses the forward and backward propagation of a neural network architecture, over several iterations, to determine the best possible character to flip to achieve the greatest change in gradient weight towards the desired result. In most cases this is used to defeat ML NLP that specifically looks for key words that would be flagged as vulgar, inappropriate, known to be intentionally misleading or spam like in nature.

In this specific area, there are several challenges being addressed. There is the challenge to the task of hand, of determining how to manipulate the models in an effective manner that goes undetected. This challenge is magnified when it comes to understanding how to manipulate a black-box model that does not have documented interworking's or provide access to directly manipulate the training. The greater challenge is then devising strategies for defending against successful attacks. Often defending is much more complex than the attack itself. Take for instance the cases of transposition and substitution at the character level on key words. It would be an overwhelming task to provide dictionaries that account for every possible permutation and if possible, would require significant compute resources. Compute resources are another challenge; it takes libraries that support GPU to efficiently train models and that compute resource that is needed increases exponentially when considering modifying algorithms beyond just training but to detect intentionally data poisoning. These high-performance resources are also needed for adversarial attacks not only for those trying to manipulate the ML systems but for those building them to test the systems vulnerability to attacks so that hardening measures can be taken.

4 Defenses Against Adversarial Attacks

Machine learning (ML) models are vulnerable and sensitive to small modifications. An attacker can tamper a model by adding just a few malicious perturbations in the input data. Therefore, it's been always an important defense goal, to develop such techniques which make these models more robust and versatile. There have been several studies so far in this area that focus on one of the two main aspects:

- Examples of what and how to change ML models with adversarial examples. Example of papers include: (Ebrahimi et al. 2017; Alzantot et al. 2018; Gao et al. 2018; Liang et al. 2017).
- Defense against the adversarial attacks: The naïve method of defensing against the adversarial attacks is to recognize and remove the unnoticeable perturbations e.g. in computer vision, the denoising auto-encoders are used to eliminate the noise which is added by the perturbations, however, this approach cannot directly be used to the NLP tasks primarily because of below reasons:
  - There are continuous pixels in the images whereas the input texts contain the discrete tokens. Because of this nature of the text, another similar or semantically matching word/token can replace a word/token which lowers the performance. Therefore, the perturbations which have a similar look with other words or get mixed up with the input text, cannot be easily identified. The other approaches capture the non-natural differences between the nearby pixels.
  - As sentences may contain words with a large vocabulary, it can be hard to iterate through all the possible perturbed sentences. As a result, the already present defense methods that heavily rely on the pixels, may not be applicable to NLP tasks.

Wang et al. proposed a Synonym Encoding Machine (SEM) to defend against adversarial perturbations (Wang et al. 2019). SEM defends against synonym substitution adversarial attacks. Text-based adversarial attacks fall into two categories: the word-level and character-level. First, word-level has defenses through the system's training data when compared against the synonym attack. Second, the character-level has a proposed defense of using a downstream classifier that will act as the defense mechanism for adversarial attacks and identify possible adversarial attack spelling mistakes. The categorization used helped cage the defense efforts and made for statistically significant results in defending against the synonym-styled attacks. As this is a small subset of the natural language processing adversarial attacks, it shows the validity in defense efforts being made to all caveats of this system. Zhou et al. proposed a randomized smoothing method to improve the training model to better defend it from adversarial substitution attacks (Zhou et al. 2019). The model would use data interpretation as its systematic central defense and placed data as the centerpiece to a holistic natural language processing defense mechanic. Rosenberg et al. evaluated 11 defense methods and found that certain ensembles and sequencing delivered different amounts of accuracy (Rosenberg et al. 2019). The effectiveness was cross correlated with the amount of effort needed for method implementation and as expected high overhead lead to the highest rate, but a low effort method with sequence squeezing was found to still produce results near 91% which shows room for growth and analysis in these methods and defense strategies. Data manipulation is not a new technique, and its employment in this research area shows machine learning modules need to be aware of security concerns just like a traditional system. To be brief, all systems are susceptible to attack, and manipulation and security techniques need to ensure these systems' confidentiality and integrity. They complete their intended functionality as machine learning cannot guarantee its safety and data surety.

4.1 Adversary Attacks and Language Comprehension

Researchers have explored many approaches related to the reading comprehension and evaluation metrics. However, its yet arguable to relate those metrics with language comprehension and semantics. It becomes even more challenging if the input text contains the sentences which have been inserted maliciously. With this adversary, the accuracy of several models can drop significantly. The existing models appear successful by standard average-case evaluation metrics, however the reading comprehension is still an appealing problem for adversarial evaluation. It seems unlikely that existing systems can truly understand languages and have reasoning capabilities.

4.2 White Versus Black-Box Attacks and Defenses

Many existing adversarial example generation algorithms require full access to target model parameters (white-box attack), which makes them impractical for many real-world attacks because in case of real-world attacks, there may not be full information available about the model. Because of this, many models are vulnerable to adversarial examples. In such cases, the inputs are crafted to make a targeted model fool into outputting an incorrect result or class. We can create the adversarial examples using white box technique, but the main problem is the discrete nature of the input text.

Research across multiple authors highlighted a trend of many subsystems being addressed as vulnerable to adversarial attacks. Systems named were deep neural networks, Dirichlet neighborhood ensemble, recurrent neural networks. Adversarial attacks were seen as credible attacks to all these methods under the natural language processing lens, and independent efforts were made to defend the specific subsystem when a combined effort to defend against adversarial attacks as a whole could prove beneficial to a more significant percentage. The trend here to specialize in a sub-topic and create a unique solution is a part of many researchers' efforts. Still, in the proposed solutions, it often was a stagnation of effects as many results proved to be along the same line of thought. Combining efforts to fight adversarial attacks might better native defense efforts with subcomponent analysis being done to prove the efficacy of efforts made by the researchers in their perspective specialties. It may be beneficial to work as a whole on problems than constantly race one another to the same solution. Similar to the above reference point, private datasets and industry-standard practices validate efforts and research consistency are difficult. The privatization of efforts in the business environment means many datasets are set to be academic as a whole. Any security defect discovered often is hidden from public purview not to hurt the company's reputation. The researchers used many academic datasets, which often had relatively robust datasets. Still, it is easy to see the lost opportunity when comparing these sets to actual data of deployed natural language processors. Company integration with the academic environment has been a significant issue in academia, and it persists as researchers are losing the ability to test and abuse real-world datasets. This point of view is easy to see when the references repeatedly used academic datasets and came to conclusions based on the data used. The significance of a topic cannot be fully understood until a product is exposed to the greater public, and in this area of review, we must stress test all defensive efforts.

5   Sequence Generative Models

A basic Generative Adversarial Network (GAN) model includes two main modules, a generator, and a discriminator. The generator and discriminator are implicit function expressions, usually implemented by deep neural networks (Creswell et al. 2018).

Applying GAN in Natural Language Processing (NLP) tasks such as text generation is challenging due to the discrete nature of the text. Consequently, it is not straightforward to pass the gradients through the discrete output words of the generator (Haidar and Rezagholizadeh 2019).

As text input is discrete, text generators model the problem as a sequential decision making process. In the model, the state is the previously generated characters, words, or sentences. The action or prediction to make is the next character/word/sentence to be generated. The generative net is a stochastic policy that maps current state to a distribution over the action space. Natural Language Generation (NLG) techniques allow the generation of natural language text based on a given context. NLG can involve text generation based on predefined grammar such as the Dada Engine, (Baki et al. 2017) or leverage deep learning neural networks such as RNN, (Yao et al. 2017) for generating text. We will describe some of the popular approaches that can be found in relevant literature in the scope of AML.

- Classical: training language models with teacher/professor forcing teacher forcing is common approach to training RNNs in order to maximize the likelihood of each token from the target sequences given previous tokens in the same sequence (Williams and Zipser 1989). In each time step, s, of training, the model is evaluated based on the likelihood of the target, t, given a groundtruth sequence. Teacher forcing is used for training the generator, which means that the decoder is exposed to the previous groundtruth token.

  RNNs trained by teacher forcing should be able to model a distribution that matches the target, where the joint distribution is modeled properly if RNN models prediction of future steps. Created error when using the model is propagated over each next or following step, resulting in low performance. A solution to this is training the model using professor forcing (Lamb et al. 2016).

  In professor forcing, RNN should give the same results when a ground truth is given as input (when training, teacher forcing) as when the output is looped back into the next step. This can be forced by training a discriminator that classifies wether the output is created with a teacher forced model or with a free running model,

- Conventional inference methods/ maximum likelihood estimation (MLE)

  MLE is conducted on real data samples, and the parameters are updated directly according to the data samples. This may lead to an overly smooth generative model. The goal is to select the distribution that maximizes the likelihood of generating the data. For practical sample scenarios, MLE is prone to over-fitting/exposure bias issues on the training set. Additionally, during the inference or generation stage, the error at each time step will accumulate through the sentence generation process (Ranzato et al. 2015).

  The following methods utilize MLE:

  - Hidden Markov Model (HMM): A Hidden Markov model (HMM) is a probability graph model that can depict the transition laws of hidden states, and mine the intentional features of data to model the observable variables. The foundation of an HMM is a Markov chain, which can be represented by a special weighted finite-state automaton. The majority of generative models require the utilization of Markov chains (Goodfellow et al. 2020; Creswell et al. 2018). The observable sequence in HMM is the participle of the given sentence in the part-of-speech PoS tag, while the hidden state is the different PoS.
  - Method of moments: The method of moments (MoM) or method of learned moments is an early principle of learning (Pearson 1893). There are situations in which MoM is preferable to MLE. One is when MLE is more computationally challenging than MoM. In the generalized method of moments (GMM), in addition to the data and the distribution class, a set of relevant feature functions is given over the instance space (Hansen 1982; Rabiner 1989). Other research contributions in AML MoM or moment matching include: (Salimans et al. 2016; Mroueh and Sercu 2017; Lewis and Syrgkanis 2018; Bennett et al. 2019).
  - Restricted Boltzmann Machine (RBM): Restricted Boltzmann Machine (RBM) is a two-layer neural network consisting of a visible layer and a hidden layer,

Hinton 2010. It is an important generative model that is capable of learning representations from data. Generative models have evolved from RBM based models, such as Helmholtz machines (HMs), (Fodor et al. 1988) and Deep Belief Nets, DBN, (Hinton et al. 2006), to Variational Auto-Encoders (VAEs), (Kingma and Welling 2013) and Generative Adversarial Networks (GANs)

- Cooperative training method In Cooperative Training Method, CTM, a language model is trained online to offer a target distribution for minimizing the divergence between the real data distribution and the generated distribution (Xie et al. 2017; Yin et al. 2020).
- RL-based versus RL-free text generation
  GAN models were originally developed for learning from a continuous, not discrete distribution. However, the discrete nature of text input handicaps the use of GANs
  In GANs, a reinforcement learning algorithm is used for policy gradient, to get an unbiased gradient estimator for the generator and obtain the reward from the discriminator (Chen et al. 2018).

  - RL-based generation
    Reinforcement learning (RL) is a technique that can be used to train an agent to perform certain tasks. Due to its generality, reinforcement learning is studied in many disciplines.
    GAN models that use a discriminating module to guide the training of the generative module as a reinforcement learning policy has shown promising results in text generation (Guo et al. 2018). Various methods have been proposed in text generation via GAN (e.g. Lin et al. 2017; Rajeswar et al. 2017; Che et al. 2017; Yu et al. 2017).
    There are several models of RL, some of which were applied to sentence generation, e.g., actor-critic algorithm and deep Q-network, (e.g. Sutton et al. 2000; Guo 2015; Bahdanau et al. 2016).
    One optimization challenge with RL-based approaches is that they may yield high-variance gradient estimates (Maddison et al. 2016; Zhang et al. 2017).
  - RL free GANs for text generation
    Examples of models that use an alternative to RL:

    Latent space based solutions
    Continuous approximation of discrete sampling

    Those models apply a simple soft-argmax operator, or Gumbel-softmax trick to provide a continuous approximation of the discrete distribution on text.
    Examples of research efforts in this category include: TextGAN, (Zhang et al. 2017) and GumbelSoftmax GAN (GSGAN), (Kusner and Hernández-Lobato 2016; Jang et al. 2016; Maddison et al. 2016), FM-GAN, (Chen et al. 2018), GSGAN, (Kusner and Hernández-Lobato 2016), and RelGAN, (Nie et al. 2018).

Generated Adversarial Networks (GAN) are a system of components that include a discriminator that attempts to distinguish between genuine samples and generated samples. The focus in this paper is the generation of text. As this paper goes into details, the resulting text generation does not belong singularly to the text generator model but rather relies upon the other components in the system as well. This will become clear as the paper explores some of the research into discriminator feature sets.

There seems to be a continued interest in selected the appropriate token size of the of the generator output. The reasoning is that the size of the output needs to contain as much meaning as possible and there is still room for improvement in that regard. The issue appears to be in reshaping the model to work with so large a variation of characters that a sub-word or full word will contain. Improvement in this area will have positive gains on both sides of the adversarial game: the discriminator will provide more meaningful feedback to the generator when both models are working from a stronger platform. Following this idea is the question of scale; how can models continue to grow and maintain adequate computational speed?

6 Adversarial Training Techniques

Adversarial training is a method to help systems be more robust against adversarial attacks. Below are examples of some adversarial training techniques reported in literature.

- Fast Gradient Sign Method FGSM
  FGSM is used to add adversarial examples to the training process (Goodfellow et al. 2014; Wong et al. 2020). During training, part of the original samples is replaced with its corresponding adversarial samples generated using the model being trained.
  Kurakin et al. suggested to use Iterative FGSM, IFGSM, FGSM-LL or FGSM-Rand variants for adversarial training, in order to reduce the effect of label leaking (Kurakin et al. 2016). Their are also other variants of FGSM such as: Momentum Iterative Fast Gradient Sign Method (MI-FGSM) (Dong et al. 2018).
- PGD-based training
  Proposed by Madry et al. (2017). At each iteration all the original samples are replaced with their corresponding adversarial samples generated using the model being trained.
  PGD was enhanced using different efforts such as: Optimization tricks such as momentum to improve adversary, (Dong et al. 2018), Combination with other heuristic defenses such as matrix estimation, (Yang et al. 2019), Defensive Quantization, (Lin et al. 2019), Logit pairing, (Mosbach et al. 2018; Kannan et al. 2018), Thermometer Encoding, (Buckman et al. 2018), Feature Denoising, (Xie et al. 2019), Robust Manifold Defense, (Jalal et al. 2017), L2 nonexpansive nets, (Qian et al. 2018), Jacobian Regularization, (Jakubovitz and Giryes 2018), Univer-

sal Perturbation, (Shafahi et al. 2020), and Stochastic Activation Pruning (Dhillon et al. 2018).

As of today, training with a PGD adversary remains empirically robust (Wong et al. 2020)

- Jacobian-based saliency map approach (JSMA)

  JSMA is a gradient based white-box method that is proposed to use the gradient of loss with each class labels with respect to every component of the input (Papernot et al. 2016b). JSMA is useful for targeted miss-classification attacks (Chakraborty et al. 2018).

- Accelerating Adversarial Training

  The cost of adversarial training can be reduced by reusing adversarial examples and merging the inner loop of a PGD and gradient updates of the model parameters (Shafahi et al. 2019; Zhang et al. 2019)

- DAWNBench competition

  Some submission projects to DAWNBench competition have shown good performance results on CIFAR10 and ImageNet classifiers in comparison with research-reported training methods (Coleman et al. 2017; Wong et al. 2020).

As the field of adversarial examples continues to expand within NLP models better systems will get put in place. However, until then the field faces many challenges, chief among them being the closed source nature of research. As pointed out by Morris et al. the need for researchers to publish there research openly is of great importance, it will ensure less time is spent reproducing and reducing errors. Another major challenge in AML text, unlike with image adversarial examples there is no one standard to detect adversarial examples within text. This lack of a standardization allows for many different ideas to exist and many different algorithms to continue to pop-up. The issue with this is while many articles reviewed utilize the same datasets and models the algorithms themselves can be flawed but that can't be determined due to lack of standardization. The third and final issue which I see is that of NLP being a recently new area of adversarial attacks in regards to research. As more research is conducted in the field their will be more advancements, this will be due to new researchers comparing their work against these ground breaking articles. This comparison will ensure algorithms which the greatest efficacy is selected and used to train deep neural networks. This training will in the future ensure a higher degree of defensive and identification of threats. This defense will allow for deep neural networks and other machine learning networks to better serve applications while protecting themselves.

7 Generation Models/Tasks/Applications

Text generation refers to the process of automatic or programmable generation of text with no or least of human intervention. The sources utilized for such generation process can also vary based on the nature of the application. The types of applications from generating text in particular are growing. We will discuss just a few in this section.

7.1 Next-Word Prediction

For many applications that we use through our smart phones, or websites, next word prediction (NWP, also called auto-completion) is a typical NLP application. From a machine-learning perspective, NWP is a classical prediction problem where previous and current text can be the pool to extract the prediction model features and other parameters and the next word to predict is the target feature. Different algorithms are proposed to approach NWP problem such as term frequencies, artificial intelligence, n-grams, neural networks, etc.

7.2 Dialog Generation

Human-machine dialog generation/prediction is an essential topic of research in the field of NLP. It has many different applications in different domains. The quality and the performance of the process can widely vary based on available resources, training/pre-training and also efficiency.

Seq2seq neural networks have demonstrated impressive results on dialog generation (Vinyals and Le 2015; Chang et al. 2019). GANs are used in dialogue generations in several research publications (e.g. Li et al. 2016; Hamilton et al. 2017; Kannan et al. 2018; Nabeel et al. 2019)

7.3 Neural Machine Translation

Neural Machine Translation (NMT) is a learning approach for automated translation, with potentials to overcome weaknesses of classical phrase-based translation systems or statistical machine learning. The main difference is that NMT is based on a model not based on some patterns. NMT tries to replicate the functions of the human brain and assess content from various sources before generating output. Further enhancements on NMT were achieved using attention based neural machine translation.

One of the popular early open source NMTs is Systran: https://translate.systran. net/, the first NMT engine launched in 2016.Other examples include those of: Google Translate, Facebook, e-bay and Microsoft.

Adversarial NMT is introduced in which training of the NMT model is assisted by an adversary, an elaborately designed 2D-convolutional neural network (CNN) (Yang et al. 2017; Wu et al. 2018; Zhang et al. 2018; Shetty et al. 2018).

8   Text Generation Metrics

One of the key issues in text generation is that there is no widely agreed-upon automated metric for evaluating the text generated output. Text generation metrics can be classified based on several categories. Here is a summary of categories and metrics:

- Document Similarity based Metrics
  One of the popular approaches to measure output TG is through comparing it with some source documents or human natural language. Some of the popular metrics in this category are Bilingual Evaluation Understudy (BLEU), (Papineni et al. 2002) and Embedding Similarity (EmbSim) (Zhu et al. 2018b).
  BLEU has several variants such as BLEU-4 and BLEU-1.
  This category can also include some of the popular classical metrics such as:Okapi BM25 (Robertson and Walker 1994), Word Mover's Distance (WMD), (Kusner et al. 2015), Cosine, Dice and Jaccard measures in addition to Term Frequency-Inverse Document Frequency (TF-IDF).
- Likelihood-based Metrics
  Log-likelihood is the negative of the training loss function, (NLL). NLL (also known as multiclass cross-entropy) outputs a probability for each class, rather than just the most likely class. The typical approach in text generation is to train the model using a neural network performing maximum likelihood estimation (MLE) by minimizing the negative log-likelihood, NLL over the text corpus. For GANs, in the standard GAN objective, the goal or objective function is to minimize NLL for the binary classification task (Goodfellow et al. 2014).
  Maximum Likelihood suffers from predicting most probable answers. This means that a model trained with maximum likelihood will tend to output short general answers that are very common in the vocabulary. The log-likelihood improves with more dimensions as it is easier to fit the hypotheses in the training step having more dimensions. Consequently, the hypothesis in the generating step have lower log-likelihood.
- Perplexity
  Perplexity measures a model's certainty of its predictions.
  There are several advantages to using perplexity (Keukeleire 2020);

  - Calculating perplexity is simple and doesn't require human interference
  - It is easy to interpret
  - It is easy to optimize a model for an improved perplexity score

  Held-out likelihood is usually presented as perplexity, which is a deterministic transformation of the log-likelihood into an information-theoretic quantity
- Inception Score (IS)
  IS rewards high confidence class labels for each generated instance (Salimans et al. 2016). IS can provide a general evaluation of GANs trained on ImageNet. However, it has limited utility in other settings (Fowl et al. 2020).

- Frechet Inception Distance (FID)
  FID is used to measure the Wasserstein-2 distance, (Vaserstein 1969) between two Gaussians, whose means and covariances are taken from embedding both real and generated data (Heusel et al. 2017; Cífka et al. 2018). FID assumes that the training data is "sufficient" and does not reward producing more diversity than the training data (Fowl et al. 2020).
- N-gram based metrics
  Distinct-n is a measure of diversity that computes the number of distinct n-grams, normalized by the number of all ngrams (Li et al. 2015).
- Sentence similarity metrics, SentenceBERT (sent-BERT)
- ROUGE metrics
  ROUGE metrics were mostly used for text generation, video captioning and summarization tasks (Lin 2004). They were introduced in 2004 as a set of metrics to evaluate machine-generated text summaries. ROUGE has several variants such as: ROUGE-1, ROUGE-2 and ROUGE-L.
- METEOR
  METEOR (Metric for Evaluation of Translation with Explicit Ordering) was proposed in 2005 (Banerjee and Lavie 2005). METEOR metric was mainly used for text generation, image and video captioning, and question answering tasks
- Embedding-based metrics
  The main approach is to embed generated sentences in latent space and then evaluate them in this space (Tevet et al. 2018). Du and Black (2019) suggest to cluster the embedded sentences with k-means and then use its inertia as a measure for diversity.
- Distributional Discrepancy (DD) is a new metric proposed by Cai et al. (2021). The new DD metric is aimed to provide a better method for estimating the quality and diversity of generated text. DD is claimed to offer significantly better performance than BLEU, self-BLEU, and Fréchet. Due to the fact that it is not possible to compute a precise DD score (the actual distribution of real text cannot be obtained directly). The author Cai et al. (2021) proposes to use a text classifier trained with real and generated text to test for discrepancies between them. This classifier uses a learning method to estimate the DD by assessing whether or not two samples are taken from the same distribution.
- Other less common metrics such as: GLEU score, edit distance, phoneme and diacritic error rate.
- Metrics for GANs, traditional probability-based LM metrics, (Tevet et al. 2018) Several papers indicated the need to use new metrics to evaluate GANs (e.g. Esteban et al. 2017; Zhu et al. 2018b; Saxena and Cao 2019). Some of the metrics proposed for GANs include:

  - Divergence based Metrics such as F-GAN, (Nowozin et al. 2016), LS-GAN (Mao et al. 2017), KL-divergence (Koochali et al. 2019), and Self-BLEU, (Zhu et al. 2018b)
  - Integral Probability Metrics such as: Wasserstein GAN (WGA)' (Arjovsky et al. 2017; Gulrajani et al. 2017).

– Domain-specific metrics, e.g. attack success rate (Gao et al. 2020).
– Random Network Distillation (RND) (Burda et al. 2018).

The clear trend is that the evaluation auto-generated text is yielding more accurate assessments which result in higher quality generated text. Trend is split between BERT based and CNN/feature based methods for determining metrics. Based on the papers, non-BERT based methods (UNION and Perception Score) seem to outperform MoverScore, BERTScore, and BLEURT. There are several ethical implications regarding trying to mimic humans. One addressed, by BOLD, which has the aim to try to gauge stereotypes and racial biases. The other is in the reward of better metrics which ultimately will make it more difficult to distinguish between human and machine text.

There are several challenges how machine to translate accurately and how to measure quality of MT evaluation. Examples of those include:

• Quality benchmarks: In order to properly evaluate text generation metrics, their is a need for a corpus of good quality human reference translations. Like the saying quantity leads to quality is an answer to improve machine translation. In case of ROUGE, it needs to research how to achieve high correlation score with human translation in single document summarization.
• Metrics value/usefulness: Once we have a common and standard metric, we can show it as an intuitive graphical representation for an easy and clear interpretation. The parameters of penalty function in METETOR are based on empirical tests of development data, this can be further trained to be optimized. BLEU is proven that it can highly correlated with human references in MT evaluation, it can be adapted in NLG tasks as a common MT evaluation. It can apply in NLG task as a next step. Labelled dependencies showed high corrections with human judgement but the comparison was only tentative. It requires more experiment with a different test set. Using f-structures require further research to enhance for potentially an accurate automated MT evaluation metrics. Above all more experiments are needed with more test sets and further research to fine tune these methods to be reliable and common in MT community.

9   Memory-Based Models

Machine learning algorithms for text generation in adversarial machine learning in text analysis and generation has become a major topic in the study of machine learning. The biggest category in this is Recurrent Neural Networks (RNN) and within RNNs there are two main contributions that are algorithms for this: Long-Short-Term Memory (LSTM) as well as Gated Recurrent Units (GRUs). Both of which address different problems and have advantages over traditional RNNs. As we mentioned earlier, vanilla RNNs do not perform well when the learning sequences have long term temporal dependence due to issues such as exploding gradients (Bengio et al. 2015).

Alternatively, Convolutional neural networks (CNNs), recurrent neural networks (RNNs), Gated recurrent unit, (GRU) and Long-short term memory (LSTM) models are effective approaches in the field of sequential modeling methods. The design of the forget gate is the essence of these models (Sun et al. 2020).

An LSTM model is a type of RNN that can remember relevant information longer than a regular RNN. As a result, they can better learn long-term patterns (Olah 2015).

LSTM models provide a mechanism that is able to both store and discard the information saved about the previous steps, limiting the accumulated error using Constant Error Carousels (Hochreiter and Schmidhuber 1997; Manzelli et al. 2018).

An RNN is a form of a neural network, one that is best represented by a graph of nodes that allows previous outputs to then be used as inputs. RNN uses a concept called back-propagation that connects each node in the network, making the network a long sequence of nodes. This is an issue because the RNN can suffer from the vanishing gradient problem and using LSTM and GRUs algorithms can help solve this problem. The gradient which is a major part of neural networks is the value that is uses to update and modify the weights in the given neural network that trains the model. The vanishing gradient problem is an issue because the gradient in the neural network shrinks as it back-propagates as time goes on, this means the gradient becomes very small which entails it does not give as much in terms of learning. These are usually in the earlier layers, which leads to the RNN to forget the earlier data in longer sequences, it has a short-term memory in other words. LSTM is an algorithm based on the use of a cell state along with numerous gates, the cell state moves the relative information through the sequence chain. There are two types of functions in the algorithm, Sigmoid which squish values between 0 and 1 and Tanh that squish values between -1 and 1. In terms of gates for this algorithm there is the input gate, the output gate and the forget gate which all use the Sigmoid function. For the forget gate, its main purpose is to decide what information, if any, should the model throw away, information from the current input gate and previous hidden state are passed through this gate. Since this is a Sigmoid function, values are evaluated to be between 0 and 1 where 0 means the model should forget and values that are closer to 1 means the model should keep them. This is the same for the input gate except it is using both Sigmoid and Tanh functions as well as for the output gate that also uses a Sigmoid function combined with a Tanh function that gets its information from the cell state that uses the output from the forget gate and input gate. GRU is very similar to the LSTM except it did away with the cell state and now uses what it calls a hidden state to transfer the relevant processed information. This algorithm features two gates: the update gate and the reset gate. For the update gate, it replaces the input and forget gate in the LSTM deciding what information it should throw away as well as what information it should add. The reset gate objective is to figure out how much past information to forget, both gates just use the Sigmoid function. By this architecture using less operations it does have an improvement in terms of training speed (Yang et al. 2020). With this it was found in "LSTM versus GRU versus Bidirectional RNN for script generation" that this improved training speed might come at a cost. When both models were trained on movie and TV series scripts, the LSTM model was found to be more efficient in the actual generation of text.

Within these two algorithms there have been numerous contributions on improving them. For example, there are models like Texar (Hu et al. 2018). Texar is a special model for LSTM and GRU architectures that aims to transform any given input into a natural language. There is also Grid LSTMs that's goal is to allow hidden layers at any step in the LSTM. With all, the results for each model vary based on the implementation and dataset used to train each model. For example, in Chung et al. (2014), both an LSTM and GRU model were trained on music and speech datasets and it was found that the GRU was superior is terms of accuracy and less loss which contradicts previous papers that state the LSTM was better in these areas. To really sum up all the research surrounding this topic it comes down to having to run both with the given datasets and see which one is better in terms of the trade-offs, are you are willing to make it such that it has faster training times, more accuracy or less loss at the cost of something else.

10 Summary and Conclusion

Adversarial Machine Learning (AML) in text analysis picked up rapidly in recent years beyond classical image-based AML applications. While their are many similarities between imaged-based AML and text-based AML, yet their are many unique issues and challenges as well. In this paper, we evaluated recent trends and issues in AML text analysis. We noticed the rapid growth of applications for AML text specially in the context of Online Social Networks (OSNs). We should notice here the difference between automatic text generation applications for malicious versus non-malicious goals. Automate, hide and mislead are three major factors that can divide text generation automation into several categories between marketing, political, social and other goals.

References

N. Akhtar, A. Mian, Threat of adversarial attacks on deep learning in computer vision: a survey. IEEE Access **6**, 14410–14430 (2018)

M. Alzantot, Y. Sharma, A. Elgohary, B.-J. Ho, M. Srivastava, K.-W. Chang, Generating natural language adversarial examples (2018). arXiv:1804.07998

M. Arjovsky, S. Chintala, L. Bottou, Wasserstein gan (2017). arXiv:1701.07875

D. Bahdanau, P. Brakel, K. Xu, A. Goyal, R. Lowe, J. Pineau, A. Courville, Y. Bengio, An actor-critic algorithm for sequence prediction (2016). arXiv:1607.07086

S. Baki, R. Verma, A. Mukherjee, O. Gnawali, Scaling and effectiveness of email masquerade attacks: exploiting natural language generation, in *Proceedings of the 2017 ACM on Asia Conference on Computer and Communications Security* (2017), pp. 469–482

S. Banerjee, A. Lavie, Meteor: an automatic metric for mt evaluation with improved correlation with human judgments, in *Proceedings of the acl Workshop on Intrinsic and Extrinsic Evaluation Measures for Machine Translation and/or Summarization* (2005), pp. 65–72

M. Barreno, B. Nelson, A.D. Joseph, J.D. Tygar, The security of machine learning. Mach. Learn. **81**(2), 121–148 (2010)

V. Behzadan, A. Munir, Vulnerability of deep reinforcement learning to policy induction attacks, in *International Conference on Machine Learning and Data Mining in Pattern Recognition* (Springer, 2017), pp. 262–275

S. Bengio, O. Vinyals, N. Jaitly, N. Shazeer, Scheduled sampling for sequence prediction with recurrent neural networks. Adv. Neural Inf. Proc. Syst. **28**, 1171–1179 (2015)

A. Bennett, N. Kallus, T. Schnabel, Deep generalized method of moments for instrumental variable analysis, in *Advances in Neural Information Processing Systems* (2019), pp. 3564–3574

B. Biggio, G. Fumera, G.L. Marcialis, F. Roli, Security of pattern recognition systems in adversarial environments (2012)

J. Buckman, A. Roy, C. Raffel, I. Goodfellow, Thermometer encoding: one hot way to resist adversarial examples, in *International Conference on Learning Representations* (2018)

Y. Burda, H. Edwards, A. Storkey, O. Klimov, Exploration by random network distillation (2018). arXiv:1810.12894

P. Cai, X. Chen, P. Jin, H. Wang, T. Li, Distributional discrepancy: a metric for unconditional text generation. Knowl.-Based Syst. **217**, 106850 (2021)

A. Chakraborty, M. Alam, V. Dey, A. Chattopadhyay, D. Mukhopadhyay, Adversarial attacks and defences: a survey (2018). arXiv:1810.00069

J. Chang, R. He, L. Wang, X. Zhao, T. Yang, R. Wang, A semi-supervised stable variational network for promoting replier-consistency in dialogue generation, in *Proceedings of the 2019 Conference on Empirical Methods in Natural Language Processing and the 9th International Joint Conference on Natural Language Processing (EMNLP-IJCNLP)* (2019), pp. 1920–1930

T. Che, Y. Li, R. Zhang, R.D. Hjelm, W. Li, Y. Song, Y. Bengio, Maximum-likelihood augmented discrete generative adversarial networks (2017). arXiv:1702.07983

L. Chen, S. Dai, C. Tao, H. Zhang, Z. Gan, D. Shen, Y. Zhang, G. Wang, R. Zhang, L. Carin, Adversarial text generation via feature-mover's distance, in *Advances in Neural Information Processing Systems* (2018), pp. 4666–4677

J. Chung, C. Gulcehre, K.H. Cho, Y. Bengio, Empirical evaluation of gated recurrent neural networks on sequence modeling (2014). arXiv:1412.3555

O. Cífka, A. Severyn, E. Alfonseca, K. Filippova, Eval all, trust a few, do wrong to none: comparing sentence generation models (2018). arXiv:1804.07972

C. Coleman, D. Narayanan, D. Kang, T. Zhao, J. Zhang, L. Nardi, P. Bailis, K. Olukotun, C. Ré, M. Zaharia, Dawnbench: an end-to-end deep learning benchmark and competition. Training **100**(101), 102 (2017)

A. Creswell, T. White, V. Dumoulin, K. Arulkumaran, B. Sengupta, A.A. Bharath, Generative adversarial networks: an overview. IEEE Signal Proc. Mag. **35**, 53–65 (2018)

G.S. Dhillon, K. Azizzadenesheli, Z.C. Lipton, J. Bernstein, J. Kossaifi, A. Khanna, A. Anandkumar, Stochastic activation pruning for robust adversarial defense (2018). arXiv:1803.01442

Y. Dong, F. Liao, T. Pang, H. Su, J. Zhu, X. Hu, J. Li, Boosting adversarial attacks with momentum, in *Proceedings of the IEEE Conference on Computer Vision and Pattern Recognition* (2018), pp. 9185–9193

W. Du, A.W. Black, Boosting dialog response generation, in *Proceedings of the 57th Annual Meeting of the Association for Computational Linguistics* (2019), pp. 38–43

J. Ebrahimi, A. Rao, D. Lowd, D. Dou, Hotflip: white-box adversarial examples for text classification (2017). arXiv:1712.06751

C. Esteban, S.L. Hyland, G. Rätsch, Real-valued (medical) time series generation with recurrent conditional gans (2017). arXiv:1706.02633

J.A. Fodor, Z.W. Pylyshyn et al., Connectionism and cognitive architecture: a critical analysis. Cognition **28**(1–2), 3–71 (1988)

L. Fowl, M. Goldblum, A. Gupta, A. Sharaf, T. Goldstein, Random network distillation as a diversity metric for both image and text generation (2020). arXiv:2010.06715

J. Gao, J. Lanchantin, M.L. Soffa, Y. Qi, Black-box generation of adversarial text sequences to evade deep learning classifiers, in *2018 IEEE Security and Privacy Workshops (SPW)* (IEEE, 2018), pp. 50–56

N. Gao, H. Xue, W. Shao, S. Zhao, K.K. Qin, A. Prabowo, M.S. Rahaman, F.D. Salim, Generative adversarial networks for spatio-temporal data: a survey (2020). arXiv:2008.08903

I.J. Goodfellow, J. Shlens, C. Szegedy, Explaining and harnessing adversarial examples (2014). arXiv:1412.6572

I. Goodfellow, J. Pouget-Abadie, M. Mirza, X. Bing, D. Warde-Farley, S. Ozair, A. Courville, Y. Bengio, Generative adversarial networks. Commun. ACM **63**(11), 139–144 (2020)

I. Gulrajani, F. Ahmed, M. Arjovsky, V. Dumoulin, A.C. Courville, Improved training of wasserstein gans, in *Advances in Neural Information Processing Systems* (2017), pp. 5767–5777

H. Guo, Generating text with deep reinforcement learning (2015). arXiv:1510.09202

J. Guo, S. Lu, H. Cai, W. Zhang, Y. Yu, J. Wang, Long text generation via adversarial training with leaked information, in *Proceedings of the AAAI Conference on Artificial Intelligence*, vol. 32 (2018)

A. Haidar, M. Rezagholizadeh, Textkd-gan: text generation using knowledge distillation and generative adversarial networks, in *Canadian Conference on Artificial Intelligence* (Springer, 2019), pp. 107–118

W. Hamilton, Z. Ying, J. Leskovec, Inductive representation learning on large graphs, in *Advances in neural information processing systems* (2017), pp. 1024–1034

L.P. Hansen, Large sample properties of generalized method of moments estimators. Econ.: J. Econ. Soc. 1029–1054 (1982)

M. Heusel, H. Ramsauer, T. Unterthiner, B. Nessler, S. Hochreiter, Gans trained by a two time-scale update rule converge to a local nash equilibrium, in *Advances in Neural Information Processing Systems* (2017), pp. 6626–6637

G.E. Hinton, S. Osindero, Y.-W. Teh, A fast learning algorithm for deep belief nets. Neural Comput. **18**(7), 1527–1554 (2006)

S. Hochreiter, J. Schmidhuber, Long short-term memory. Neural Comput. **9**(8), 1735–1780 (1997)

Z. Hu, H. Shi, B. Tan, W. Wang, Z. Yang, T. Zhao, J. He, L. Qin, D. Wang, X. Ma, et al., Texar: a modularized, versatile, and extensible toolkit for text generation (2018). arXiv:1809.00794

R. Huang, B. Xu, D. Schuurmans, C. Szepesvári, Learning with a strong adversary (2015). arXiv:1511.03034

D. Jakubovitz, R. Giryes, Improving dnn robustness to adversarial attacks using jacobian regularization, in *Proceedings of the European Conference on Computer Vision (ECCV)* (2018), pp. 514–529

A. Jalal, A. Ilyas, C. Daskalakis, A.G. Dimakis, The robust manifold defense: adversarial training using generative models (2017). arXiv:1712.09196

E. Jang, S. Gu, B. Poole, Categorical reparameterization with gumbel-softmax (2016). arXiv:1611.01144

H. Kannan, A. Kurakin, I. Goodfellow, Adversarial logit pairing (2018). arXiv:1803.06373

P. Keukeleire, Correspondence between perplexity scores and human evaluation of generated tv-show scripts (2020)

D.P. Kingma, M. Welling, Auto-encoding variational bayes (2013). arXiv:1312.6114

A. Koochali, P. Schichtel, A. Dengel, S. Ahmed, Probabilistic forecasting of sensory data with generative adversarial networks-forgan. IEEE Access **7**, 63868–63880 (2019)

A. Kurakin, I. Goodfellow, S. Bengio, Adversarial machine learning at scale (2016). arXiv:1611.01236

M.J. Kusner, J.M. Hernández-Lobato, Gans for sequences of discrete elements with the gumbel-softmax distribution (2016). arXiv:1611.04051

M.J. Kusner, Y. Sun, N.I. Kolkin, K.Q. Weinberger, From word embeddings to document distances, in *Proceedings of the 32nd International Conference on International Conference on Machine Learning*, vol. 37, JMLR.org (2015), pp. 957–966

A.M. Lamb, A.G.A.P. Goyal, Y. Zhang, S. Zhang, A.C. Courville, Y. Bengio, Professor forcing: a new algorithm for training recurrent networks, in *Advances in Neural Information Processing Systems* (2016), pp. 4601–4460

G. Lewis, V. Syrgkanis, Adversarial generalized method of moments (2018). arXiv:1803.07164

J. Li, M. Galley, C. Brockett, J. Gao, B. Dolan, A diversity-promoting objective function for neural conversation models (2015). arXiv:1510.03055

J. Li, W. Monroe, A. Ritter, M. Galley, J. Gao, D. Jurafsky, Deep reinforcement learning for dialogue generation (2016). arXiv:1606.01541

B. Liang, H. Li, M. Su, P. Bian, X. Li, W. Shi, Deep text classification can be fooled (2017). arXiv:1704.08006

C.-Y. Lin, Rouge: a package for automatic evaluation of summaries, in *Text Summarization Branches Out* (2004), pp. 74–81

J. Lin, C. Gan, S. Han, Defensive quantization: when efficiency meets robustness (2019). arXiv:1904.08444

K. Lin, D. Li, X. He, Z. Zhang, M.-T. Sun, Adversarial ranking for language generation, in *Advances in Neural Information Processing Systems* (2017), pp. 3155–3165

C.J. Maddison, A. Mnih, Y.W. Teh, The concrete distribution: a continuous relaxation of discrete random variables (2016). arXiv:1611.00712

A. Madry, A. Makelov, L. Schmidt, D. Tsipras, A. Vladu, Towards deep learning models resistant to adversarial attacks (2017). arXiv:1706.06083

R. Manzelli, V. Thakkar, A. Siahkamari, B. Kulis, An end to end model for automatic music generation: combining deep raw and symbolic audio networks, in *Proceedings of the Musical Metacreation Workshop at 9th International Conference on Computational Creativity, Salamanca, Spain* (2018)

X. Mao, Q. Li, H. Xie, R.Y.K. Lau, Z. Wang, S. Paul Smolley, Least squares generative adversarial networks, in *Proceedings of the IEEE International Conference on Computer Vision* (2017), pp. 2794–2802

M. Mosbach, M. Andriushchenko, T. Trost, M. Hein, D. Klakow, Logit pairing methods can fool gradient-based attacks (2018). arXiv:1810.12042

Y. Mroueh, T. Sercu, Fisher gan, in *Advances in Neural Information Processing Systems* (2017), pp. 2513–2523

M. Nabeel, A. Riaz, W. Zhenyu, *Cas-Gans: An Approach of Dialogue Policy Learning Based on Gan and Rl Techniques* (Int. J. Adv. Comput. Sci, Appl, 2019)

W. Nie, N. Narodytska, A. Patel, Relgan: relational generative adversarial networks for text generation, in *International Conference on Learning Representations* (2018)

S. Nowozin, B. Cseke, R. Tomioka, f-gan: training generative neural samplers using variational divergence minimization, in *Advances in Neural Information Processing Systems* (2016), pp. 271–279

C. Olah, Understanding lstm networks (2015)

N. Papernot, P. McDaniel, X. Wu, S. Jha, A. Swami, Distillation as a defense to adversarial perturbations against deep neural networks, in *2016 IEEE Symposium on Security and Privacy* (IEEE, 2016a), pp. 582–597

N. Papernot, P. McDaniel, S. Jha, M. Fredrikson, Z.B. Celik, A. Swami, The limitations of deep learning in adversarial settings, in *2016 IEEE European symposium on security and privacy (EuroS&P)* (IEEE, 2016b), pp. 372–387

K. Papineni, S. Roukos, T. Ward, W.-J. Zhu, Bleu: a method for automatic evaluation of machine translation, in *Proceedings of the 40th Annual Meeting of the Association for Computational Linguistics* (2002), pp. 311–318

K. Pearson, Asymmetrical frequency curves. Nature **48**(1252), 615–616 (1893)

Q. Qian, M. Huang, H. Zhao, J. Xu, X. Zhu, Assigning personality/profile to a chatting machine for coherent conversation generation, in *IJCAI* (2018), pp. 4279–4285

L.R. Rabiner, A tutorial on hidden markov models and selected applications in speech recognition. Proc. IEEE **77**(2), 257–286 (1989)

S. Rajeswar, S. Subramanian, F. Dutil, C. Pal, A. Courville, Adversarial generation of natural language (2017). arXiv:1705.10929

M. Ranzato, S. Chopra, M. Auli, W. Zaremba, Sequence level training with recurrent neural networks (2015). arXiv:1511.06732

S.E. Robertson, S. Walker, Some simple effective approximations to the 2-poisson model for probabilistic weighted retrieval, in *Proceedings of the 17th Annual International ACM SIGIR Conference on Research and Development in Information Retrieval* (1994), pp. 232–241

I. Rosenberg, A. Shabtai, Y. Elovici, L. Rokach, Defense methods against adversarial examples for recurrent neural networks (2019). arXiv:1901.09963

T. Salimans, I. Goodfellow, W. Zaremba, V. Cheung, A. Radford, X. Chen, Improved techniques for training gans (2016). arXiv:1606.03498

D. Saxena, J. Cao, D-gan: deep generative adversarial nets for spatio-temporal prediction (2019). arXiv:1907.08556

A. Shafahi, M. Najibi, M.A. Ghiasi, Z. Xu, J. Dickerson, C. Studer, L.S. Davis, G. Taylor, T. Goldstein, Adversarial training for free!, in *Advances in Neural Information Processing Systems* (2019), pp. 3358–3369

A. Shafahi, M. Najibi, Z. Xu, J.P. Dickerson, L.S. Davis, T. Goldstein, Universal adversarial training, in *AAAI* (2020), pp. 5636–5643

R. Shetty, B. Schiele, M. Fritz, A4nt: author attribute anonymity by adversarial training of neural machine translation, in *27th {USENIX} Security Symposium {USENIX}* Security, vol. 18 (2018), pp. 1633–1650

A. Sun, J. Wang, N. Cheng, H. Peng, Z. Zeng, L. Kong, J. Xiao, Graphpb: graphical representations of prosody boundary in speech synthesis (2020). arXiv:2012.02626

R.S. Sutton, D.A. McAllester, S.P. Singh, Y. Mansour, Policy gradient methods for reinforcement learning with function approximation, in *Advances in Neural Information Processing Systems* (2000), pp. 1057–1063

C. Szegedy, W. Zaremba, I. Sutskever, J. Bruna, D. Erhan, I. Goodfellow, R. Fergus, Intriguing properties of neural networks (2013). arXiv:1312.6199

G. Tevet, G. Habib, V. Shwartz, J. Berant, Evaluating text gans as language models (2018). arXiv:1810.12686

F. Tramèr, A. Kurakin, N. Papernot, I. Goodfellow, D. Boneh, P. McDaniel, Ensemble adversarial training: attacks and defenses (2017). arXiv:1705.07204

D. Tsipras, S. Santurkar, L. Engstrom, A. Turner, A. Madry, Robustness may be at odds with accuracy (2018). arXiv:1805.12152

L.N. Vaserstein, Markov processes over denumerable products of spaces, describing large systems of automata. Problemy Peredachi Informatsii **5**(3), 64–72 (1969)

O. Vinyals, Q. Le, A neural conversational model (2015). arXiv:1506.05869

E. Wallace, T.Z. Zhao, S. Feng, S. Singh, Concealed data poisoning attacks on nlp models (2021)

X. Wang, H. Jin, K. He, Natural language adversarial attacks and defenses in word level (2019). arXiv:1909.06723

R.J. Williams, D. Zipser, A learning algorithm for continually running fully recurrent neural networks. Neural Comput. **1**(2), 270–280 (1989)

E. Wong, L. Rice, J.Z. Kolter, Fast is better than free: revisiting adversarial training (2020). arXiv:2001.03994

E. Wong, F. Schmidt, J.H. Metzen, J.Z. Kolter, Scaling provable adversarial defenses, in *Advances in Neural Information Processing Systems* (2018), pp. 8400–8409

L. Wu, Y. Xia, F. Tian, L. Zhao, T. Qin, J. Lai, T.-Y. Liu, Adversarial neural machine translation, in *Asian Conference on Machine Learning* (PMLR, 2018), pp. 534–549

C. Xie, J. Wang, Z. Zhang, Z. Ren, A. Yuille, Mitigating adversarial effects through randomization (2017). arXiv:1711.01991

C. Xie, Y. Wu, L. van der Maaten, A.L. Yuille, K. He, Feature denoising for improving adversarial robustness, in *Proceedings of the IEEE Conference on Computer Vision and Pattern Recognition* (2019), pp. 501–509

Z. Yan, Y. Guo, C. Zhang, Deep defense: training dnns with improved adversarial robustness. Adv. Neural Inf. Proc. Syst. **31**, 419–428 (2018)

Z. Yang, W. Chen, F. Wang, B. Xu, Improving neural machine translation with conditional sequence generative adversarial nets (2017). arXiv:1703.04887

S. Yang, X. Yu, Y. Zhou, Lstm and gru neural network performance comparison study: taking yelp review dataset as an example, in *2020 International Workshop on Electronic Communication and Artificial Intelligence (IWECAI)* (IEEE, 2020), pp. 98–101

Y. Yang, G. Zhang, D. Katabi, Z. Xu, Me-net: towards effective adversarial robustness with matrix estimation (2019). arXiv:1905.11971

Y. Yao, B. Viswanath, J. Cryan, H. Zheng, B.Y. Zhao, Automated crowdturfing attacks and defenses in online review systems, in *Proceedings of the 2017 ACM SIGSAC Conference on Computer and Communications Security* (2017), pp. 1143–1158

H. Yin, D. Li, X. Li, P. Li, Meta-cotgan: a meta cooperative training paradigm for improving adversarial text generation, in *AAAI* (2020), pp. 9466–9473

F. Yu, Z. Xu, Y. Wang, C. Liu, X. Chen, Towards robust training of neural networks by regularizing adversarial gradients (2018). arXiv:1805.09370

L. Yu, W. Zhang, J. Wang, Y. Yu, Seqgan: sequence generative adversarial nets with policy gradient, in *Proceedings of the AAAI Conference on Artificial Intelligence*, vol. 31 (2017)

Y. Zhang, Z. Gan, K. Fan, Z. Chen, R. Henao, D. Shen, L. Carin, Adversarial feature matching for text generation (2017). arXiv:1706.03850

Z. Zhang, S. Liu, M. Li, M. Zhou, E. Chen, Bidirectional generative adversarial networks for neural machine translation, in *Proceedings of the 22nd Conference on Computational Natural Language Learning* (2018), pp. 190–199

D. Zhang, T. Zhang, Y. Lu, Z. Zhu, B. Dong, You only propagate once: accelerating adversarial training via maximal principle, in *Advances in Neural Information Processing Systems* (2019), pp. 227–238

Z. Zhou, H. Guan, M.M. Bhat, J. Hsu, Fake news detection via nlp is vulnerable to adversarial attacks (2019). arXiv:1901.09657

J. Zhu, R. Kaplan, J. Johnson, L. Fei-Fei, Hidden: hiding data with deep networks. In *Proceedings of the European conference on computer vision (ECCV)* (2018a), pp. 657–672

Y. Zhu, S. Lu, L. Zheng, J. Guo, W. Zhang, J. Wang, Y. Yu, Texygen: a benchmarking platform for text generation models, in *The 41st International ACM SIGIR Conference on Research & Development in Information Retrieval* (2018b), pp. 1097–1100

Multi-agent Systems for Distributed Data Mining Techniques: An Overview

Mais Haj Qasem, Amjad Hudaib, Nadim Obeid,
Mohammed Amin Almaiah, Omar Almomani, and Ahmad Al-Khasawneh

Abstract The term "multi-agent systems" (MAS) refers to a mechanism that is used to create goal-oriented autonomous agents in a shared environment and have communication and coordination capabilities. This goal-oriented mechanism supports distributed data mining (DM) to implement various techniques for distributed clustering, classification, and prediction. Different distributed DM (DDM) techniques, MASs, the advantages of MAS-based DDM, and various MAS-based DDM approaches proposed by researchers are reviewed in this study.

Keywords Distributed data mining · FIPA standards · Multi-agent system

1 Introduction

Data mining (DM) is a process that is used to analyze data, determine data patterns, and predict future trends from previously analyzed data. Training data for DM techniques is useful only when data can be accessed by the running algorithm and therefore should be located physically or virtually in a single unit to increase computation cost and maintain data privacy. Computation of data from different geographical locations is a future perspective in DM and is referred to as distributed DM (DDM).

M. H. Qasem · A. Hudaib · N. Obeid
King Abdullah II School for Information Technology, The University of Jordan, Amman, Jordan

M. A. Almaiah (✉)
Department of Computer Networks and Communications, King Faisal University, Al-Ahas, Saudi Arabia
e-mail: malmaiah@kfu.edu.sa

O. Almomani
Network Computer and Information Systems Department, the World Islamic Sciences & Education University, Amman, Jordan

A. Al-Khasawneh
Department of Computer Information Systems, Hashemite University, Zarqa, Jordan

© The Author(s), under exclusive license to Springer Nature Switzerland AG 2022
Y. Baddi et al. (eds.), *Big Data Intelligence for Smart Applications*,
Studies in Computational Intelligence 994,
https://doi.org/10.1007/978-3-030-87954-9_3

As distributed data, cloud computing, and multiagent systems (MASs) increase and microprocessor devices, such as mobiles and sensors, become used widely, the transfer of data acquired from multiple parties into a centralized data warehouse becomes difficult (Tsoumakas and Vlahavas 2009). The transfer of some secure data, such as financial records, is impossible due to exposure risk. The sharing or transfer of other data, such as medical data, to other locations is also impossible. The data mining task consequently turns into distribution in nature.

An option for the aforementioned cases is to share learned elements into a DDM environment. DDM analyzes distributed data and provides algorithmic solutions to implement different data analysis and mining operations in a distributed manner in consideration of resource constraints. Patters are determined and multiple distributed data sources are utilized to predict. DM problems that require placing data in a single unit are also solved by DDM. The literature has presented various DDM techniques, such as distributed clustering, distributed frequent pattern mining and distributed classification. MASs are included in distributed systems; thus, DDM is combined with MASs for data-intensive applications.

MASs are efficient in distributed problem solving. Their communication and coordination capabilities allow the creation goal-oriented autonomous agents that operate in a shared environment. The agents have their own problem-solving capabilities and can interact among one another to achieve a general goal. They also demonstrate flexible and pro-active behavior.

MASs maintain a pool of agents on a global level and deploy such agents to each distributed data site. Thus, they reduce computation time and memory cost and are suitable for DDM. Agents can also adapt to errors and faults in the entire system by connecting with all distributed data sites. Tasks that are difficult for a single agent to achieve are completed by the multiple agents in MASs. The success of MASs, as an artificial intelligence technology, depends on how it coordinates the behavior of individual agents (Obeid and Moubaiddin 2010).

The MAS technology in DDM reduces network traffic to enhance efficiency and scalability (only codes, not the entire data, are shared among distributed sites to reduce bandwidth) and advances the system without increasing its complexity; agents react dynamically, adapt easily to the changing environment, operate asynchronously (they disconnect from and reconnect to the network automatically after their tasks), and are fault-tolerant (they avoid a fault-distributed data site and approach a reliable distributed data site) (Sawant and Shah 2013).

Agents are autonomous, not needing guidance when operating, and are able to control their internal state and actions. They also make the decision on whether to perform a requested action. They fulfill their objectives with their flexible problem-solving characteristic. Autonomous agents can interact with other agents by using a specific communication language. Accordingly, they may participate in large social networks and respond to changes and/or achieve goals with the help of other agents.

Agents typically operate in a dynamic, nondeterministic complex environment. In MAS environments, no assumptions on global control, data centralization, and synchronization are established. Agents operate with incomplete information or capabilities to solve problems. They collect information, share such information, coordinate their actions, and enhance interoperation via communication. Agents interact for such reasons as requests for information, particular services or actions to be performed by others, and issues on cooperation, coordination, and/or negotiation to organize interdependent activities.

Such interactions occur mainly to complete individual objectives and take place at the knowledge level (Obeid and Moubaiddin 2010). The interactions depend on which goals should be followed, at what time, and by which agent. Flexibility is required, considering the agents have only partial control and knowledge of the environment where they operate. Runtime decisions on whether to initiate interactions and the nature and scope of such interactions are crucial.

The communication protocols used in MASs control the interactions and communications of agents. The external behavior of agents is determined by the protocols through their inside structure without the need to focus on their internal behavior. These specifications and the task implemented by the underlying MAS define the specification of the internal behavior of agents.

Various MAS-based DDM approaches have been proposed. The capabilities of agents to build their own learned model and control the transfer of their learned information or output results to other agents are among the advantages of such approaches. A combination of the results of multiple classifiers may yield an excellent output, and the accuracy of each agent can be maintained.

This paper discusses different DDM techniques and reviews MASs. The combination of DDM with MASs for data-intensive applications, the advantages of MAS-based DDM, and the various MAS-based DDM approaches presented by other researchers are elaborated. The remainder of this paper is divided into several sections. Section 2 provides an overview of the DDM approach and its related works. Section 3 discusses the MAS approach and its related works. Section 5 presents works on agent-based DDM.

2 Distributed Data Mining

DM identifies hidden data patterns and categorizes them to obtain useful information. The information is then collected in common areas, such as data warehouses, and analyzed using DM techniques, including pattern matching, clustering, rule association, regression, and classification (Chiang et al. 2011). DM algorithms are utilized in business decision-making and other processes that require information to decrease costs and increase revenues. Data are assumed to be centralized, memory-resident, and static in traditional DM algorithms (Gan et al. 2017).

The collection of distributed data in a data warehouse is required in the application of the classical knowledge discovery process in distributed environments for central processing (Tsoumakas and Vlahavas 2009). This process, however, is usually either ineffective or infeasible because of the following reasons:

1. **High storage cost**: A large data warehouse is required to store all data in a central system.
2. **High communication cost**: The transfer of large data volumes over networks requires considerable time and financial resources. Wireless network environments with limited bandwidth may encounter problems when transferring even a small volume of data.
3. **High computational cost**: Mining a central data warehouse incurs a higher cost than the sum of the cost of analyzing small parts of data.
4. **Risk of exposure of private and sensitive data**: DM applications handle sensitive data, such as people's medical and financial records. Central collection may jeopardize the privacy of data.

The advancement of the Internet has led to the rapid generation of a large amount of data, which are difficult to process even by supercomputers. Such data are stored in multiple locations, requiring costly centralization. Data centralization is also affected by bandwidth limitation and privacy concerns. DDM has gained considerable attention to solve such problems (Liu et al. 2011).

DDM mines data sources regardless of their physical locations. This characteristic is crucial, given that locally produced data may not be shared across the network because of their large amount and privacy issues. As the Internet advances, the data from businesses increase rapidly. Such data may not be shared or transferred because of security and privacy issues, which involve legislation of particular locales and system architecture (Moemeng et al. 2009). The decentralized architecture of DDM can reach every networked business; hence, DDM has become a critical component of knowledge-based systems. The business intelligence market is one of the fastest growing and most profitable areas in the software industry, which consequently leads to the popularization of DDM.

The need to extract information from decentralized data sources has given rise to DDM. DM techniques must be dynamic because changes in the system can influence its overall performance. DDM is a rapidly expanding area that aims to find data patterns in an environment with distributed data and computation. Current data analysis systems require centralized storage of data; nevertheless, the combination of computation with communication needs DDM environments that can utilize distributed computation (Li et al. 2003).

DDM has evolved mainly due to factors such as privacy of sensitive data, transmission cost, computation cost, and memory cost. DDM aims to extract useful information from data located at heterogeneous sites. Distributed computing comprises distributed sites that have computing units situated at individual heterogeneous points. A decentralized mining strategy that makes the entire working system scalable by distributing workload across heterogeneous sites is followed by DDM (Sawant and Shah 2013). Figure 1 shows the typical DM architecture versus the DDM working architecture used to construct a data warehouse.

The advances in computing and communication, including the Internet, intranets, local area networks, and wireless networks, have resulted in many pervasive distributed environments, which comprise various sources of large data and multiple

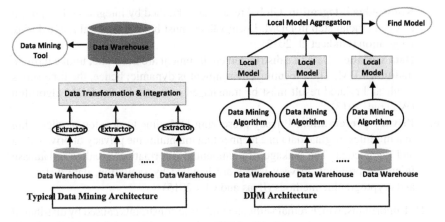

Fig. 1 Typical data mining architecture versus DDM architecture

computation nodes. Therefore, the analysis and management of these distributed data sources require DDM technology that is designed for distributed applications (Ali et al. 2018).

The DDM is applicable to the following conditions:

1. Systems have multiple independent data and computation sites and communicate exclusively through messages.
2. Communication among sites is expensive.
3. Sites include resource constraints, such as battery power issues.
4. Sites involve privacy concerns.

Communication is assumed to be conducted solely by message passing. Several DDM methods thus aim to minimize the number of messages sent. Load balance is also proposed to prevent performance from being dominated by the time and space usage of any individual site. Reference indicated that "building a monolithic database, in order to perform non-distributed DM, may be infeasible or simply impossible" in many applications. The transferring of considerable data may incur a high cost and result in inefficient implementations.

The performance of DDM is affected by the following aspects:

1. **Heterogeneous DM**: Heterogeneous data lead to contradiction among the data attributes. In case of heterogeneous data, a local data management model needs to be integrated into a global model (Chen and Huang 2019)
2. **Data consistency**: Data from different sites are inconsistent. Modification of the local data model affects the global data model and, consequently, the final output produced after DM (Li et al. 2017).
3. **Communication cost**: Communication cost is dependent on network bandwidth and the amount of transferred information, thereby requiring a cost model to be built in DDM (Clifton et al. 2002).

4. **Knowledge integration**: Global results are obtained by integrating local ones; this is a crucial step in DDM. Local values must be in the global range during integration (Ouda et al. 2012).
5. **Data variance**: Data in a distributed environment are not static, unlike those in traditional DM. The executing environment is dynamic; hence, the time-series result and related result must be transferred correctly by the DDM algorithm (Wu et al. 2013).
6. **Privacy preservation**: Privacy-preserving DM aims to develop an algorithm that modifies original data in a manner that maintains the privacy of private data and knowledge. The leakage of confidential information is called a database inference problem. Privacy issues are handled by sanitation, data distortion, and cryptographic methods (Ilyas and Chu 2015).

DM algorithms, which remain efficient under the constraints caused by distributed datasets, have been developed by the DM community in recent years. The DDM field has been an active area of study. The DDM methods in the literature work in an abstract architecture that consists of several sites with independent computing power and storage capability. Each site conducts local computation. Either a central site interacts with each distributed site or a peer-to-peer (P2P) architecture is adopted to compute global models. In the latter case, individual nodes may interact with a centralized node that is rich in resources, but they communicate with neighboring nodes by message passing over an asynchronous network to perform most tasks. For example, independent sensor nodes that connect to one another in an ad hoc manner may be represented by sites.

Cuzzocrea (2013) stated that a methodology for DDM is difficult to establish because of not only the distributed environment but also the requirement for efficient resource sharing and minimized computational complexity. Two variations of DDM are data- and computation-distributed methods. The former method handles data distributed in heterogeneous sites at the local level and computation at the global level. The latter method conducts computation in heterogeneous sites at the local level and hosts data at the global level. Useful, unknown information are hosted on the database of heterogeneous sites. DDM algorithms are applied to data from heterogeneous sites as a local model, and the computed results obtained by DM are combined to form a global model (Gan et al. 2017).

Kargupta et al. (1999) and Zaki et al. (Almaiah and Al-Khasawneh 2020) indicated that several researchers have analyzed the effective and efficient usage of computational resources in individual distributed data sites, performed knowledge discovery in such sites (local level), and aggregated the discovered knowledge at the global level to solve the complexities in constructing DDM methodologies. Fu et al. (Adil et al. 2020) discussed the formulation of suitable DDM algorithms for heterogeneous datasets, the minimization of computational and space complexities, the enhancement of data privacy at distributed sites, and the preservation of the autonomy of local datasets as issues in constructing DDM algorithms. These issues are interrelated, and many researchers have been working in this field. Park et al. (Adil et al. 2020 Mar) proposed an architecture for DDM, in which data are processed at the local level and accumulated at the global level. Tsoumakas et al. (Adil et al. 2020) presented a similar architecture.

2.1 DDM Approach

One of the two assumptions in the DDM literature is about how data are distributed across sites, namely, homogeneously (horizontally partitioned) and heterogeneously (vertically partitioned). In both viewpoints, data tables at each site are partitioned into a single global table. Classifiers for DDM include homogeneous and heterogeneous approaches.

Vertical partitioning of data refers to the collection of different information on the same set of entities or people by different sites or organizations; for example, hospitals and insurance companies collect data on a set of people that can be jointly linked. The data mined at such sites are similar (Ouda et al. 2012). In horizontal partitioning, the same information on different entities or people is collected; for example, supermarkets collect the transaction information of their clients. In this case, the mined data are a combination of the data at the sites.

- **Homogeneous Classifier**

A global table is horizontally partitioned in this case. The tables at each site are subsets of the global table and have similar attributes, as illustrated in Fig. 2. The same set of attributes over all participating nodes is mined in homogeneous DDM. This case occurs when databases are owned by the same organization (e.g., local stores of a chain). Bandyopadhyay et al. (2006) proposed a technique based on the principles of the k-means algorithm to cluster homogeneously distributed data in a P2P environment, including sensor networks. The technique operates in a localized, asynchronous manner by interacting with neighboring nodes. They theoretically analyzed the algorithm and found that it binds errors in a distributed clustering process, unlike the centralized approach, which requires all observed data to be downloaded to a single site.

Li et al. (2017) identified dense regions as clusters and removed clutter at the coarse level by using a proposed density-based clustering mechanism. The "cannotlink" (CL) constraint was applied at the fine level, and overlapping clusters were determined. A multisource homogeneous data-clustering model was also investigated. The model poses CL constraints on the data from the same source. High

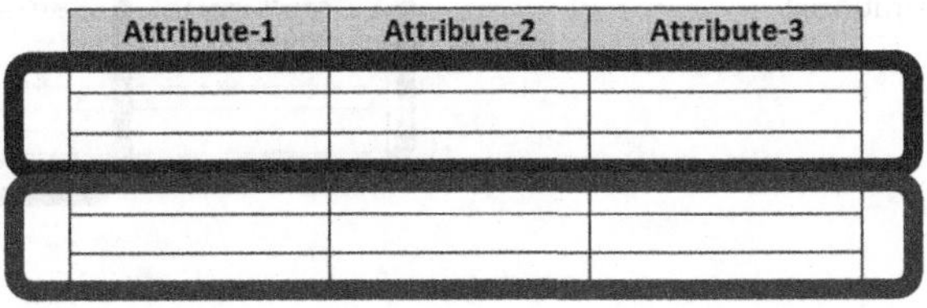

Fig. 2 Homogeneous partitioned

level of clutter, misdetection, and unknown number of potential clusters affect the dataset. Coarse-level clustering was based on density for fast computing, whereas the fine-level one was based on CL-constrained distance to divide closely connected clusters.

Montero-Manso et al. (2018) proposed a novel learning algorithm that combines the outputs of classifiers trained at each node. Such outputs have different weights. A distributed framework with training and test samples that belonged to the same distribution was considered. The training instances spread across disjoint nodes. The distributional distance between each node and the test set in the feature space was used as a reference for the weights. Per-node weighting (pNW) and per-instance weighting (pIW) were introduced. The same weight was assigned to all test instances at each node in the former approach, whereas distinct weights were allowed for test instances that are differently represented at the node in the latter approach. Experiments showed that pIW was superior to pNW and standard unweighted approaches in classification accuracy.

Amir et al. (2018) used the distributed associative memory tree (DASMET) algorithm and presented a fully distributed pattern recognition system within P2P networks to detect spam. Unlike server-based systems, the presented system is cost-efficient and avoids a single point of failure. The algorithm can be scaled for large and frequently updated datasets and is designed specifically for datasets with similar existing patterns. DASMET used a relatively small amount of resources but performed best in spam detection among all distribution methods.

- **Heterogeneous Classifier**

The table is vertically partitioned in this case. A collection of columns exist in each site (sites do not have the same attributes), as illustrated in Fig. 3. However, a unique identifier is present in each tuple at each site for matching. The global table viewpoint is strictly conceptual. Physical realization and partitioning of such a table to form tables at each site are unnecessarily assumed. Different sets of attributes in each participating node are mined in heterogeneous DDM.

Stahl et al. (2012) proposed a pocket DM (PDM) that analyzes data streams in mobile ad hoc distributed environments. Various applications can be run in such environments given the advances in mobile devices, including smartphones and tablet computers. Classification techniques for data streams for PDM were proposed for adoption. An experiment verified that the application of classification techniques for heterogeneous or homogeneous data streams to vertically partitioned data (data

Fig. 3 Heterogeneous partitioned

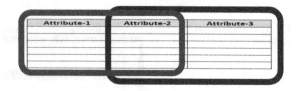

partitioned according to the feature space) results in comparable performance to that of batch and centralized learning techniques.

Iddianozie and McArdle (2019) used a transfer learning methodology to evaluate the relatedness of street networks. Such methodology is included in the formal contexts of inherently dependent and heterogeneous spatial data. They analyzed street networks from eight cities by adopting a statistical multimeasure to determine their similarities. Random forest was used to predict street types, and accuracy was evaluated as a function of transfer polarity. A positive transfer occurs when the transferred models perform better than the parent model, whereas a negative one occurs when they perform worse.

Ranwa et al. (2019) proposed a framework to classify congestion into its components. VANET was used as an alternative cost-effective and flexible solution to monitor effectively road traffic in heterogeneous networks. The objectives of the proposed framework were to exchange traffic flow data and to embed reasoning machinery in vehicles to infer the cause of NRC. Machine-learning methods for classifying congestion into its components in consideration of traffic features for the inference were evaluated; an algorithm that can detect, identify, and propagate via VANET the cause of NRC was provided; and the inference methods were validated through simulation scenarios extended from the real-world Cologne scenario.

2.2 DDM Information Sharing

DDM is implemented by multiple methods with a unified goal. The roles of data analysis at multiple distributed data sources are established. The contribution of distributed methods is facilitated by result sharing. DDM can generally be classified into low- and high-level DDM according to the level of information sharing. Low-level DDM integrates the voting results of multiple independent decision-making units, and high-level DDM combines learning results by meta-learning.

- **Low-level DDM**

Each method (i.e., decision-maker) is trained on the basis of its own data in low-level DDM. The same task, which can be classifying, clustering, or analyzing an instance, is given to all decision-makers. An output is acquired by combining all results. DBDS (Januzaj et al. 2004) is a distributed clustering approach that is implemented on dense clustering algorithms and operates locally. Cluster centers are produced locally at each distributed source and transformed with a small number of data elements to a decision-making center, which then recalculates the cluster centers in accordance with the received centers and elements. Examples of such approach are bagging and boosting, in which multiple classifiers operate on different datasets (Al Hwaitat et al. 2020). Bayesian classifiers are operated in a distributed environment; the local model of distributed sources is averaged to obtain a global one (Adil et al. 2020).

Yan et al. (2019) discussed a distributed ensemble technique for mining health care data under privacy constraints. A novel privacy-based distributed ensemble classifier technique called adaptive privacy-based boosting was proposed to predict the model for EHR data. Each distributed site was allowed to learn data distribution effectively and share medical data without the need to reveal patients' sensitive data. Minimal computational complexity and communication cost were achieved accordingly.

- **High-level DDM**

The learned model of each source is shared with the global model to yield a single learning model for mining input data in high-level DDM. This DDM type is called metalearning or meta-DM. Various tools, such as JAM (Almaiah et al. 2020) and BODHI (Qasem et al. 2021), were proposed for classification. Da Silva et al. (Almaiah) discussed distributed data clustering inferences and proposed the kernel-based distributed clustering scheme algorithm. A local-level estimate was established by a helper site for each local-level distributed site and then forwarded to peer distributed sites. A global-level estimate was then determined by a local-density-based clustering algorithm, and data confidentiality was achieved. Aouad et al. (Almaiah et al.) described a lightweight clustering technique that minimizes communication overhead by minimum variance formation of clusters.

Dong et al. (Yuan et al. 2019) introduced a clustering algorithm based on data partitioning for unevenly distributed datasets. The proposed algorithm was used on uneven datasets. They developed a fuzzy connectedness graph algorithm (PFHC) on the basis of partitioning uneven datasets into similar datasets with equal density. Clusters with equal density were obtained in the local level of distributed sites. Several local clusters were integrated into global clusters, and uneven datasets were mined efficiently.

As indicated above, several DDM techniques, such as distributed clustering, distributed frequent pattern mining, and distributed classification, were provided in the literature. Table 1 summarizes the DDM literature and highlights the characteristics of such techniques.

3 Multi-agent System

Multi-agent systems (MASs) are a computerized environment that ensures that intelligent communication and interactions among agents can achieve certain tasks by integrating involved agents. Such systems are mostly created by the agent-based model, which is an intelligent software that features states and behaviors. Protocols that control the communication and interaction among the involved agents manage the interaction in such MASs (Shen et al. 2006).

A MAS is a group of agents that each have their own problem-solving capabilities and can interact with one another to achieve an overall goal (Obeid and Moubaiddin 2010). Agents are specialized problem-solving entities that have clear boundaries, and they agents fulfill a clearly defined purpose and demonstrate flexible and proactive behavior (Obeid and Moubaiddin 2010).

Table 1 DDM literature

References	Machine learning techniques	DDM approach	Information sharing	Application	Algorithm	Publish year
Bandyopadhyay et al. (2006)	Clustering	Homogenous	Low-Level DDM	Sensor Network	K-Means	2006
Stahl et al. (2012)	Classification	Homogenous Heterogeneous	Low-Level DDM	Mobile Sensor	Naive Bayes	2004
Iddianozie et al. (2019)	Ensemble	Heterogeneous	High-Level DDM	Street Networks	Random Forests	2019
Li et al. (2017)	Clustering	Homogenous	Low-Level DDM	Sensor MODE	Density-Based Clustering	2017
Ranwa et al. (2019)	Classification	Heterogeneous	High-Level DDM	Nonrecurrent Congestion NRC	• Classification Tree • Random Forest • Bayesian Network	2019
Montero-Manso et al. (2018)	Classification	Homogenous	Low-Level DDM	Intrusion Detection	• Random Forest • Support Vector Machine • Logistic Regression	2018
Amir et al. (2018)	Classification	Homogenous	Low-Level DDM	Image Spam Detection	Distributed Associative Memory Tree	2018
Jeong et al. (2019)	Clustering	Heterogeneous	Low-Level DDM	UAV System	Raft Algorithm	2019
Chen and Huang (2019)	Clustering	Heterogeneous	High-Level DDM	Sensorless Drive Diagnosis Cartographic	K-Means, SOM, And Spectra	2019
Januzaj et al. (2004)	Clustering	Homogenous	Low-Level DDM	–	K-Means Density-Based Clustering	2004
Chen and Huang (2019)	Ensemble	Homogenous	Low-Level DDM	Healthcare	ADABOOST	2019

(continued)

Table 1 (continued)

References	Machine learning techniques	DDM approach	Information sharing	Application	Algorithm	Publish year
Al Hwaitat et al. (2020)	Ensemble	Homogenous	Low-Level DDM	–	Decision Trees	2001
Khan et al. (2020)	Classification	Heterogeneous	Low-Level DDM	–	*Bayesian Networks*	2004
Almaiah et al. (2020)	Meta-Learning	Homogenous	High-Level DDM	Fraud And Intrusion Detection	JAM	1997
Qasem et al. (2021)	Regression	Heterogeneous	High-Level Ddm	–	CMD BODHI	1999
Almaiah et al.	Clustering	Homogeneous	High-Level DDM	Internet Of Things	LPDA	2007
Bandyopadhyay et al. (2006)	Clustering	Homogenous	High-Level DDM	Attacks	KDEC	2006
Tong et al. (2018)	Clustering	Homogenous	High-Level DDM	Felony Incidents Of New York City	• Density Based • Affinity Propagation • Spectral	2018

The literature provides many definitions of agents due to their diverse application-specific features. According to the authors in Russell et al. (1995), an agent is "a flexible autonomous entity capable of perceiving the environment through the sensors connected to it." Other researchers have corroborated this definition (Jain and Srinivasan 2010). Another point of view was expressed in Ye et al.; according to the authors, an agent is "an encapsulated computational system that is situated in some environment and this is capable of flexible, autonomous action in that environment in order to meet its design objective."

The above definitions, among others, frame agents in the context of a specific field of study. However, the concept of agents is generic and applicable to many disciplines. Dorri et al. (2018) proposed a generalized definition that considers the fundamental abilities and features of agents and defines them as an entity that is placed in an environment and senses different parameters used to make a decision on the basis of the entity's goal. The entity then conducts the necessary action on the environment on the basis of this decision.

Agents in MASs are autonomous in that they can function independently without guidance. Such agents have control over their internal state and actions. An agent can independently whether to perform a requested action. Agents exhibit flexible problem-solving behavior to achieve their objectives, which they are designed to fulfill. Autonomous agents can interact with other agents by using a specific communication language. Thus, they can take part in large social networks and respond to changes and/or achieve goals with the assistance of other agents.

Agents typically operate in a dynamic, non-deterministic complex environment. In MAS environments, no assumption exists with regard to global control and data centralization and synchronization. Thus, the assumption is that agents in a MAS operate with partial information or limited capabilities to solve the problem. Through communication, agents share the information that they gathered for coordinating their actions and increasing interoperation. The interactions between the agents can be requests for information, particular services or an action to be performed by other agents, and issues related to cooperation, coordination, and/or negotiation to arrange related activities.

Agents must interact with one another to achieve their individual objectives. However, such interaction is conceptualized as occurring at the knowledge level (LiviuPenait 2005). Specifically, interaction can be conceived in terms of the goals that should be followed, at a specified time, and by the assigned agent. Given that agents are flexible problem solvers that have only partial control and knowledge of the environment where they operate, interactions must be handled flexibly, and agents must make runtime decisions as to whether or not to initiate interactions and the nature and scope of these interactions. Through the following features, agents can have broad applicability and solve complex tasks (Garcia et al. 2010):

- **Sociability**: Agents can share their knowledge and request information from other agents so they can perform better in achieving their objective.
- **Autonomy**: Each agent is able to execute the decision-making process independently and implement suitable actions.

- **Proactivity**: To predict possible future actions, each agent uses its history, sensed parameters, and information of other agents, thereby enabling agents to take effective actions that meet their objectives. From this ability, we can assume that the same agent may resort to disparate actions when it is placed in different environments.

Although agents working by themselves can take actions on the basis of autonomy, their real benefit can be harnessed only when they collaborate with other agents. Multiple agents that collaborate to solve a complex task are known as MASs (McArthur et al. 2007).

Agent interactions and communications in MASs are controlled by the utilized communication protocols. MAS protocols determine the external agent behavior along with the internal agent structure without focusing on the internal behavior of the agent. These specifications, along with the task implemented by the underlying MAS, can determine the specification of the internal agent behavior. The existing protocols for agent communication can be classified into two categories, namely, centralized and decentralized protocols as illustrated in Fig. 4.

- **Centralized protocols**: In this protocol, the agent that is requesting to interact with others, which is called an initiator, forwards its request to a center agent, called broker. The center agent then selects the intended agent, which is called the participant, on the basis of the specification given by the initiator and forwards the initiator request accordingly.

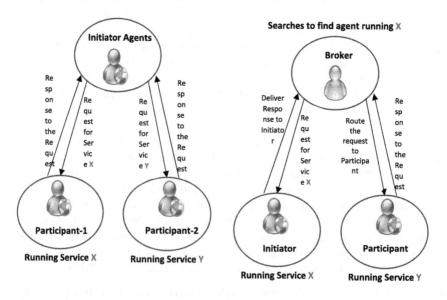

Fig. 4 Centralized versus Decentralized Protocol

The advantage of this type is the freedom of the broker to select a suitable participant. However, the disadvantages include the need to record, manage, and exchange the history of the participant to avoid an incorrect selection.

- **Decentralized protocols**: In decentralized protocols, the agent that is requesting to directly interact with the participant obtains the address and status information from the registry agent, which functions as a "yellow page" that contains no information about the agent activities.

Through decision-making, or planning and control in MAS, an agent can accomplish its goals by determining the actions that it should perform. The decision-making problem can either be episodic or sequential (Claes et al. 2011). The output of the episodic problem is a single action, while the sequential problem produces a sequence of actions or policy. The decision-making algorithm is evaluated on the basis of policy optimality, search completeness, and time and space complexities.

3.1 MAS Features

This section presents the seven important features of MAS and discusses the different categorizations that emerge when each feature is considered.

1. **Heterogeneity**: MAS can be divided into two categories, namely, homogeneous and heterogeneous, on the basis of agent heterogeneity. Homogeneous MASs include agents that have the same characteristics and functionalities, while heterogeneous MASs are composed of those with diverse features (Wen et al. 2013).
2. **Topology**: Topology is the location and relations of agents. MAS topology can be categorized as either static or dynamic. In a static topology, the position and relations of an agent remain unchanged over the agent's lifetime. In a dynamic MAS topology, the position and relations of an agent change as the agent moves, leaves or joins the MAS, or establishes new communications, such as relations, with other agents (Li et al. 2014).
3. **Data transmission frequency**: Agents sense the environment and share the data they sensed data with other agents either in a time-triggered or an event-triggered way. In the time-triggered approach, the agent continuously senses the environment, collects data, and transmits all newly sensed data to other agents in pre-defined time intervals. In the event-triggered approach, the agent senses the environment only when a particular event takes place. Subsequently, the agent sends the collected data to other agents.
4. **Agreement parameters**: In certain MAS applications, agents need to agree on particular parameters referred to as metrics. MASs are classified as first, second, or higher order on the basis of the number of metrics. In the first order, agents collaborate to agree on one metric (Du et al. 2013).
5. **Mobility**: Agents can be classified as static or mobile on the basis of their dynamicity. Static agents are always located in the same position in the environment.

By contrast, mobile agents can move around the environment. A mobile agent can be hosted by other agents, that is, it uses the resources of other agents, monitors them, or senses the environment from the position of other agents so that it can perform actions (Ma et al. 2015).

6. **Delay consideration**: Agents might face multiple sources of delay when performing tasks. Such delay sources include those communication media, including wireless or wired, which are used by agents to exchange data, or scheduling resources for each agent. MASs can be classified into two groups, namely, with or without delay, depending on whether the delays are substantial and relevant (Olfati-Saber and Murray 2004). The former considers the delay sources, while the latter assumes that no delay sources exist. The latter instance is simplified because no communication and processing delays exist. However, most real-world applications always encounter non-trivial delays.

7. **Leadership**: We consider the existence of a leader, that is, an agent that defines goals and tasks for the other agents on the basis of one global goal. MAS can be categorized as either as leaderless or leader follow depending on the presence or absence of such a leader (Du et al. 2013).

3.2 MAS Applications

MASs have been applied to various fields, including cloud computing, robotics, smart grid, modeling complex systems, and social network. In cloud computing, the complexity in computer networks increases considerably as a result of the emergence of new technologies and proliferation of Internet-connected devices. Agents are primarily used to address this complexity given the broad range of applications of MAS in various networks.

Bouchemal (2018) proposed an approach that is based on MASs and the cloud computing concept to assist users, such as employees, managers, and customers, in being permanently connected to the ERP system. The aforementioned author proposed to assign a cloud manager agent at the cloud level to serve as a mediator between the user and the ERP manager agents. The author initiated a preliminary implementation by using the JADE-LEAP framework. Januário et al. (2018) presented an architecture that is dependent on a hierarchical MAS to enhance resilience, with particular attention paid to wireless sensor and actuator networks. An IPv6 test bed that comprised several distributed devices was used to evaluate the proposed framework; this evaluation involved the analysis of performance and communication link health. The test bed results show the relevance of the proposed approach.

Over the past two decades, the use of agents for robotics has been studied. Duan et al. (2012) proposed an agent-based soccer robot to study such complexity. Agents are grouped into teams. The agents in a team learn knowledge about the opponent team and determine possible actions by interacting with the environment. These agents then share the policies they have learned with other agents in their team.

Reinforcement learning is used along with probabilistic neural networks to ensure that the final decision made by the agent is accurate. According to implementation results, the agents predict the correct actions. As a result, the ball possession percentage (a key performance measure in soccer) in the agent team is higher than that of the non-agent one.

In smart grid literature, agents address multiple challenges posed by smart grids, including balancing the generated and demanded energy, negotiating between the energy consumer and the producer over the energy price, storing energy in home storages, and restoring energy. Santos et al. (2019) proposed a centralized hierarchical MAS that coordinates the different monitoring stages and decision-making processes. The main contribution of this MAS is that it improves traditional contingency response algorithms, such as load shedding schemes, thereby maximizing future smart grid infrastructure. The results showed that improving the traditional load shedding philosophy schemes and using advanced communication infrastructure, monitoring, and embedded processing capabilities to provide enhanced stability and reduce the unnecessary load disconnections from the system are difficult.

In modeling complex dynamic systems, this field is costly and results in a significant processing overhead because of the demand for powerful modeling platforms and high complexity. Given the flexibility, autonomy, and scalability afforded by agents, agent-based modeling (ABM) has become a low-cost and low-resource solution for modeling complex systems. Unlike other modeling methods that use equations, ABM uses a rule-based methodology to model the environment. The key advantages of ABM are as follows (Domínguez et al. 2015):

1. Ability to be aggregated and combined with other modeling methods
2. Flexibility in assumptions for modeling a MAS
3. Flexibility in pre-defined knowledge given that agents can acquire knowledge by learning from the environment
4. Possibility of parallel execution, which can accelerate the modeling process
5. Ability to explore emergent behaviors owing to agent proactivity.

Wangapisit et al. (2014) proposed an ABM to model supply chains. Each entity in the supply chain is separately modeled with its own policies and can define its own interactions with other entities. The proposed method comprises two groups of agents, namely, planning and physical agents. Customers and suppliers use six agents in the planning group to negotiate and reach an agreement over product price. The agents in the planning group are the demand fulfillment agents who manage the customer demand, the material resource planning agents who communicate with the producers and purchases products, the demand forecast agents that predict the demand of the customers on the basis of the current and history of demands, the master planning agents who aggregate production planning, production planning agents who disaggregate the aggregated planning by the master planning agent, and the scheduling agents who schedule jobs in multiple agents. Three agents in the physical group perform physical tasks, such as receiving and storing raw materials, manufacturing, and delivering products to the customer.

The popularity of social networks has exponentially increased with the rise in the number of Internet users. A social network comprises various actors, such as users, groups, and services (Jiang and Jiang 2014). The complexity of social networks originates from its dynamicity, that is, a large number of participants joining or leaving the network or establishing new connections, and its broad range of applications and services. MAS may potentially overcome the complexity of social networks. Gatti et al. (2013) proposed an agent-based method for predicting user behavior, including likes, posts, and follows, in social networks, such as Twitter. Such authors proposed the use of multiple agents, such as actors, which are distributed throughout the social network to gather a dataset of the user behavior. The agents then conduct topic and sentiment classification on the data of each particular to create a user profile. Finally, the user profile is input into a prediction system that predicts future user behavior, including likes, topics, replies, posts, and shares.

In their work, Ma and Zhang (2014) considered a school as a social network. They used MAS to help school managers finding the relations between the allocation of funds to different school programs, such as athletic, academic, and cultural activities, and school performance. In this approach, students, teachers, and school departments are agents who collaboratively form the social network, that is, the school. A hierarchical structure is developed to organize the interactions of each student with other students or teachers, the fund distribution policy from previous years, and the performance of students in the current and past years. This structure is then input into a learning function that appraises the relation between the fund distribution policy and the academic performance of the students.

3.3 MAS Platform

A MAS platform is a generic framework that provides the support required to perform agent-based simulations. MAS systems are designed to make the creation and evaluation of multi-agent models easier. MASs have been used in different fields as they provide logical, scientific, realistic, and efficient methods for deciphering and solving challenges (Der et al. 2007). In addition to their ability to communicate with one another and with their alternative behavior, MASs also feature self-healing defense systems (LiviuPenait 2005; Charlton and Cattoni 2001). MASwill adapt to situational variations and communicate and share information to achieve a common goal without external interventions (LiviuPenait 2005).

During the last two decades, different Agent Platforms (APs) have been built by the research community; these APs are either general-purpose or address a specific domain of usage. Several platforms are no longer updated, although the introduction of new versions of other platforms continues. In the meantime, a growing number of new initiatives continue to be introduced by the agent-oriented research community. The nature of these sites competes with that of their user base. Because of the large number of platforms, people usually rely on word of mouth, past experiences, and platform ads to decide the most appropriate one for their intent. Recently, however, people have become interested in using MASs to take advantage of agent technologies that are increasingly dependent on solid survey products.

Several tools to produce agents have grown, and several more are being developed. Choosing an effective agent is a critical factor in agent growth. Java Agent Development Environment (JADE) and Smart Python Agent Development Environment (SPADE) are the common modeling and assessment methods for evaluating performance metrics in lectures, which are better suited for agent development. There are also numerous agent structures, such as GAMA and PANGEA. Implementation of agents in this framework is based on requirements set by the Foundation of Intelligent Physical Agents (FIPA) (FIPA 2002a). FIPA an association of standards for the IEEE Computer Society, it is supporting agent-based technology and the interoperability of its standards with other technologies. FIPA establishes standards for computer software, interacting agents, and agent-based systems and defines an agent as a computational process that implements the autonomous functionality of an application. Agents are autonomous, that is, they are independent, self-governing, and exercise control over their actions. Such agents are reactive and respond to variations in the environment. Agents learn from previous experiences, make impromptu tactical adjustments, and have the initiative to achieve their goals as they are goal-directed, thereby demonstrating their proactivity and flexibility (Liviu-Penait 2005). Also, agents are social, able to interact with humans and other agents, mobile, and move easily from one machine to the other. Given their robust nature, they can achieve continuity and vigorous fault recovery to meet their goals (Charlton and Cattoni 2001). The FIPA reference model comprises the following (The FIPA Specifications):

- **Agent Management System:** This system registers and deregister agents, provides a directory of agent descriptions in alphabetical order, supervises and controls access to the AP, and delivers life cycle management facilities.
- **Directory Facilitator**: This facilitator provides an indexed almanac to the services given by other agents and registers and deregisters the agents.
- **Agent Communication Channel**: It facilitates inter-agent communication internally and to agents traversing through platforms.
- **Internal Platform Message Transport**: It ensures that message forwarding services take place within a platform.

FIPA Open Source (FIPA-OS) is used to develop agents that conform to FIPA benchmarks. Figure 5 shows the FIPA-OS Agent Platform. FIPA-OS has the following abilities: maintenance of agent shells from various categories; message and conversation management; dynamic alignments to support multiple IPMTs; diagnostics and virtualization tools; abstract interfaces; and software designs (FIPA 2002b). FIPA Agent Communication Language (ACL) is used by the agents to communicate. The four FIPA ACL components are conversation, ACL message, syntax, and semantics. The message syntax is domain-independent and does not need to be defined for all addressed domains.

The semantics/ontologies help define the names and data types to be used in messages and their meanings. All involved participants interpret the messages in the

Fig. 5 FIPA-OS agent
platform

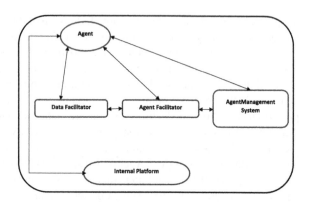

colloquy explicitly (Charlton et al. 2002). The FIPA standards have played a major role in the development of MASs (Nicholas 1998).

Next, we discuss different agent frameworks:

A. **Java Agent Development Framework (JADE)**

JADE is a software framework that is fully implemented in the Java language. This framework simplifies the implementation of MASs through middleware that is compliant with FIPA specifications and by using a set of graphical tools that support the debugging and deployment phases. A JADE-based system can be distributed across machines, which do not require the same OS to be shared, and a remote GUI can be used to control the configuration, which can be changed during runtime by transferring agents from machine to machine as necessary. The minimumsystem requirement for JADE is the Java 5 runtime environment or the JDK (Jain and Srinivasan 2010).

Hudaib et al. (2017)proposed a semi-centralized protocol to address the limitations of existing protocols and enable the development of a robust and adaptable solution in a highly dynamic environment. The outcome of the JADE framework implementation proves that the proposed protocol is highly capable. The proposed protocol has better time and communication overhead under various conditions than the Contract Net Protocol.

B. **Smart Python Multi-agent Development Environment (SPADE)**

SPADE is an instant messaging-based MAS platform written in Python. In SPADE, agents can chat with other agents and humans. A SPADE agent comprises a procedure for connecting to the platform, message disseminator, and several behaviors to which the messages are transmitted. Each agent establishes a platform link by using a unique Jabber identifier and valid platform.

A SPADE agent has built-in jabber technology. Through this technology, an agent who is connected to the SPADE platform is automatically registered and remains active, constantly communicating with the agent platform.

- **Message disseminator**: This component ensures the delivery of a message to the correct recipient or agent.
- **Agent behavior**: This task is performed by the agent.

An agent can exhibit several behaviors, each of which is attached with a message template. This template determines the type of message to be received by a behavior.

Goryashchenko (2019) described the group formation problem in MASs. In this study, SPADE MAS is selected after its scalability and performance for several hundreds of agents were evaluated. A greedy algorithm for agent group formation is proposed and implemented using SPADE MAS. The characteristics of the implemented algorithm are studied. The results showed that the performance of the proposed SPADE-based framework allows the implementation of additional complex algorithms for the formation of an agent group, which may provide close-to-optimal results.

C. **GAma Modeling Language(GAMA)**

GAMA is a simulation platform that offers a complete modeling and simulation development environment where modelers, field experts, and computer scientists can build multi-agent simulations that are spatially explicit. The platform enables capabilities related to the right combination of multilevel modeling, 3D visualization, and GIS data management. For support in rapid design and prototyping, large models written in the GAML agent-oriented language can be built in GAMA with an optional graphical modeling technique. Agents can be instantiated by users from GIS data, files, or databases; large-scale simulations reaching millions of agents can be run, and rich user interfaces (UIs) can be designed to support in-depth inspections on agents, multiple multi-layer 2D/3D displays, and aspects, and user-controlled panels and actions (Kravari et al. 2010).

A methodology was developed by Ruiz-Chavez et al. (2018) that identifies the influencing factors of solid waste management in the Centro Histórico de Quito. This method involves simulating with georeferenced multi-agents that improve waste handling. A simulation model was built using GAMA with three types of agents, namely, people, collectors, and waste containers, in an environment modeled with geographic information systems (GIS) layers. According to the results, the complexity of waste handling increases with the population in the area. Modifying the routers is insufficient. Thus, the number of collectors must be increased.

D. **EMERALD**

EMERALD is a novel knowledge-based implementation framework wherein trusted third-party reasoning services can be used for interoperable reasoning among agents in the Semantic Web. The benefit of this framework is that each agent can trade its position justification arguments even with agents that do not follow the same rule logic or paradigm. EMERALD is fully Foundation of Intelligent Physical Agents (FIPA)-compliant and established on top of JADE.

It is the only agent platform that supports reputation and trust mechanisms that enable efficient decision-making and trustworthiness in MASs. Thus far, it has been employed to examine how agents behave on behalf of their users in trading and similar cases.

Kravari et al. (2010) took advantage of EMERALD to develop a MAS on top of JADE. They used a reusable agent prototype for knowledge-customizable agent behavior, trusted third-party reasoning services, and a reputation mechanism for ensuring trust in the framework. The authors presented a use case scenario to show the proposed framework's viability.

Based on the MAS status (open or closed system) and communication type (symmetric or asymmetric), Kravari et al. (2012) defined the space of potential methods of interoperation for heterogeneous MASs. Their study showed how heterogeneous MASs can interoperate and eventually automate collaboration between communities using one of these techniques. To exemplify the method, the authors used two SW-enabled MASs, namely, Rule Responder and EMERALD, which help user communities according to declarative SW and multi-agent standards, including OWL, RDF, FIPA, and RuleML. This interoperation depends on high-quality rules and facts and utilizes a declarative, knowledge-based method that allows information agents to generate consistent and smart decisions. Finally, this work demonstrated the interoperation's added value by presenting multi-step interaction use cases between agents from both MAS communities. Table 2 present a compersion between agent frameworks.

MASs are applied in diverse fields to resolve and decipher challenges through the technical, logical, efficient, and practical methods of the systems. MASs have self-healing protection systems and the capacity to interact with one another and their alternative actions. These intelligent agent systems can interact and exchange knowledge to attain common objectives without external intervention as well as adjust to situational changes. A MAS may involve a combination of agents and humans. Agents in a MAS have only a local system view and are autonomous. No agent in a MAS is granted supervisory or governing authority. Table 3 is summarizedMAS literature.

3.4 MAS with Distributed Data Mining

MASs provides an architecture for distributed problem solving, and Distributed Data Mining (DDM) algorithms focus on one category of such problem-solving tasks, namely, distributed datamodeling and analysis. Here we present a viewpoint on DDM algorithms in the MAS context and offer a general discussion on the relationship between MASs and DDM..

MASs frequently handle complex applications that need distributingproblem-solving. Meanwhile, DDM is a complex system focused on data mining processes and resource distribution over networks. Scalability lies at the core of DDM systems.

Table 2 Comparative analysis of different agent frameworks

Framework	Performance	Stability	Robustness	Programming language	Operating systems	Popularity	Platform security	Standard compatibility	Communication
JADE	Very High	High	High	JAVA	Any with JVM	The most popular	Strong	FIPA, CORBA	ACL
SPADE	Very High	High	High	Python	Any	High	Strong	FIPA	ACL
GAMA	Good	Good	Good	GMAL	Mac OS X, Windows, Linux	Low	Average	FIPA, (GIS, 3D capabilities)	ACL
EMERALD	High (on JADE)	High	High	Java, JESS, RuleML, Prolog (plus use of XML, RDF)	Any with JVM	Low	Strong	FIPA, Semantic Web standards	ACL

Table 3 MAS literature

References	Application	Platform	Communication protocol	Data transmission frequency	Mobility	Topology	Heterogeneity
Bouchemal and Bouchemal (2018)	Cloud	JADE	Centralized	Event Triggered	Mobile Agent	Dynamic	Heterogeneous
Duan et al. (2012)	Robotics	–	Decentralized	Time-Triggered	Mobile Agents	Dynamic	Homogenies
Santos et al. (2019)	Smart Grids	–	Centralized	Event Triggered	Mobile Agents	Dynamic	Homogenies
Ren et al. (2019)	Smart Grids	–	Decentralized	Time-Triggered	Static	Static	Heterogeneous
Domínguez et al. (2015)	Car Parking	–	Decentralized	Event Triggered	Static	Static	Homogenies
(Hudaib et al. (2017)	–	JADE	Semi- Centralized	Event Triggered	Static	Static	Homogenies
Kannan et al. (2019)	E-Commerce	Jade	Centralized	Event Triggered	Static	Static	Heterogeneous
Goryashchenko (2019)	–	SPADE	Decentralized	Event Triggered	Mobile Agents	Dynamic	Heterogeneous
Louati et al. (2018)	Traffic Signal Control Systems	SPADE	Centralized	Time-Triggered	Mobile Agents	Dynamic	Homogenies
	Smart City	Gama	Decentralized	Event Triggered	Mobile Agents Static	Static, Dynamic	Heterogeneous
Olszewski et al. (2019)	Geographic Information Systems	Gama	Centralized	Time-Triggered	Mobile Agents	Dynamic	Heterogeneous
Ruiz-havez et al. (2018)	Semantic Web	Emerald	Centralized	Event Triggered	Static	Static	Homogenies
Kravari et al. (2010)	Semantic Web	Emerald	Centralized	Event Triggered	Static	Static	Homogenies

Given that system configurations may sometimes change, DDM system design looks at many details regarding software engineering, such as extensibility, reusability, and robustness. Therefore, the characteristics of agents are favorable for DDM systems (Sawant and Shah 2013).

MAS is appropriate for distributed problem solving because it allows the creation of autonomous, goal-oriented agents that operate in shared environments with coordination and communication capabilities. This mechanism is beneficial for DDM in implementing different distributed clustering, prediction, and classification methods. For example, an agent may seek assistance from other agents in classifying instances that cannot be easily classified locally. The agent communicates its beliefs or outcomes, and the other agents decide whether using such beliefs will be advantageous for classifying instances and adjusting their prior assumptions about each data class.

In MASs, several agents mine and coordinate together. These systems are particularly appropriate for DDM because memory cost and computation time can be decreased by having agents at the global level and deploying numerous agents to each site of the distributed data. Agents can adjust to the faults and errors arising in the entire system by cooperating with all distributed data sites. A MAS consists of several agents that can perform tasks that a single agent cannot easily carry out. MAS technology is classified under artificial intelligence; thus, its success depends on the coordination of individual agent behaviors (Januário et al. 2018).

Agent technology in DDM improves scalability and efficiency by decreasing network traffic, that is, only the code and not the data is shared among distributed sites, thereby decreasing bandwidth and enabling efficient system scaling without increasing system complexity (Gan et al. 2017). Agents adapt easily and react vigorously to changing environments and operate asynchronously; that is, an agent disconnects from the network and automatically reconnects after completing its tasks (Gan et al. 2017). Finally, agents are fault-tolerant; they bypass faults in distributed data sites and shelter on reliable distributed data sites. MAS agents obtain the following basic properties.

- Autonomy
- Decentralization
- Limited system view.

Security is a vital concern. Enterprises will not adopt systems that cannot secure their data. The concern is also called data security and trust-based agents. However, a rigid security model that aims to guarantee security can degrade the scalability of systems. Agents that can travel through system networks provide alternative methods. Also, scalability is an important concern in distributed systems. Added human interventions or highly complex mechanisms, which may compromise performance, are needed to inform every unit in the system about any system configuration update, such as during the joining of a new data site. Concerning this issue, collaborative learning agents can exchange and disseminate information (about modifications in system configuration in this case) with one another, thereby enabling the system to adapt at the individual agent level. Furthermore, as previously discussed, these agents can assist in decreasing DM application server and network load, as in state-of-the-art systems (Liu et al. 2011).

Using DDM include intrusion detection, credit card fraud detection, security-related systems, health insurance, distributed clustering, sensor networks, market segmentation, customer profiling, credit risk analysis, and retail promotion assessment. A growing number of sophisticated applications are being proposed along with the advancement of the Internet. Consequently, the number of free services being offered is also increasing. Service subscribers (i.e., customers) are persuaded to complete their profiles, which are adapted to match them with affiliate advertisements delivered side-by-side with the services, thereby displaying the best advertisements for online users. Learning from user profiles is a perfect illustration of the use of data mining methods. This task can be performed fully through database operations, but autonomous agents can improve numerous DDM features.

3.5   Key Characteristics of DDM with MAS

DDM with MAS considers data mining a basic foundation and is improved with agents; thus, this novel data-mining method will inherit all the powerful properties of agents and acquire favorable features. Constructing a DDM with MAS involves three key characteristics: dynamic system configuration, interoperability, and performance aspects.

- **Interoperability**: This characteristic involves the collaboration of agents both inside and outside a system, thus enabling new agents to integrate seamlessly into the system. The system architecture should be flexible and open to dealing with such interaction, which includes service directory, communication protocol, and integration policy.

  Communication protocol involves message encryption, encoding, and transportation between agents. These operations are standardized by the FIPA and are accessible to the public. Most agent platforms are FIPA-compliant, such as SPADE and JADE, and thus support interoperability among them. Integration policy stipulates the behavior of a system when an external component, such as a data site or agent, seeks to leave or join.

  The policy is known by all system components and is maintained by a version that can be updated periodically. Document updating (e.g., policy) requires that the present version of the document be known by all system components, thereby resulting in system inconsistency. A service directory publishes system documents, such as system configuration and policy; all components may validate the document version to guarantee system consistency. Further adoption may consider how a service directory can be published, propagated, and maintained.
- **Dynamic system configuration**: Concerning interoperability, handling dynamic system configurations is difficult because of the complex nature of mining algorithms and planning. A mining task may include several data sources and agents,

with the latter being equipped with an algorithm and dealing with datasets. Any modification in data affects the mining task because an agent may still be running the algorithm during the update. Therefore, the task priority should be whether it can be restarted. The ideal algorithm is preemptive; it can pause for an update in the dataset and continue the execution after the change is completed. Adding an agent expecting to increase the speed of the mining task is likewise ideal because the agents must share the instance of agent memory.

- **Performance**: Performance can be enhanced or degraded depending on the major constraint posed by data distribution. Tasks can be executed in parallel in distributed environments; however, parallelism leads to concurrency concerns. Thus, ADDM systems must ensure acceptable (if not improved) system performance. Other performance issues may be considered. For instance, when the system enhances user experience, it may also optimize the mining performance. An agent may represent a user with a profile. ADDM must determine the user profile before data analysis. A user profile presents a pattern of queries; ADDM must be able to manage query caches or some relevant performance improvement methods.

In this manner, ADDM may ignore certain repetitive queries by replying with a previous query result from the query cache. An agent may also describe a data source with a profile. Heterogeneous data sources lead to various processing and query specifications. A less powerful data source may impair the performance of the entire MAS; thus, an ADDM must consider alternative data sources to acquire input data for a query. Agents may share information with peers by initiating communications in the MAS for a costor searching the service directory to identify alternative data sources. This process is the extent of the cost–benefit analysis between communication and query processing. As shown in Fig. 6, an ADDM system can be generalized into a set of components. System activities may be simplified into request and response, each involving distinct components.

The following are the basic components of an ADDM.

- **Data**: Data isthe foundation layer of interest. In distributed environments, data can be hosted in different forms, such as data streams, online relational databases, and web pages, where data have varied purposes.

Fig. 6 Overview of ADDM systems

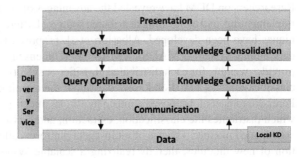

- **Communication**: The system selects the associated resources from the directory service, which keeps a list of data sources, data schemas, mining algorithms, and data types, among others. Communication protocols, such as client–server and peer-to-peer, may change by system execution.
- **Presentation**: The UI cooperates with the user by receiving and responding. It simplifies complicated distributed systems into user-friendly messages, including visual reporting tools and network diagrams.

When a user requests data mining through the UI, the following components are involved.

- **Query optimization**: A query optimizer analyzes the request to identify the mining task type and then selects resources appropriate for the request. The query optimizer also decides whether the tasks can be parallelized, given that the data are distributed and can be mined in parallel.
- **Discovery plan**: A planner allocates sub-tasks with related resources. Mediating agents play an important role during this stage by coordinating numerous computing units, given that mining sub-tasks are accomplished asynchronously, and the results from such tasks.

When a mining task is executed, the following components are involved.

- **Local knowledge discovery (KD)**: Mining may occur locally at each data site depending on the individual implementation to transform data into patterns that sufficiently represent the data and can be reasonably transferred over the network.
- **KD:** Also called mining, KD runs the algorithm as entailed by the task to acquire knowledge from the specified data source.
- **Knowledge consolidation**: The knowledge obtained from different sources should be normalized to offer a meaningful and compact mining result to the user. This component includes a complex technique that incorporates patterns or knowledge from distributed sites. Combining homogeneous knowledge or patterns is probable but challenging for heterogeneous cases.

3.6   DDM with MAS Approaches

Researchers on DDM have revealed the advantages of applying MAS in organizing, implementing, and controlling distributed sources. Different MAS-based DDM techniques have been developed. Albashiri et al. proposed the extendible multi-agent data mining system (EMADS), a MAS based on ensemble classification. EMADS provides weights for each distributed classifier or selects one of them to execute the classification task based on knowledge about the learning model at each classifier. Pérez-Marcos et al. proposed a music multi-classification platform based on MASs which distributes the classification, extraction, and service tasks among agents in a coordinated manner. The use of multi-agent platforms allows for the dynamic addition of new modules, thereby realizing a scalable system. Each sub-process that comprises the global system is distributed. The platform implements emotional and musical genre classification and then gives context information about songs from social networks, such as Twitter and Facebook. The selected techniques base on meta-classifiers to conduct single- and multi-label classification and obtain superior results. The proposed platform performed better in multi-label classification than did platforms from previous works. Moghadam and Ravanmehr proposed a new DDM method named multi-agent hierarchical data mining (MHDM) to classify meteorology data gathered from various sites that were broadly distributed around Iran. The REPTree classification algorithm is executed in local agents to obtain the needed information from raw data kept in local databases, which are optimized to operate in MAS. The researchers assessed the proposed method by executing it on 20 million pieces of meteorology data. Experimental results indicated that MHDM can substantially enhance performance compared with parallel and centralized methods for KD in cases with large volumes of data.

A MAS-based clustering framework capable of enhancing the initial cluster centers at each agent was recently developed. Chaimontree et al. developed a collaborative clustering framework that presented better performance over non-collaborative agent-based clustering. This framework retained the particularity of each agent and enabled information exchange, including private and non-shareable data, among local agents to improve the capabilities of autonomous agents.

A clustering system based on cooperative agents through a common and centralized ERP database was presented by Mesbahi et al. to enhance decision support in ERP systems. The authors used a MAS paradigm to distribute the complexity of the k-means algorithm to numerous autonomous entities called agents, whose objective is to assemble observations on or records of similar object classes. This process will assist business decision-makers in making sound decisions and giving superior response time using the MAS. The JADE platform is a convenient means of implementing the proposed architecture while offering a complete set of services and having agents comply with the FIPA specifications. Qayumi and Norta proposed a MAS-based business intelligence (BI) system model called BI-MAS with comprehensive designing steps as a running case. Given the complexity of BI generation in distributed environments, the use of such a system is diverse because of integrated DDM and MAS technologies. Incorporating these frameworks in the content of BI systems poses difficulties during the design, analysis, and testing stages in the development life cycle. To demonstrate the proposed approach, the authors first considered an evaluation process of identifying appropriate agent-oriented techniques. Second, the researchers adopted the chosen methodologies in designing and analyzing concepts for BI-MAS life cycles. Finally, they demonstrate a novel verification and validation approach for BI-MAS life cycles. Sun et al. examined distributed deep learning as a MAS problem and designed a MAS approach for distributed training of deep neural networks. In their approach, the agents assess the utility of each action and then implement the best response strategy based on their estimated utilities mixed with ϵ-random exploration. The proposed approach, called Slim-DP, differs from the typical data parallelism in that it only communicates a subset of the gradient or the global model. The experimental results demonstrated that Slim-DP can decrease communication cost more and obtain better speedup than the typical data parallelism and its quantization version without loss of accuracy. Table 4 summarizes the advantages of the methods in the literature.

4 Conclusion

A MAS is a process for creating goal-oriented autonomous agents in shared environments with coordination and communication capabilities. DDM benefits from this goal-oriented mechanism by executing different distributed clustering, prediction, and classification techniques. MASs are appropriate for distributed problem solving and allows for the development of goal-oriented autonomous agents that operate in shared environments with coordination and communication capabilities. They can be defined as a collection of agents that possess problem-solving functions and can interact with one another to achieve an overall objective. Agents are specialized problem-solving entities with well-defined boundaries and the ability to communicate with other agents. They are flexible, pro-active, and fulfill a distinct objective.

No assumptions are made about MAS environments in terms of data centralization, global control, and synchronization. Therefore, agents in MAS are assumed to operate with incomplete information or problem-solving capabilities. Communication is the key factor for agents to exchange information, coordinate their actions, and increase interoperation. Interactions between agents can be sought for particular services, information, or actions to be performed by other agents; these interactions are likewise necessary for issues concerning coordination, cooperation, and/or negotiation needed to organize interdependent activities.

The study presents a comprehensiveoverview ofvarious DDM methods, MASs, the advantages of MAS-based DDM, and the different MAS-based DDM methods that have been proposed in existing research.

Table 4 MAS with DDM literature

References	Mutual collaboration	Information sharing	Information hiding	Agent framework	Algorithm	Application	Publish year
MAS-Based Classification	No	Low-Level	No	JADE	Random Decision Tree	–	2008
MAS-Based Classification	No	Low-Level	No	JADE	Reptree	Weather Stations	2018
MAS-Based Classification	Yes	Low-Level	No	PANGEA	SVM	Musical Genre	2018
MAS-Based Classification	No	High-Level	No	JADE	Neural Network	Smart Grid	2015
MAS-Based Clustering	Yes	Low-Level	Yes	JADE	K-Means	–	2011
MAS-Based Clustering	No	High-Level	No	JADE	K-Means	ERP	2019
MAS-Based Clustering	No	High-Level	No	JADE	K-Means	ERP	2018
MAS-Based DDM	No	Low-Level	No	ROADMAP	Colored Petri Nets	Business-Intelligence	2018
MAS-Based Deep learning	No	High-Level	No	–	Plump-DP	Image	2018

References

A.A. Ali, P. Vařacha, S. Krayem, P. Žáček, A. Urbanek, Distributed data mining systems: techniques, approaches and algorithms, in *MATEC Web of Conferences*, vol. 210. (EDP Sciences, 2018), p. 04038

A. Amir, B. Srinivasan, A.I. Khan, Distributed classification for image spam detection. Multimedia Tools Appl. **77**(11), 13249–13278 (2018)

M.A. Almaiah, A. Al-Khasawneh, Investigating the main determinants of mobile cloud computing adoption in university campus. Educ. Inf. Technol. **25**(4), 3087–3107 (2020)

M. Adil, R. Khan, M.A. Almaiah, M. Al-Zahrani, M. Zakarya, M.S. Amjad, R. Ahmed, MAC-AODV based mutual authentication scheme for constraint oriented networks. IEEE Access **4**(8), 44459–44469 (2020)

M. Adil, R. Khan, M.A. Almaiah, M. Binsawad, J. Ali, A. Al Saaidah, Q.T.H. Ta, An efficient load balancing scheme of energy gauge nodes to maximize the lifespan of constraint oriented networks. IEEE Access **8**, 148510–148527 (2020)

M. Adil, M.A. Almaiah, A. Omar Alsayed, O. Almomani, An anonymous channel categorization scheme of edge nodes to detect jamming attacks in wireless sensor networks. Sensors **20**(8), 2311 (2020)

A.K. Al Hwaitat, M.A. Almaiah, O. Almomani, M. Al-Zahrani, R.M. Al-Sayed, R.M. Asaifi, K.K. Adhim, A. Althunibat, A. Alsaaidah, Improved security particle swarm optimization (PSO) algorithm to detect radio jamming attacks in mobile networks. Quintana **11**(4), 614–624 (2020)

M. Adil, R. Khan, J. Ali, B.H. Roh, Q.T. Ta, M.A. Almaiah, An energy proficient load balancing routing scheme for wireless sensor networks to maximize their lifespan in an operational environment. IEEE Access **31**(8), 163209–163224 (2020)

M.A. Almaiah, Z. Dawahdeh, O. Almomani, A. Alsaaidah, A. Al-khasawneh, S. Khawatreh, A new hybrid text encryption approach over mobile ad hoc network. Int. J. Electr. Comput. Eng. (IJECE) **10**(6), 6461–6471 (2020)

M.A. Almaiah, A. Al-Zahrani, O. Almomani, A.K. Alhwaitat, Classification of cyber security threats on mobile devices and applications. Artif. Intell. Blockchain Future Cybersecur. Appl. **107**

M.A. Almaiah, A new scheme for detecting malicious attacks in wireless sensor networks based on blockchain technology. Artif. Intell. Blockchain Future Cybersecur. Appl. **217**

M.A. Almaiah, M. Al-Zahrani, Multilayer neural network based on MIMO and channel estimation for impulsive noise environment in mobile wireless networks. Int. J. Adv. Trends Comput. Sci. Eng. **9**(1), 315–321 (2020)

M.A. Almaiah, M.M. Alamri, Proposing a new technical quality requirements for mobile learning applications. J. Theore. Appl. Inf. Technol **96**(19) (2018)

S. Bandyopadhyay, C. Giannella, U. Maulik, H. Kargupta, K. Liu, S. Datta, Clustering distributed data streams in peer-to-peer environments. Inf. Sci. **176**(14), 1952–1985 (2006)

N. Bouchemal, N. Bouchemal, Intelligent ERP based multi agent systems and cloud computing. In *International Conference on Machine Learning for Networking* (Springer, Cham, Nov. 2018), pp. 378–386

D. Chiang, C. Lin, M. Chen, The adaptive approach for storage assignment by mining data of warehouse management system for distribution centres. Enterp. Inf. Syst. **5**(2), 219–234 (2011)

C.Y. Chen, J.J. Huang, Double deep autoencoder for heterogeneous distributed clustering. Information **10**(4), 144 (2019)

C. Clifton, M. Kantarcioglou, X. Lin, M. Zhu, Tools for privacy preserving distributed data mining. ACM SIGKDD Exp. **4**(2) (2002)

A. Cuzzocrea, Models and algorithms for high-performance distributed data mining. Elsevier J. Parallel Distrib. Comput. **73**(93), 281–283 (2013)

R. Claes, T. Holvoet, D. Weyns, A decentralized approach for anticipatory vehicle routing using delegate multiagent systems. IEEE Trans. Intell. Transp. Syst. **12**(2), 364–373 (2011)

P. Charlton, R. Cattoni, Evaluating the deployment of FIPA standards when developing application services. Int. J. Pattern Recogn. Artif. Intell. **15**(03) (2001)

P. Charlton, R. Cattoni, A. Potrich, E. Mamdani, Evaluating the FIPA standards and their role in achieving cooperation. In *Multi-Agent Systems* (IEEE Xplore, Aug. 2002)

A. Dorri, S.S. Kanhere, R. Jurdak, Multi-agent systems: A survey. *IEEE*. Access **6**, 28573–28593 (2018)

G. Dudek, M.R. Jenkin, E. Milios, D. Wilkes, A taxonomy for multi-agent robotics. Auton. Robot. **3**(4), 375–397 (1996)

Y. Duan, B.X. Cui, X.H. Xu, A multi-agent reinforcement learning approach to robot soccer. Artif. Intell. Rev. **38**(3), 193–211 (2012)

R. Domínguez, S. Cannella, J.M. Framinan, Scope: a multi-agent system tool for supply chain network analysis, in *EUROCON 2015-International Conference on Computer as a Tool (EUROCON), IEEE* (IEEE, 2015), pp. 1–5

H. Du, S. Li, S. Ding, Bounded consensus algorithms for multiagent systems in directed networks. Asian J. Control **15**(1), 282–291 (2013)

FIPA, FIPA Abstract Architecture Specification, SC 00001L (2002a). http://www.fipa.org/specs/fip a00001/SC000011.pdf

FIPA, SC00067F (2002b). http://www.fipa.org/specs/fipa00067/SC00067F.pdf

I.E. Foukarakis, A.I. Kostaridis, C.G. Biniaris, D.I. Kaklamani, I.S. Venieris, *Webmages: An Agent Platform Based on Web Services*

W. Gan, J.C.W. Lin, H.C. Chao, J. Zhan, Data mining in distributed environment: a survey. Wiley Interdiscip. Rev.: Data Mining Knowl. Dis. **7**(6), e1216 (2017)

A.P. Garcia, J. Oliver, D. Gosch, An intelligent agent-based distributed architecture for smart-grid integrated network management, in *2010 IEEE 35th Conference on Local Computer Networks (LCN)* (IEEE, 2010), pp. 1013–1018

A. González-Briones, F. De La Prieta, M. Mohamad, S. Omatu, J. Corchado, Multi-agent systems applications in energy optimization problems: a state-of-the-art review. Energies **11**(8), 1928 (2018)

M. Gatti, P. Cavalin, S.B. Neto, C. Pinhanez, C. dos Santos, D. Gribel, A.P. Appel, Large-scale multi-agent-based modeling and simulation of microblogging-based online social network, in *International Workshop on Multi-Agent Systems and Agent-Based Simulation* (Springer, 2013), pp. 17–33

A. Goryashchenko, Algorithm and application development for the agents group formation in a multi-agent system using SPADE system, in *Future of Information and Communication Conference.* (Springer, Cham, Mar. 2019), pp. 1136–1143

A. Hudaib, M.H. Qasem, N. Obeid, FIPA-Based semi-centralized protocol for negotiation, in *Proceedings of the Computational Methods in Systems and Software* (Springer, Cham, Sept. 2017), pp. 135–149

D. Helbing, Agent-based modeling, in *Social self-organization* (Springer, 2012), pp. 25–70

C. Iddianozie, G. McArdle, A transfer learning paradigm for spatial networks, in *Proceedings of the 34th ACM/SIGAPP Symposium on Applied Computing* (ACM, Apr. 2019), pp. 659–666

I.F. Ilyas, X. Chu, X. Trends in cleaning relational data: Consistency and deduplication. Found. Trends® Databases **5**(4), 281–393 (2015)

T. Ishida, H. Yukoi, Y. Kakazu, Self-organized norms of behaviour under interactions of selfish agents, in *IEEE SMC '99 Conference Proceedings and IEEE Xplore Systems, Man and Cybernetics, 1999*, Aug. 2002

E. Januzaj, H.P. Kriegel, M, Pfeifle, Dbdc: density based distributed clustering, in *International Conference on Extending Database Technology.* (Springer, Berlin, Heidelberg, Mar. 2004), pp. 88–105

S. Jeong, U. Choi, J. Ahn, Distributed clustering algorithm for UAV systems, in *AIAA Scitech 2019 Forum*, p. 1795 (2019)

L.C. Jain, D. Srinivasan, *Innovations in Multi-agent Systems and Applications* (Springer, 2010)

F. Januário, A. Cardoso, P. Gil, Multi-Agent framework for resilience enhancement over a WSAN, in *2018 15th International Conference on Electrical Engineering/Electronics, Computer, Telecommunications and Information Technology (ECTI-CON)* (IEEE, July 2018), pp. 110–113

Y. Jiang, J. Jiang, Understanding social networks from a multiagent perspective. IEEE Trans. Parallel Distrib. Syst. **25**(10), 2743–2759 (2014)

N.R. Jennings, K. Sycara, M. Wooldridge, *A Roadmap of Agent Research and Development, Springer: Autonomous Agents and Multi-Agent Systems*, vol. 1, Issue 1, pp 7–38 (1998)

K. Kasemsap, Multifaceted applications of data mining, business intelligence, and knowledge management, in *Intelligent Systems: Concepts, Methodologies, Tools, and Applications*, (IGI Global, 2018), pp. 810–825

K. Kargupt, Chan, Distributed and parallel data mining: emergence, growth and future directions, in *Advances in Distributed Data Mining*, ed. by H. Kargupta, P. Chan (AAAI Press, 1999), pp. 407–416

M.N. Khan, H.U. Rahman, M.A. Almaiah, M.Z. Khan, A. Khan, M. Raza, M. Al-Zahrani, O. Almomani, R. Khan, Improving energy efficiency with content-based adaptive and dynamic scheduling in wireless sensor networks. IEEE Access **25**(8), 176495–176520 (2020)

K. Kannan, K. Raja, A. Rajakumar, P.K. Nizar Banu, *E-Business Decision Support System for Online Shopping using MAS with Ontology and JADE Methodology* (2019)

K. Kravari, E. Kontopoulos, N. Bassiliades,. EMERALD: a multi-agent system for knowledge-based reasoning interoperability in the semantic web, in *Hellenic Conference on Artificial Intelligence* (Springer, Berlin, Heidelberg, May 2010), pp. 173–182

K. Kravari, N. Bassiliades, H. Boley, Cross-community interoperation between knowledge-based multi-agent systems: A study on EMERALD and Rule Responder. Expert Syst. Appl. **39**(10), 9571–9587 (2012)

H. Li, L. Xu, J. Wang, Z. Mo, Feature space theory in data mining: transformations between extensions and intensions in knowledge representation. Expert Syst. **20**(2), 60–71 (2003)

B. Liu, S. Cao, W. He, Distributed data mining for e-business. Inf. Technol. Manag. **12**(2), 67–79 (2011)

T. Li, F. De la Prieta Pintado, J.M. Corchado, J. Bajo, Multi-source homogeneous data clustering for multi-target detection from cluttered background with misdetection. Appl. Soft Comput. **60**, 436–446 (2017)

R. Lu, K. Heung, A.H. Lashkari, A.A. Ghorbani, A lightweight privacy-preserving data aggregation scheme for fog computing-enhanced IoT. IEEE Access **5**, 3302–3312 (2017)

O. Lopez Ortega, F. Castro Espinoza, O. Perez-Cortes, *An Intelligent Multiagent System to Create and Classify Fractal Music* (Springer-Verlag GmbH Austria, Jan. 2018)

A. Louati, S. Elkosantini, S. Darmoul, H. Louati, Multi-agent preemptive longest queue first system to manage the crossing of emergency vehicles at interrupted intersections. Eur. Transp. Res. Rev. **10**(2), 52 (2018)

H. Li, C. Ming, S. Shen, W. Wong, Event-triggered control for multi-agent systems with randomly occurring nonlinear dynamics and time-varying delay. J. Franklin Inst. **351**(5), 2582–2599 (2014)

C. Moemeng, V. Gorodetsky, V., Z. Zuo, Y. Yang, C. Zhang, Agent-based distributed data mining: a survey, in *Data Mining and Multi-Agent Integration*, (Springer, Boston, MA, 2009), pp. 47–58

R. Mendes, J.P. Vilela, Privacy-preserving data mining: methods, metrics, and applications. IEEE Access **5**, 10562–10582 (2017)

P. Montero-Manso, L. Morán-Fernández, V. Bolón-Canedo, J.A. Vilar, A. Alonso-Betanzos, Distributed classification based on distances between probability distributions in feature space. Inf. Sci. (2018)

L.S. Melo, R.F. Sampaio, R.P.S. Leão, G.C. Barroso, J.R. Bezerra, Python-based multi-agent platform for application on power grids. Int. Trans. Electr. Energy Syst. e12012 (2019).

S.D. McArthur, E.M. Davidson, V.M. Catterson, A.L. Dimeas, N.D. Hatziargyriou, F. Ponci, T. Funabashi, Multi-agent systems for power engineering applicationsâ Tpart i: concepts, approaches, and technical challenges. IEEE Trans. Power Syst. **22**(4), 1743–1752 (2007)

L. Ma, H. Min, S. Wang, Y. Liu, S. Liao, An overview of research in distributed attitude coordination control. IEEE/CAA J. Automatica Sinica **2**(2), 121–133 (2015)

L. Ma, Y. Zhang, Hierarchical social network analysis using multiagent systems: a school system case, in *2014 IEEE International Conference on Systems, Man and Cybernetics (SMC)* (IEEE, 2014), pp. 1412–1419

L. Niu, N. Feng, Research on cooperation control of chassis multi-agent, in *2010 International Conference on Computer, Mechatronics, Control and Electronic Engineering*, vol. 2 (IEEE, Aug. 2010), pp. 464–467

M.A. Ouda, S.A. Salem, I.A. Ali, E.S.M. Saad, Privacy-preserving data mining (PPDM) method for horizontally partitioned data. Int. J. Comput. Sci. **9**(5), 339–347 (2012)

N. Obeid, A. Moubaiddin, A. Towards a formal model of knowledge sharing in complex systems, in *Smart Information and Knowledge Management* (Springer, Berlin, Heidelberg, 2010), pp. 53–82

R. Olszewski, P. Pałka, A. Turek, B. Kietlińska, T. Płatkowski, M. Borkowski, Spatiotemporal modeling of the smart city residents' activity with multi-agent systems. Appl. Sci. **9**(10), 2059 (2019)

R. Olfati-Saber, R.M. Murray, Consensus problems in networks of agents with switching topology and time-delays. IEEE Trans. Autom. Control **49**(9), 1520–1533 (2004)

A. Patel, W. Qi, C. Wills, A review and future research directions of secure and trustworthy mobile agent-based e-marketplace systems. Inf. Manag. Comput. Secur. **18**(3), 144–161 (2010)

L. Penait, S. Luke, *Co-operative Multi-Agent Learning: The State of the Art, Springer Science + Business Media, Netherlands: Autonomous Agents and Multi-Agent Systems*, vol. 11, pp. 387–434 (2005)

M.H. Qasem, N. Obeid, A. Hudaib M.A Almaiah, A. Al-Zahrani, A. Al-khasawneh, *Multi-Agent System Combined with Distributed Data Mining for Mutual Collaboration Classification* (IEEE Access, 20 Apr. 2021)

A.M. Ranwa, F. Bilal, F., Q. Alejandro, *Distributed Classification of Urban Congestion Using VANET (2019).* arXiv:1904.12685.

Russell, A.P. Norvig, Intelligence, "A modern approach", vol. 25 (Artificial Intelligence. Prentice-Hall, Egnlewood Cliffs, 1995), p. 27

Y. Rizk, M. Awad, E.W. Tunstel, Decision making in multiagent systems: a survey. IEEE Trans. Cogn. Dev. Syst. **10**(3), 514–529 (2018)

H. Rezaee, F. Abdollahi, Average consensus over high-order multiagent systems. IEEE Trans. Autom. Control **60**(11), 3047–3052 (2015)

Y. Ren, D. Fan, Q. Feng, Z. Wang, B. Sun, D. Yang, Agent-based restoration approach for reliability with load balancing on smart grids. Appl. Energy **249**, 46–57 (2019)

Z. Ruiz-Chavez, J. Salvador-Meneses, S. Díaz-Quilachamín, C. Mejía-Astudillo, (, October). Solid Waste Management using Georeferenced Multi-agent Systems. In *2018 IEEE Third Ecuador Technical Chapters Meeting (ETCM)* (IEEE, Oct. 2018), pp. 1–6

V. Sawant, K. Shah, A review of distributed data mining using agents. Int. J. Adv. Technol. Eng. Res. (IJATER) **3**(5), 27–33 (2013)

F. Stahl, M.M. Gaber, P. Aldridge, D. May, H. Liu, M. Bramer, S.Y. Philip, Homogeneous and heterogeneous distributed classification for pocket data mining, in *Transactions on Large-Scale Data-and Knowledge-Centered Systems V* (Springer, Berlin, Heidelberg, 2012), pp. 183–205

S. Sharmila, S. Vijayarani, Association rule hiding using firefly optimization algorithm, In *International Conference on Intelligent Systems Design and Applications* (Springer, Cham, Dec. 2018), pp. 699–708

W. Shen, et al. Applications of agent-based systems in intelligent manufacturing: an updated review. Adv. Eng. Inf. **20.4**, 415–431 (2006)

C.S. Shih, *Cooperative Adaptive Control for Multi-Agent Systems* (2018)

A.Q. Santos, R.M. Monaro, D.V. Coury, M. Oleskovicz, M., A new real-time multi-agent system for under frequency load shedding in a smart grid context. Electric Power Syst. Res. **174**, 105851 (2019)

S. Seng, K.K. Li, W.L. Chan, Z. Xiangjun, D. Xianzhong, Agent-based Self-healing Protection System, in *IEEE transactions on Power Delivery*, vol. 21, Issue 02, Apr. 2006

G. Tsoumakas, I. Vlahavas, Distributed data mining, in *Database Technologies: Concepts, Methodologies, Tools, and Applications* (IGI Global, 2009), pp. 157–164

Q. Tong, X. Li, B. Yuan, Efficient distributed clustering using boundary information. Neurocomputing **275**, 2355–2366 (2018)

The FIPA Specifications. www.fipa.org

S. Uppoor, M. Fiore, Large-scale urban vehicular mobility for networking research. in *Proceedings of the IEEE Vehicular Networking Conference(VNC)*, Nov. 2011, pp. 62–69

W. Van Der, M. Woolridge, Multi-Agent systems. *Handbook of Knowledge Representation.Elsevier B.V.* 2007.M (2007)

J. Vrancken, M.D.S. Soares, A real-life test bed for multi-agent monitoring of road network performance. Int. J. Crit. Infrastruct. **5**(4), 357–367 (2009)

X. Wu, X. Zhu, G.Q. Wu, W. Ding, Data mining with big data. IEEE Trans. Knowl. Data Eng. **26**(1), 97–107 (2013)

F. Wang, J. Sun, Survey on distance metric learning and dimensionality reduction in data mining. Data Min. Knowl. Disc. **29**(2), 534–564 (2015)

T.Y. Wu, J.C.W. Lin, Y. Zhang, C.H. Chen, A grid-based swarm intelligence algorithm for privacy-preserving data mining. Appl. Sci. **9**(4), 774 (2019)

O. Wangapisit, E. Taniguchi, J.S. Teo, A.G. Qureshi, Multi-agent systems modelling for evaluating joint delivery systems. Procedia Soc. Behav. Sci. **125**, 472–483 (2014)

G. Wen, G. Hu, W. Yu, J. Cao, G. Chen, Consensus tracking for higher-order multi-agent systems with switching directed topologies and occasionally missing control inputs. Syst. Control Lett. **62**(12), 1151–1158 (2013)

M. Wooldridge, *An Introduction to Multiagent Systems* (Wiley, NJ, 2008)

G. Weiss, *Multiagent Systems: A Modern Approach to Distributed Artificial Intelligence* (MIT Press, Cambridge, 1999)

D. Yuan, A. Proutiere, A., G. Shi, *Distributed Online Linear Regression* (2019). arXiv:1902.04774.

D. Ye, M. Zhang, A.V. Vasilakos, A survey of self-organization mechanisms in multiagent systems (IEEE)

N.-P. Yu, C.-C. Liu, Multiagent systems, in *Advanced Solutions in Power Systems: HVDC, FACTS, and artificial intelligence* (Wiley, Hoboken, NJ, 2016), pp. 903–930

Time Series Data Analysis Using Deep Learning Methods for Smart Cities Monitoring

Giuseppe Ciaburro

Abstract A time series is a sequence of empirical data ordered as a function of time. Time series analysis models exploit forecasting techniques based solely on the history of the variable of interest. They work by capturing patterns in historical data and extrapolating them into the future. The Times Series features recurring structures that can be captured through careful and precise analysis of its performance. Machine Learning-based methods are able to identify these recurring structures fully automatically. In this chapter we have faced the problem of the elaboration of forecasting models based on Deep Learning algorithms for data with time series characteristics. First, we introduced the Time Series, and we analyzed the most popular forecast models based on the traditional methodologies of classical Statistics. Next, we introduced Deep Learning-based methodologies that are inspired by the structure and function of the brain and which have proven effective in capturing the recurring characteristics of time series. In this context, we developed a model based on Recurrent Neural Networks for the prediction of equivalent noise levels produced by a road infrastructure. The model based on the LSTM was able to memorize the recurring structures present in the trend of the noise values and in the forecast, it preserved the daily and weekly trend characteristics already verified through a visual analysis.

Keywords Time Series analysis · Deep learning · Smart Cities · Time Series forecasting

1 Introduction

A time series is a finite set of observations, ordered in time, of a certain phenomenon that evolves randomly. It is a sequence of empirical data ordered as a function of time, this allows us to represent them in a time diagram and be able to understand any trends. Time, therefore, represents the crucial parameter for the detection of this

G. Ciaburro (✉)
Department of Architecture and Industrial Design, Università degli Studi della Campania Luigi Vanvitelli, 81031 Borgo San Lorenzo, Aversa, CE, Italy
e-mail: giuseppe.ciaburro@unicampania.it

© The Author(s), under exclusive license to Springer Nature Switzerland AG 2022   93
Y. Baddi et al. (eds.), *Big Data Intelligence for Smart Applications*,
Studies in Computational Intelligence 994,
https://doi.org/10.1007/978-3-030-87954-9_4

type of data, which may present characteristics of periodicity on a daily, weekly, monthly, half-yearly or yearly basis (Hamilton 2020). The data then follow one another maintaining a link between the subsequent observations that characterizes their behavior. It is precisely this link that is interesting to capture in order to be able to use it to elaborate future predictions of the evolution of the system over time (Wei 2006).

Time series analysis models exploit forecasting techniques based solely on the history of the variable of interest. They work by capturing patterns in Time data and extrapolating them into the future. Time series models can be applied when a reasonable amount of data and continuity in the near future of past conditions can be assumed (Chatfield and Xing 2019). These models are more suitable for making short-term forecasts, while they lose validity when the required forecast is long-term. This is due to the assumption that past patterns and current trends resemble patterns and trends in the future. The purpose of the analysis of the Time series consists in the study of the past evolution of the phenomenon with respect to time: The forecast is obtained by assuming that these regularities of behavior are repeated in the future. It is therefore assumed that the observation times are equidistant. This is also supported by the fact that many physical phenomena are recorded at intervals of the same amplitude (Fuller 2009).

The past dependence of time series is only partially dependent on the past, which is why these structures are treated as stochastic processes. A stochastic process can be interpreted as a family of signals defined on a common set of times or as a family of random variables defined on a common probability space (Christ et al. 2018). Stochastic processes are a tool for probabilistically representing uncertain quantities that depend on time in the same way that random variables are used to describe uncertain quantities that do not depend on time (Fulcher 2017).

Conversely, a deterministic signal is characterized by a predetermined trajectory that crosses time and space. In other words, the fluctuations of the signal are completely described by a functional relationship that allows to obtain the exact value at every instant of past, present, and future time (Moritz and Bartz-Beielstein 2017). Unfortunately, the realization of deterministic signals requires to be able to explain parameters whose knowledge is very expensive or even impossible (Ren et al. 2019).

In real life, the trend of the observed variables is probabilistic in the sense that it is not possible to accurately identify its future values. Therefore, to a deterministic contribution it is necessary to add a random disturbance component that characterizes its stochastic behavior (Folgado et al. 2018). The Times Series, however, have recurring structures that can be captured through a careful and precise analysis of its performance (Abanda et al. 2019). However, these structures are not easily identifiable, at least they are not simply through a visual approach. In the past years, researchers have attempted to capture these characteristics through methods based on traditional statistics. To these were then added the methods based on Machine Learning able to identify the recurring structures of the Time Series in a completely automatic way (Goodfellow et al. 2016). The term Machine Learning encompasses a set of techniques belonging to the world of artificial intelligence that allow a machine to improve its performance capabilities over time (Jordan and Mitchell 2015). The peculiarity of this family of techniques is the fact that the machines, as the name suggests, can automatically learn with experience to perform certain tasks, improving their performance more and more over time (Ciaburro et al. 2020a). Specifically, in recent years the use of methodologies based on Deep Learning has become widespread. It is a sub-category of Machine Learning and indicates that family of methods belonging to artificial intelligence that refers to algorithms inspired by the structure and function of the brain, or artificial neural networks (Mohri et al. 2018). These architectures are applied in different contexts, such as in Computer Vision, audio and spoken language recognition, natural language processing and bioinformatics (Mohri et al. 2018).

In this chapter we will analyze and describe some methods based on Deep Learning to capture the recurring structures from Times Series Data in order to be able to elaborate forecast scenarios. The remainder of the chapter is structured as follows: Sect. 2 introduces the Times Series, examining their characteristics and describes some of the traditional statistical techniques used for their analysis. Section 3 introduces deep learning-based methods for time series analysis. Section 4 reports the results of a study on a model for predicting the noise levels produced near a road infrastructure using a model based on recurrent neural networks. Finally, Sect. 5 summarizes the results obtained in applying these methods to real cases, highlighting their merits, and listing their limits.

2 Times Series Basic Concepts

Time series modeling has been an area of growing interest for many disciplines and in which much effort has been devoted to the development of new methods and techniques. Its objective is to provide the researcher with the mathematical representation of a time series, which allows to grasp in whole or in part the most relevant characteristics of the real phenomenon, from the information contained in the data (Granger and Newbold 2014). Although various models oriented to the representation of time series have been proposed in the literature, their usefulness depends on the degree of similarity between the dynamics of the series generation process and the mathematical formulation of the model with which it is represented. A time series is a stochastic process in which the set of observations are taken sequentially over time, usually at equal intervals. Therefore, a time series model for observed data is a specification of the joint distribution, possibly only the means and covariances, of a sequence of random variables (Tsay 2005).

It is commonly accepted that many time series exhibit non-linear dynamic behaviors, the complexity of which makes it impossible to formulate a model based on physical or economic laws that adequately represent their evolution (Lütkepohl 2013). The problem of formulating the model is aggravated by the presence of atypical observations and structural changes, for which there are no mathematical models that allow their representation in the non-linear case (Brockwell et al. 2016). One of the characteristics that distinguishes time series from other types of statistical data is the fact that, in general, the values of the series at different times are correlated.

In this way, the fundamental problem in the analysis of time series is the study of the correlation pattern of the values of the series and the elaboration of a statistical model that explains its structure (Lütkepohl 2005).

Traditionally, time series have been considered as the sum of two contributions: a deterministic component and a random disturbance contribution. We can therefore represent a Time Series through Eq. (1):

$$Y_t = f(t) + w(t), \tag{1}$$

The terms present in Eq. (1) are defined as follows,

- Y_t is the Time Series
- $f(t)$ is the deterministic component
- $w(t)$ is the random noise component

The deterministic part determines the temporal development of the event through a specific law, on the contrary the random contribution adds information of negligible entities which, however, modify its progress (Hannan 2009). The random contribution is due to random phenomena, attributable to a series of accidental errors: It assumes stochastic characteristics similar to a white noise signal, therefore a sequence of independent and identically distributed random variables with zero mean and constant variance. Traditional statistical methods assume the deterministic contribution is relevant, leaving out the random contribution that is assumed with unrelated components and therefore negligible (Weigend 2018).

In these data, time becomes crucial as it specifies the sequence of events and determines the position of observation along the time dimension. To carry out a first visual analysis of the data, it is possible to represent the pair of values (time, observation) in a Cartesian diagram. Figure 1 shows the trend in the number of airline passengers from 1949 to 1960 (Datasets and Package 2021).

Figure 1 shows regular trends and other systematic trends over time: We can indeed see a systematically increasing trend in the long run, while a zigzag trend is observed and since the data are monthly, we can distinguish the phenomenon called seasonality. In fact, high peaks occur in the summer months when more air travel is expected due to summer holidays.

In the trends relating to Time series, we can see oscillations around a long-term trend called components of the series. We can define four main types of components (Durbin and Koopman 2012):

- Trend component (T) (Fig. 2a): it is the result of factors that influence the data in the long run, which generate a gradual and consistent pattern of variations in the same series. In the case of the data shown in Fig. 1, the trend of airline passengers has been growing over the years, due to the greater number of trips for work and leisure that social and technological progress has brought about. According to traditional methodologies, the trend can be exponential, hyperbolic, or harmonic, but other trends, such as linear, quadratic, or cubic, can be considered in the data analysis.

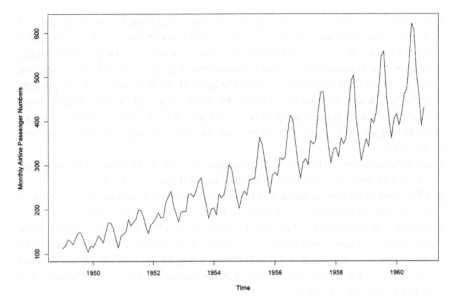

Fig. 1 Monthly airline passenger numbers from 1949 to 1960

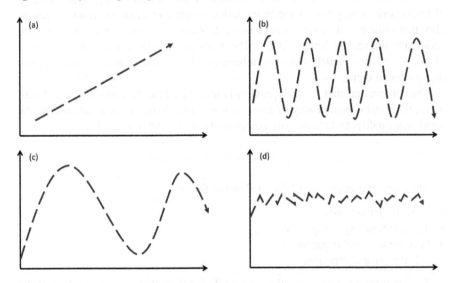

Fig. 2 Components of a time series: **a** Trend component (T); **b** Seasonal Components (S); **c** Cyclical component (C); **d** Irregular component (E)

- Seasonal Components (S): Both cyclical and seasonal changes can be understood as the undulating movement of the data above and below the trend line. But the difference between one and the other has to do with the range of data being analyzed. In the case of the cyclical components (Fig. 2c) we refer to processes that occur more widely within the observation interval, while the seasonal components occur with less separation over time. For example, when analyzing the data shown in Fig. 1, a seasonal component (Fig. 2b) may refer to changes in the number of passengers in the summer when more air travel is expected due to summer holidays.
- Cyclical component (C): These are fluctuations with a periodicity of between one and a half and up to ten years, depending on the definition of the cycle used. They tend to be less frequent and less systematic than seasonal ones. A cyclical component may present expansion or contraction over a longer period.
- Irregular component (E) (Fig. 2d): it relates to data that are very deviating from the trend, as occurs when an aircraft is stopped. In this case, decisions need to be made on how to handle these outliers, as including them in the calculations would add a lot of noise to their description. For this reason, a value is usually estimated to replace anomalous data.

The methods of time series analysis consider the fact that the data collected in different periods may have some characteristics of autocorrelation, trend or seasonality that must be taken into consideration to model the behavior of the observed series (Baydogan and Runger 2016). The application of these methods serves two purposes: to understand the forces of influence in the data and to discover the structure that produced them.

The components that we have previously introduced can be combined in different ways, the most common way of representing a time series is as a function of the trend, seasonality, and irregularity components, that is through Eq. (2).

$$Y_t = f(T_t, S_t, \varepsilon_t) = T_t + S_t + \varepsilon_t, \tag{2}$$

The terms present in Eq. (2) are defined as follows,

- Y_t is the Time Series
- T_t is the trending component
- S_t is the seasonal component
- ε_t is the error component

The error term ε_t represents the random fluctuations that cause the values of the series to deviate from the average level, these fluctuations are a little more complex to explain within the analysis of the time series in most cases. For this reason, it is convenient to introduce a random component that allows greater flexibility within the analysis (Costa et al. 2002).

An alternative model predicts only the presence of the trend component and the irregular component, as indicated in Eq. (3).

$$Y_t = T_t + \varepsilon_t, \qquad (3)$$

The terms present in Eq. (3) are defined as follows,

- Y_t is the Time Series
- T_t is the trend component
- ε_t is the error component

There are several widely used mathematical forms for trends, for example linear, exponential, and logistic.

The simplest trend is the linear trend represented by Eq. (4).

$$T_t = \alpha + \beta * t, \qquad (4)$$

where α, β are constant, the latter is also called the trend component. The coefficient β represents the average increase in Y per unit of time. Other authors prefer to call it the trend of the parameter β, which represents the slope of the linear model. According to the latter interpretation, the trend is the variation of the average level per unit of time. The trend in Eq. (4) is a deterministic function of time and is known as the global linear trend. If β is greater than zero there is a linear growth, while if β is less than zero there is a decline over a long period. In the case where $\beta = 0$, it means that there is no growth or decline for a long time in the series, and it is said to be a trendless series. A convenient way to estimate α and β is with the least square's method (Ng and Perron 2005).

The exponential trend shape is used when the trend reflects an almost constant percentage of growth. Since it poses a risk to assume that exponential growth will continue indefinitely, then one can adapt the logistic trend form that provides S-like trends.

In case a seasonal component is detected, it must be added to the model. This addition can be made in different ways, the first and simplest is the additive model which provides a simple sum between the different contributions, as already reported in Eq. (2). Another model is the multiplicative one in which the three components (trend, seasonality, and error) are multiplied by each other. Finally, a third model provides a mixed mode between the two just seen, then we can multiply the trend and seasonality components together, and then add the term due to the error to the result. In all cases, the component due to the error must have the characteristics of white noise, therefore an average of zero and constant variance.

An additive model is suitable, for example, when the seasonal component does not depend on other components, such as for the trend, such as, it can be used for series of time models that show constant seasonal variation. If, on the contrary, the seasonality varies with the trend, such as, the time series shows an increasing or decreasing seasonal variation, the most appropriate model is the multiplicative one: However, the multiplicative model can be transformed into additive using logarithms.

After having analyzed in detail the components of a time series, it is necessary to introduce the most widespread models that have allowed researchers over time to elaborate predictions on time series data.

2.1 Autoregressive Model (AR)

In the time series, it is possible to notice a strong correlation between consecutive values, when this correlation is verified for adjacent values, it is called first order autocorrelation (Wei 2006). If this correlation occurs between the values of the series after two periods, then it is defined as second order autocorrelation, and if this characteristic is highlighted after p periods it will be defined as pth order autocorrelation. The Autoregressive models exploit these correlations between the time series data to obtain predictions on the future trend of the series: This is a linear predictive modeling methodology. The procedure involves the estimation of specific parameters, called AR parameters, which will then be used as coefficients for the estimation of future values. The future value of the series is estimated as a linear combination of the past values of the variables. The AR model of order p is represented by Eq. (5).

$$Y_t = c + \phi_1 * Y_{t-1} + \cdots + \phi_p * Y_{t-p} + \varepsilon_t \tag{5}$$

The terms present in Eq. (5) are defined as follows,

- Y_t are the components of the Time Series
- c is a constant
- ϕ_i are the parameters of the AR model
- ε_t is the error component (white noise)

As you can see, this is a normal linear regression model in which the response variable is the present value of the process while the explanatory variables are the past values of the process itself.

2.2 Moving Average Model

The moving average (MA) model models time series, based on the moving average of their past terms. It is an alternative tool for estimating the trend and cyclical components (Said and Dickey 1984). The moving average acts by reducing the amplitude of the oscillations, so the greater the number of observations used for calculating the average, the greater the flattening. Recall that the t-term moving average is equal to the arithmetic mean of the series.

The MA model of order p is represented by Eq. (6).

$$Y_t = c + \theta_1 * \epsilon_{t-1} + \cdots + \theta_p * \epsilon_{t-p} + \varepsilon_t \tag{6}$$

The terms present in Eq. (6) are defined as follows,

- Y_t are the components of the Time Series
- c is a constant
- θ_i are the parameters of the MA model
- ε_t is the error component (white noise)

If the series consists only of trends and the irregular component, the moving average eliminates the effects of noise. If in the original series there is also the seasonal phenomenon of period p, then a moving average of amplitude p is able to eliminate also the seasonality.

2.3 Autoregressive Moving Average Model

The autoregressive moving average model (ARMA) is a linear mathematical model that provides instant-by-instant an output value based on previous input and output values (Said and Dickey 1984). Each internal parameter, therefore, will at each instant be set equal to a linear combination of all internal parameters of the previous instant and of the input value, and the output value will in turn be a linear combination of the internal parameters and in rare cases also than the incoming one. They combine the two classes of models just seen.

The MA model of order p is represented by Eq. (7).

$$Y_t = c + \phi_1 * Y_{t-1} + \cdots + \phi_p * Y_{t-p} + \theta_1 * \varepsilon_{t-1} + \cdots + \theta_p * \varepsilon_{t-p} + \varepsilon_t, \quad (7)$$

The terms present in Eq. (7) are defined as follows,

- Y_t are the components of the Time Series
- c is a constant
- ϕ_i are the parameters of the AR model
- θ_i are the parameters of the MA model
- ε_t is the error component (white noise)

The autocorrelation function can have very different trends. The general rule is that the first p coefficients are substantially arbitrary, after the autocorrelation function converges towards zero, in the same way, starting from zero of the autocorrelation functions of an AR model.

3 Machine Learning-Based Methods

Machine Learning, or automatic learning, is a rapidly expanding field of artificial intelligence that allows machines to perform specific tasks in the light of experience and mistakes made, without the need for a continuous input into programming by the man. In this way it is possible to acquire greater independence and efficiency. In fact, these machines are able to autonomously reach the goal for which they were programmed by extracting recurring information from a set of data with the aid of Machine Learning methods. However, it has been observed that some activities of immediate resolution for humans are complicated for a machine if only the classical methods of Machine Learning are applied. To address this problem, new types of algorithms called Deep Learning have been developed, capable of providing the machine with the ability to extract simple information from a set of data and then combine them to obtain more elaborate information. The final result is achieved by processing, in a hierarchical way, different levels of representation of the data corresponding to characteristics, factors or concepts.

3.1 Artificial Neural Networks

Artificial Neural Networks—ANNs are mathematical models formed by artificial neurons that take inspiration from the biological functioning of the human brain. To function properly they need considerable hardware resources, and it is for this reason that they have become very popular lately despite the fact that the technology was born many years ago (Hassoun 1995). The key feature of these systems lies in the fact that they learn by exploiting mechanisms similar to those of human intelligence, giving them performances that are impossible to achieve for other algorithms (Ciaburro et al. 2020b).

Artificial neural networks are inspired by the functioning of the human brain: The neural networks of the human brain are in fact the seat of the ability to understand the environment and its changes and are able to provide adaptive responses specifically tailored to the needs that arise. They consist of highly interconnected sets of nerve cells (Gevrey et al. 2003). Within them we find a series of fundamental elements:

- Neuronal soma: represents the body of neurons. Their job is to receive and process information.
- Axons: they constitute the way of communication in output from a neuron.
- Neurotransmitter: these are biological compounds of different categories, synthesized in the somes and responsible for the modulation of any nerve impulse.
- Synapses: these are highly specialized functional sites where information is passed between neurons.
- Dendrites: they are the main input communication route.

A fundamental concept of neurons, from which the technology has taken its cue having also been implemented in artificial networks, is that a single neuron can simultaneously receive signals from different synapses. One of its intrinsic capacity is to measure the electrical potential of these signals globally, thus establishing whether the activation threshold has been reached to generate a nerve impulse in turn (Abiodun et al. 2018). This property is also implemented in artificial networks. Another key

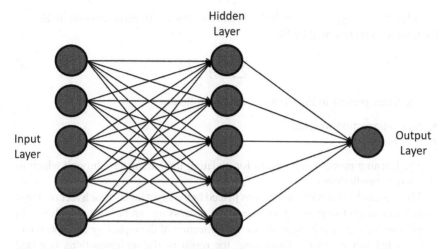

Fig. 3 Architecture of a three-layer Artificial Neural Network

feature is the dynamism of the synaptic configuration: The number of synapses can increase or decrease depending on the stimuli that the network receives.

A biological network receives external data and signals that are perceived through the senses. These are processed into information through a massive number of interconnected neurons in a non-linear and variable structure in response to those external data and stimuli. Likewise, artificial neural networks are non-linear structures of statistical data organized as modeling tools. The ANNs receive external signals on a layer of nodes, each of these input nodes is connected to various internal nodes of the network which, typically, are organized on several levels so that each single node can process the received signals by transmitting to subsequent levels the result of his elaborations (Anitescu et al. 2019).

Typically, neural networks consist of three layers (Fig. 3):

- Input Layer: it is the first layer of the network and is the one that has the function of receiving and processing the input signals, adapting them to the demands of the neurons within the network.
- Hidden Layer: it is the intermediate layer that is in charge of the actual processing process.
- Output Layer: it is the final layer where the processing results of the Hidden layer are collected and adapted to the requests of the next block-level of the neural network.

An artificial neuron represents the processing unit of the structure and its action can therefore be represented by the following equation:

$$y_i = f(x_i) \tag{8}$$

In Eq. (8), f is typically non-linear and is called the activation function. In Eq. (8) the term xi is represented by Eq. (9).

$$x_i = \sum (w_{ji} + b_i)$$ (9)

The terms present in Eq. (9) are defined as follows:

- w_{ji} = connection weights
- b_i = bias

The learning process is intended for updating the parameters through which to obtain data predictions.

The process begins with the neurons placed in the input layer, which receive input stimuli that are subsequently processed. The processing can be very complex or in simple cases it can be represented as a multiplication of the inputs for an appropriate value called weight. In the latter case, the result of the multiplications is added and if the sum exceeds a certain threshold, the neuron is activated, consequently activating its output. The weight therefore has the function of an indicator of the synaptic efficacy of the entry line and has the function of quantifying its importance (Ciaburro and Iannace 2021).

The tasks to which neural networks are applied can be classified into the following broad categories of applications:

- Approximation or regression functions including time series analysis and modeling.
- Classification, including the structuring of sequences of generic data, the identification of novelties within a dataset with the consequent facilitation of the decision-making process.
- Data processing, including filtering, clustering, signal separation and compression.

Finally, it should be noted that the models produced by neural networks, even if very efficient, cannot be explained in human symbolic language: the results must in fact be accepted as they are, hence the definition of neural networks as black boxes.

3.2 Big Data Analytics

The term Big Data defines a large set of data in terms of volume and management complexity that requires different processing methods and technologies than traditional software and computer architectures. To understand the difficulties inherent in managing Big Data, we can think of a traditional database consisting of tables with many records but a few dozen fields: In the case of Big Data, tools are required that can manage the same number of records, but with thousands of fields. The problem of Big Data management arises because the amount of data that modern infrastructures are able to generate has become really high: Web traffic, that generated by modern smartphones, as well as the data detected by the countless quantities of sensors installed for city monitoring represent only a few sources of data generation (Tsai et al. 2015).

Huge volumes of heterogeneous data by source and format, which can be analyzed in real time, constitute what are defined as Big Data. In recent years, the number of data produced in the world has increased dramatically and companies are expected to be able to produce Zettabytes of data by considering sources such as sensors, data, financials, satellites, etc. The need for analysis on a single set of data is the cause of the progressive increase in the size of the data sets. The goal is to extract patterns and useful information compared to those that could be obtained by analyzing small series as in the past. Another distinction between Big Data and traditional datasets is that Big Data mainly includes data from heterogeneous sources, therefore they are not formed only by structured data, such as databases, but also unstructured, such as images, data collected by GPS, email, and information taken from social networks (Rajaraman 2016).

To analyze huge amounts of data, different types of algorithms have been born over the years. In addition to differing in the technologies used and the logic they use to bring results, these algorithms are distinguished by the type of analysis they conduct on the data. In particular, there are three types of analysis that can be carried out on Big data:

- Descriptive analysis: It is the simplest type of analysis; these are studies that aim to synthesize and describe raw data. These analyzes are conducted on large databases with the aim of finding hidden correlations between the data, so that useful information emerges for those conducting the analysis. In order to summarize and clarify the dynamics and performance of pre-established metrics and to obtain indications on the best way to approach future activities, it deals with analyzing past events from different points of view. This type of analysis is of vital importance for companies, which thanks to the results obtained are able to understand what is happening in a specific domain of interest and make targeted choices accordingly (Ristevski and Chen 2018). Although it is not necessary to use sophisticated technologies to conduct these research, without the help of computational power, humans would not be able to conduct these types of analyzes.
- Predictive analysis: It aims to predict what will happen in the future in a given application field. To conduct these analyzes it is necessary to have previously carried out a descriptive analysis on the data of interest, in order to have information regarding the temporal evolution of the data and any correlations between data sets. Subsequently, algorithms will be able to project this information in the more or less near future and highlight what may be the trends and the results that will be reached. It should be emphasized, however, that no statistical algorithm is able to predict the future with accuracy, but it is still possible to arrive at very good absolute probabilities which are often satisfactory for companies (Choi et al. 2018).

- Prescriptive analysis: This analysis goes beyond the prediction of future results that predictive analysis performs, automatically providing recommendations on the best actions to take. This is possible thanks to the synthesis of data and the joint use of mathematical sciences, business rules and Machine Learning technologies. The similarity with predictive analysis is very high, both provide the user with information about what the future is likely to be. The difference with the previous type of analysis, however, is that the prescriptive analysis explains how to act to achieve the desired goal. Prescriptive analytics are complex to administer, and most companies are not using them yet. However, when implemented correctly, they can have a big impact on how companies make decisions and, in doing so, help them deliver the right products at the right time, thus optimizing the customer experience.

3.3 Deep Learning Algorithms

These are algorithms that use various levels of non-linear units. These levels are used in cascade to perform tasks that can be classified as problems of transforming characteristics extracted from the data. Each level uses the output of the previous level as an input (Sejnowski 2018). These algorithms fall within the broader class of data representation learning algorithms within machine learning. They are formed by multiple levels of representation that can be understood as different levels of abstraction, capable of forming a hierarchy of concepts. The mechanisms by which the human brain works are then simulated by these algorithms, in order to teach machines not only to learn autonomously as in Machine Learning, but to do it in a deeper way as the human brain would also do, whereby deep we mean on multiple conceptual levels. Although the demand for enormous computational capabilities may represent a limit, the scalability of Deep Learning to the increase in available data and algorithms is what differentiates it from Machine Learning (Ciaburro 2020).

Deep Learning systems, in fact, improve their performance as data increases, while Machine Learning applications, once a certain level of performance is reached, are no longer scalable even by adding a greater number of examples and training data to the neural network. To train a Deep Learning system usually the employees use to label the data. For example, in the field of visual recognition, the bird meta tag can be inserted into images containing a bird and, without explaining to the system how to recognize it, the system itself, through multiple hierarchical levels, will guess what characterizes a bird. (the wings, the beak, the feathers), and therefore to learn to recognize it (Esteva et al. 2019).

In order for the final output to be satisfactory, huge amounts of data are needed, but thinking about Big Data right away is a mistake, as at least in the initial part of training, the system needs labeled and structured data, therefore the heterogeneity of Big Data cannot be considered a solution. Unstructured data can be analyzed by a deep learning model once it is formed and reached an acceptable level of accuracy, but not for the initial training phase (Voulodimos et al. 2018). Deep Learning today is already applied in various fields: self-driving cars, but also drones and robots used for the delivery of packages or for emergency management are excellent examples. Finally, speech recognition and synthesis for chatbots and service robots, facial recognition for surveillance and predictive maintenance are among the most relevant emerging applications (Ciaburro and Iannace 2020).

4 Implementing an LSTM to Forecast the Traffic Noise

4.1 Introduction

Knowledge of the noise levels that characterize a given area represents a fundamental step for the description of the acoustic state and for the definition of remediation interventions, but also an important basis for territorial and urban planning and programming. The possible objectives of a monitoring activity for the estimation of the population index exposed to noise are such as to require a separate assessment of the noise coming from the different sources (Ko et al. 2011). In fact, both in order to define the mitigation interventions and to establish impacts on the population, the distinction of the share of environmental noise to be attributed to the different sources is indispensable.

It is evident that the main source of noise to which large sections of the population are systematically exposed is the urban road network (Smith et al. 2017). The acoustic state of the city can be defined through a representation of data relating to an existing noise situation as a function of an acoustic descriptor, which indicates the exceeding of relevant limit values, the number of people exposed, or the number of houses exposed to certain values of the chosen acoustic descriptor (Can et al. 2008). But it is equally useful to have models for forecasting the noise produced by vehicular traffic starting from the data collected during the monitoring period (Cirianni and Leonardi 2012; Kumar et al. 2011).

The noise generated by road traffic varies according to the volume of traffic, the type of vehicles that constitute it, the speed, the driving behavior. The resulting sound field depends on numerous external conditions that affect its propagation.

By analyzing the traces of the noise emitted by vehicular traffic, it is possible to identify trends that distinguish it as a data with time series characteristics (Murthy et al. 2007).

Road traffic noise is caused by heavy vehicles, light vehicles, and motorcycles. The noise produced by vehicles originates from various components: engine, air resistance, tire rolling, accessory engines, as well as the actuation of the brakes (Kulauzović et al. 2020). The engine is home to compressions, pops and decompressions that produce a quantity of noise that is directly related to the number of revolutions.

The rolling of the tires on the asphalt is a source of noise as a result of the trapping and subsequent release of air from the cavities, as well as vibrations on the bodywork.

The noise deriving from the air resistance is generally detected only at speeds above 200 km/h, therefore in a field that is extraneous to the normal flow of urban road traffic. Finally, the action of the brakes that manifests itself through the friction between the lining and the disc. If the pressure between the two elements is high, it can cause the tire to drag on the asphalt; the combined action of the two phenomena causes high noise levels. The noise produced by the motor vehicle engine is, at low speeds, higher than that produced by the rolling of the tires on the asphalt. As the speed increases, the rolling noise becomes more intense until it prevails over that produced by the engine.

On the other hand, as far as heavy vehicles are concerned, the engine component always predominates over the pneumatic component. It should also be considered that particular aspect of urban noise consisting of the sound of sirens that inform the activity of various public utility services (Ögren et al. 2018).

4.2   Materials and Methods

The models for predicting noise from vehicular traffic are elaborated starting from a database of sound levels detected in standard conditions when single vehicles of predefined types and at pre-established speeds pass (Steele 2001). In this work we will apply a different approach, we will make predictions of the noise generated by road traffic on the basis of the data detected by appropriate sensors using algorithms based on recurrent neural networks. The data was collected in a measurement campaign carried out near a road artery. The area in which the measurement was carried out is characterized by intense human activity due to productive settlements and widespread urbanization, with a population density of about 1500 inhabitants/sq. Km. These activities contribute significantly to traffic volumes, particularly during peak hours.

The measurement methods, the position of the microphone and the acquisition parameters are linked to the nature and purpose of the survey (ISO 2017). The microphone was positioned in a place representative of the sound energy to which the disturbed are exposed, at a distance of at least one meter from the facades of the buildings exposed to the highest noise levels and at an altitude of 4 m from the ground.

The monitoring was carried out over sufficiently long detection times, in order to allow an overall view of the sound energy present in the investigated place. The monitoring was performed with special instrumentation, with the microphone placed in correspondence with the position occupied by the sensitive receptors.

The noise measurement was performed for a measurement time of not less than one week. In this period, the A-weighted equivalent continuous level was measured for each hour over the entire 24-h period using a mobile monitoring station. A mobile station consists of devices equipped with a sufficient amount of memory to store the acoustic monitoring data for several days, periodically downloading them manually or with the aid of GSM transmitters.

These stations provide for the use of simplified outdoor microphones and an autonomous power supply system that allows operation without being connected to the mains. The instruments are normally placed inside mobile vehicles, specially set up, or in suitable suitcases, to be used in short and medium-term monitoring. The following instrumentation was used:

- Only 01 dB integrating sound level meter model of "Class 1"
- Larson Davis CAL 200 Calibrator
- Tripod
- Windproof headphones

The monitoring period lasted three weeks, which was sufficient to analyze the daily and weekly trends.

4.3 Recurrent Neural Network (RNN)

Recurrent neural networks are one of the possible models for dealing with time series. Its main advantage lies in the possibility of storing a representation of the recent history of the series, which allows, unlike what happens with artificial neural networks, that the output of a given input vector can vary according to the current internal configuration of the network (Li et al. 2018). A recurrent neural network (RNN) is a type of ANN that provides feedback connections, or loops or loops within the network structure. This type of backward connection allows you to store the status from one iteration to the next, using your output as an input for the next step (Arras et al. 2017). The network receives inputs, produces an output, and returns the latter back to itself (Fig. 4).

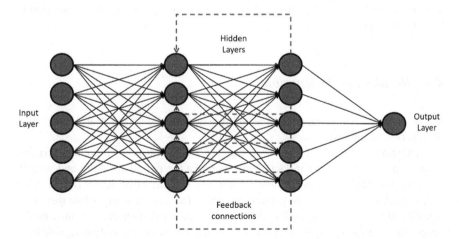

Fig. 4 Architecture of a Recurrent Neural Network with indication of the feedback connections

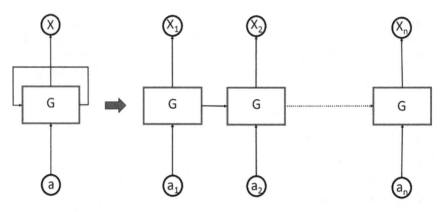

Fig. 5 Simplified diagram of the learning mechanisms of an RNN

To better understand the structure of an RNN, we can unfold the network which gives us a feedforward version of the structure (Fig. 5). In this structure the weights and biases of block G are shared, each output Xt depends on the processing by the network of all inputs ai with i ∈ [0; t]. The number of blocks of the unfolded version essentially depends on the length of the sequence to be analyzed (Lin et al. 2021).

Among the many RNN architectures developed over the years, Long Short-Term Memory (LSTM) (Hochreiter and Schmidhuber 1997) has met with considerable success and has proposed itself as a structure capable of reducing learning problems in RNNs. The LSTMs have a structure that allows them to grasp long-term dependencies in the data, thanks to the memory cell, capable of storing the state outside the normal flow of the recurring network (Fischer and Krauss 2018). The state is calculated as the sum between the previous state and the current input processing modulated through masks so that only part of them is remembered. The network has, in fact, the ability to remove, add or read the information contained therein at any time, respectively through the gates.

4.4   Results and Discussion

In Fig. 6 are shown the equivalent sound pressure level detected near a road artery for a period of three weeks.

Analyzing the graph of Fig. 6 we can see the typical recurring characteristics of the time series, we can in fact clearly identify the periods of daily and weekly observations. During a day, in fact, we can see that the noise levels rise in value during peak periods when the population moves for work reasons, while they drop significantly during the night. But this is not the only trend we can see, moving the observation period from day to week we can see that noise levels drop significantly over the weekend. In fact, both Saturdays and Sundays there are values that are

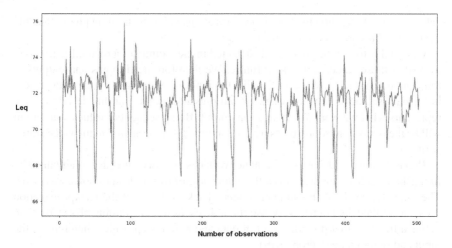

Fig. 6 Trend of vehicular traffic noise for an observation period of three weeks (number of observations versus Leq (dBA))

significantly lower than working days. This demonstrates that the percentage of the population that goes to work at the weekend is much lower.

From this first visual analysis we can deduce that road traffic noise, being dependent on traffic flows, essentially depends on the volume of the population that uses this means of transport to go to work. If we had extended the observation period to one year, we could have seen that this descriptor is characterized by seasonal components. In fact, in the summer period as a large part of the population moves to the holiday resorts, then the cities record a considerable reduction in noise levels, which instead increase precisely in the holiday resorts.

In order to be able to adequately manage a traffic plan in order to keep the population's exposure to noise within the limits imposed by the legislation, it is crucial to have forecast models that can provide us with estimates of road traffic noise levels. In this section I present a methodology for the elaboration of such forecast scenarios based on algorithms that use recurrent neural networks. Specifically, a Long Short-Term Memory (LSTM) is used, implemented with the use of the Python Keras library (Hochreiter and Schmidhuber 1997).

To begin with, the data in our possession, which refer to a monitoring of the noise levels introduced by vehicular traffic, have been appropriately separated into two subsets that will be used for training and subsequently for testing the network. Given a dataset of approximately 504 data records, we divide it into a 70:30 ratio, meaning that 70% of the data (353) will be used in the network training phase, while the remaining 30% (151) will be used for model testing. Usually, the data subdivision is performed randomly in order to guarantee an effective network training procedure. With data with time series characteristics this procedure is not feasible as the sequence of values is important. A simple method we can use is to divide the dataset sequentially, first the first 353 values are used for the training phase, then the next 151 values are used for the test.

Our goal is to use the noise levels detected in the vicinity of a road infrastructure, to predict these values in the near future. We therefore need inputs and outputs to train and test our network. It is clear that the input is represented by the data present in the data set, hence the equivalent sound pressure levels. But this descriptor also represents what we want to obtain as an output: Being sequential values, our output will be the equivalent sound pressure level at time t + 1 compared to the value stored at time t.

Previously, we said that a recurrent network has memory, and this is maintained by fixing the so-called time step, this is the main reason why this design choice was made. The time step is about the number of steps back in time that the backpropagation uses to calculate the gradients for weight updates during the training phase. We then define a function that provides a set of data and a time step, which then returns the input and output data (Géron 2019).

At each level of a LSTM an input must be sent in the form of a 3D tensor with these features:

- Batch: Number of noise levels detected
- TimeSteps: Equivalent to the amount of time steps performed on the recurring neural network
- Features: they are simply the number of dimensions that we feed in each time step

Once trained, we use the model to make a prediction on vehicle noise levels Fig. 7. From the analysis of the results, we can see that the model returns predictions that are very close to the values detected in the monitoring phase. This result is guaranteed both in the training phase and in the test phase. From a more precise control we can see that the model underestimates the peak values of the noise, on the contrary we can

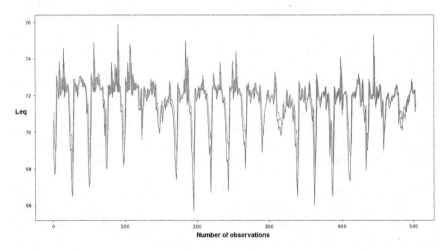

Fig. 7 Comparison of measured and predicted values from the LSTM-based model

Table 1 RMSE for the LSTM model

Subset	RMSE
Training	1.07
Test	1.00

see that the lowest noise values detected are instead overestimated. Therefore, the difficulties encountered by the model are in predicting the limit values (Sherstinsky 2020).

To evaluate the performance of the model, we calculated the Root Mean Square Error, which represents an accuracy measure based on the scale of the analyzed data and is therefore not very suitable for comparing time series with data with different scales. This measure metric is defined by the Eq. (10).

$$RMSE = \sqrt{\frac{1}{N} \sum_{i=1}^{N} (y_i - \hat{y}_i)^2} \tag{10}$$

Here,

- y_i is the actual value.
- $\hat{y}_i$ is the forecasted value.
- N is the number of the sample.

Table 1 shows the RMSE values returned by the forecasts made with the model based on the LSTMs.

The obtained RMSE values confirm that the forecasts obtained with the LSTM-based model are very close to the data collected through monitoring. This is what we had already deduced from the visual analysis of the data shown in Fig. 7.

5 Conclusions

In this chapter we have faced the problem of elaborating forecast models for data with time series characteristics. A time series is a finite set of observations, ordered in time, of a certain phenomenon that evolves randomly. It is a sequence of empirical data ordered as a function of time, this allows us to represent them in a time diagram and be able to understand any trends. Time series analysis models exploit forecasting techniques based solely on the history of the variable of interest. They work by capturing patterns in historical data and extrapolating them into the future. The Times Series present recurring structures that can be captured through a careful and precise analysis of its performance.

However, these structures are not easily identifiable, at least they are not simply through a visual approach. The methods based on Machine Learning are able to identify these recurring structures in a completely automatic way: These techniques allow a machine to improve its performance capabilities over time, automatically learning with experience to perform certain tasks, improving more and more its performance over time. Deep Learning-based methodologies are inspired by the structure and function of the brain and have proven effective in capturing the recurring characteristics of Time Series. In this context, the Recurrent Neural Networks with their memory capacity have proved to be very efficient in predicting the equivalent levels of noise produced by a road infrastructure. The model based on the LSTM was able to memorize the recurring structures present in the trend of the noise values and in the forecast, it preserved the daily and weekly trend characteristics already verified through a visual analysis.

The methodology studied in this work will be able to enrich the technologies supporting smart cities. The architectures of the cities of the future will be full of interconnected sensors that will return a large amount of data. These data will be able to provide all the information necessary to guarantee the needs of citizens and their safety. Furthermore, all the information necessary to ensure high health standards can be extracted from these data. Among these models based on machine learning coupled with a correct installation of acoustic sensors will provide useful tools for effective noise control.

The design choices on the network architecture are essential in order to ensure that the network learns well and is able to correctly approximate the function you intend to learn. Future progress of this work will focus on possible improvements in forecasting capacity. A possible approach to improve the convergence of an LSTM consists in starting from a sufficiently high learning-rate that does not lead to divergences, then decreasing it over time. Another advance can be made with the adoption of regularization techniques that aim to increase the generalization capacity of an NN by reducing overfitting and therefore the difference between the error in the training-set and in the test-set.

References

A. Abanda, U. Mori, J.A. Lozano, A review on distance based time series classification. Data Min. Knowl. Disc. **33**(2), 378–412 (2019)

O.I. Abiodun, A. Jantan, A.E. Omolara, K.V. Dada, N.A. Mohamed, H. Arshad, State-of-the-art in artificial neural network applications: a survey. Heliyon **4**(11), e00938. (2018)

C. Anitescu, E. Atroshchenko, N. Alajlan, T. Rabczuk, Artificial neural network methods for the solution of second order boundary value problems. Comput. Mater. Continua **59**(1), 345–359 (2019)

L. Arras, G. Montavon, K.R. Müller, W. Samek, Explaining recurrent neural network predictions in sentiment analysis (2017). arXiv:1706.07206

P.J. Brockwell, R.A. Brockwell, R.A. Davis, R.A. Davis, *Introduction to Time Series and forecasting* (Springer, 2016)

M.G. Baydogan, G. Runger, Time series representation and similarity based on local autopatterns. Data Min. Knowl. Disc. **30**(2), 476–509 (2016)

C. Chatfield, H. Xing, *The Analysis of Time Series: An Introduction with R* (CRC Press, 2019)

M. Christ, N. Braun, J. Neuffer, A.W. Kempa-Liehr, Time series feature extraction on basis of scalable hypothesis tests (tsfresh–a python package). Neurocomputing **307**, 72–77 (2018)

G. Ciaburro, G. Iannace, J. Passaro, A. Bifulco, D. Marano, M. Guida, ... F. Branda, Artificial neural network-based models for predicting the sound absorption coefficient of electrospun poly (vinyl pyrrolidone)/silica composite. Appl. Acoustics **169**, 107472 (2020)

M. Costa, A.L. Goldberger, C.K. Peng, Multiscale entropy analysis of complex physiologic time series. Phys. Rev. Lett. **89**(6), 068102 (2002)

G. Ciaburro, G. Iannace, M. Ali, A. Alabdulkarem, A. Nuhait, An artificial neural network approach to modelling absorbent asphalts acoustic properties. J. King Saud Univ.-Eng. Sci. (2020)

G. Ciaburro, G. Iannace, Acoustic characterization of rooms using reverberation time estimation based on supervised learning algorithm. Appl. Sci. **11**(4), 1661 (2021)

T.M. Choi, S.W. Wallace, Y. Wang, Big data analytics in operations management. Prod. Oper. Manag. **27**(10), 1868–1883 (2018)

G. Ciaburro, Sound event detection in underground parking garage using convolutional neural network. Big Data Cogn. Comput. **4**(3), 20 (2020)

G. Ciaburro, G. Iannace, Improving smart cities safety using sound events detection based on deep neural network algorithms, in *Informatics*, vol. 7, No. 3 (Multidisciplinary Digital Publishing Institute, Sep. 2020) ,p. 23

A. Can, L. Leclercq, J. Lelong, J. Defrance, Capturing urban traffic noise dynamics through relevant descriptors. Appl. Acoust. **69**(12), 1270–1280 (2008)

F. Cirianni, G. Leonardi, Environmental modeling for traffic noise in urban area. Am. J. Environ. Sci. **8**(4), 345 (2012)

J. Durbin, S.J. Koopman, *Time Series Analysis by State Space Methods* (Oxford University Press, 2012)

Esteva, A. Robicquet, B. Ramsundar, V. Kuleshov, M. DePristo, K. Chou, ... J. Dean, A guide to deep learning in healthcare. Nat. Med. **25**(1), 24–29 (2019)

W.A. Fuller, *Introduction to Statistical Time Series*, vol. 428. (Wiley, 2009)

B.D. Fulcher, *Feature-based Time-series Analysis* (2017). arXiv:1709.08055

D. Folgado, M. Barandas, R. Matias, R. Martins, M. Carvalho, H. Gamboa, Time alignment measurement for time series. Pattern Recogn. **81**, 268–279 (2018)

T. Fischer, C. Krauss, Deep learning with long short-term memory networks for financial market predictions. Eur. J. Oper. Res. **270**(2), 654–669 (2018)

I. Goodfellow, Y. Bengio, A. Courville, Mach. Learn. Basics. Deep Learn. **1**, 98–164 (2016)

C.W.J. Granger, P. Newbold, *Forecasting Economic Time Series* (Academic Press, 2014)

M. Gevrey, I. Dimopoulos, S. Lek, Review and comparison of methods to study the contribution of variables in artificial neural network models. Ecol. Model. **160**(3), 249–264 (2003)

A. Géron, Hands-on machine learning with Scikit-Learn, Keras, and TensorFlow: Concepts, tools, and techniques to build intelligent systems. O'Reilly Media (2019)

J.D. Hamilton, *Time Series Analysis* (Princeton University Press, 2020)

Hannan, E. J. (2009). Multiple time series (Vol. 38). John Wiley & Sons.

M.H. Hassoun, *Fundamentals of Artificial Neural Networks* (MIT Press, 1995)

S. Hochreiter, J. Schmidhuber, Long short-term memory. Neural Comput. **9**(8), 1735–1780 (1997)

ISO 1996–2: 2017. Description, measurement and assessment of environmental noise. Part 2: determination of sound pressure levels. Switzerland: International Organization for Standardization (2017)

M.I. Jordan, T.M. Mitchell, Machine learning: trends, perspectives, and prospects. Science **349**(6245), 255–260 (2015)

J.H. Ko, S.I. Chang, M. Kim, J.B. Holt, J.C. Seong, Transportation noise and exposed population of an urban area in the Republic of Korea. Environ. Int. **37**(2), 328–334 (2011)

K. Kumar, V.K. Katiyar, M. Parida, K. Rawat, Mathematical modeling of road traffic noise prediction. Int. J. Appl. Math Mech, **7**(4), 21–28 (2011)

B. Kulauzović, T. Pejanović Nosaka, J. Jamnik, Relationship between weight of the heavy trucks and traffic noise pollution in the viewpoint of feasibility of fines for exceeded noise–a case study, in *Proceedings of 8th Transport Research Arena TRA* (2020)

H. Lütkepohl, *Introduction to Multiple Time Series Analysis* (Springer Science & Business Media, 2013)

H. Lütkepohl, *New Introduction to Multiple time Series Analysis* (Springer Science & Business Media, 2005)

J.C.W. Lin, Y. Shao, Y. Djenouri, U. Yun, ASRNN: a recurrent neural network with an attention model for sequence labeling. Knowl.-Based Syst. **212**, 106548 (2021)

S. Li, W. Li, C. Cook, C. Zhu, Y. Gao, Independently recurrent neural network (indrnn): Building a longer and deeper rnn, in *Proceedings of the IEEE Conference on Computer Vision and Pattern Recognition*, pp. 5457–5466 (2018)

M. Mohri, A. Rostamizadeh, A., A. Talwalkar, *Foundations of Machine Learning* (MIT Press, 2018)

V.K. Murthy, A.K. Majumder, S.N. Khanal, D.P. Subedi, Assessment of traffic noise pollution in Banepa, a semi urban town of Nepal. Kathmandu Univ. J. Sci. Eng. Technol. **3**(2), 12–20 (2007)

S. Moritz, T. Bartz-Beielstein, imputeTS: time series missing value imputation in R. R J. **9**(1), 207 (2017)

S. Ng, P. Perron, A note on the selection of time series models. Oxford Bull. Econ. stat. **67**(1), 115–134 (2005)

M. Ögren, P. Molnár, L. Barregard, Road traffic noise abatement scenarios in Gothenburg 2015–2035. Environ. Res. **164**, 516–521 (2018)

H. Ren, B. Xu, Y. Wang, C. Yi, C. Huang, X. Kou, ... Q. Zhang, Time-series anomaly detection service at Microsoft, in *Proceedings of the 25th ACM SIGKDD International Conference on Knowledge Discovery & Data Mining*, pp. 3009–3017, July 2019

V. Rajaraman, Big data analytics. Resonance **21**(8), 695–716 (2016)

B. Ristevski, M. Chen, Big data analytics in medicine and healthcare. J. integr. Bioinf. **15**(3) (2018)

S.E. Said, D.A. Dickey, Testing for unit roots in autoregressive-moving average models of unknown order. Biometrika **71**(3), 599–607 (1984)

T.J. Sejnowski, *The Deep Learning Revolution* (Mit Press, 2018)

R.B. Smith, D. Fecht, J. Gulliver, S.D. Beevers, D. Dajnak, M. Blangiardo, ... M.B. Toledano, Impact of London's road traffic air and noise pollution on birth weight: retrospective population based cohort study. Bmj, **359** (2017)

C. Steele, A critical review of some traffic noise prediction models. Appl. Acoust. **62**(3), 271–287 (2001)

A. Sherstinsky, Fundamentals of recurrent neural network (RNN) and long short-term memory (LSTM) network. Physica D: Nonlinear Phenomena, **404**, 132306 (2020)

R.S. Tsay, *Analysis of Financial Time Series*, vol. 543 (Wiley, 2005)

The R Datasets Package (2021). https://stat.ethz.ch/R-manual/R-devel/library/datasets/html/00Index.html Accessed 15 April 2021

C.W. Tsai, C.F. Lai, H.C. Chao, A.V. Vasilakos, Big data analytics: a survey. J. Big Data **2**(1), 1–32 (2015)

A. Voulodimos, N. Doulamis, A. Doulamis, E. Protopapadakis, Deep learning for computer vision: a brief review. Comput. Intell. Neurosci. (2018)

A.S. Weigend, *Time Series Prediction: Forecasting the Future and Understanding the Past* (Routledge, 2018)

W.W. Wei, Time series analysis, in *The Oxford Handbook of Quantitative Methods in Psychology*, vol. 2 (2006)

A Low-Cost IMU-Based Wearable System for Precise Identification of Walk Activity Using Deep Convolutional Neural Network

Amartya Chakraborty⊕ and Nandini Mukherjee⊕

Abstract The increasing popularity of smart wearable devices can be attributed to progress in research towards human activity recognition. With the use of tiny sensing units, the different human activities of day-to-day life can be identified with a high accuracy. However, as a fitness tracking product, smart wearables are not always accurate in determining actual physical motion details. For example, recognizing walk activity may not always be possible when the monitoring device is held in hand, or kept in pockets. For precise recognition of walk activity, movement of legs needs to be monitored. However, the movement of legs while walking must be distinguished from simple leg swing activities. The present work designs an IMU based sensing system that can prevent false identification of a mimicked walk or leg swing in sitting posture as a real walk activity, using conventional and convolutional deep learning algorithms. The system shows remarkable capability of identifying an actual walk from a mimicked walk activity using CNN 95% of the time.

1 Introduction

Rapid advancement in sensing technologies has ushered in a new, smart era. Coupled with the strive for a better life, these smart solutions have been adapted as products of daily use in almost every sector of our life, starting from education to healthcare. The recognition of human activities from a wider aspect to an intrinsic level, lies at the centre of such smart products. Simultaneously, as a research problem, the recognition of human Activities of Daily Living (ADL) has been addressed for many years now. This finds applications in elderly monitoring, indoor tracking, fall detection, assisted living etc. Depending on the target activity set, the sensing can be performed using ambient sensors or wearable devices individually, or a combination of both. While

A. Chakraborty (✉) · N. Mukherjee
Department of Computer Science and Engineering, Jadavpur University, Kolkata, India
e-mail: nandini.mukhopadhyay@jadavpuruniversity.in

© The Author(s), under exclusive license to Springer Nature Switzerland AG 2022   117
Y. Baddi et al. (eds.), *Big Data Intelligence for Smart Applications*,
Studies in Computational Intelligence 994,
https://doi.org/10.1007/978-3-030-87954-9_5

the environment-deployed ambient sensing systems have been extensively used for determining human activity related information, wearable systems perform on a per-person basis, at a more intrinsic level.

One popular and frequently used product is a smart fitness tracking device, generally used for monitoring human walk activity. Recent studies show that such type of wearable devices are projected to ship over ten times more in 2023, starting with less than 30 million units in 2014 (Statista 2014). This reflects how people have become more conscious about their bodily fitness related statistics. These fitness trackers are generally worn on the wrist or as anklets, and are capable of alerting the users about their vital physical statistics such as duration of walk, heart rate, step count etc. With increasing dependence on such systems in recent times, researchers have tried to determine their ineffectiveness in estimating fitness characteristics (Chen et al. 2016). The development of miniature, smart health sensing wearable devices is also a very popular field of research. Most of the human motion data captured by such wearable devices are generated by the Inertial Measurement Unit (IMU) based accelerometer and gyroscope sensors. Such sensors generate time-variant data corresponding to the subjects' linear acceleration and angular velocity in a 3-axis co-ordinate system. Coupled with a micro-controller chip, these sensors can be used to determine different sets of human ADLs with high precision. Their presence is also noted in smartphones nowadays, and are utilized by multiple application programs for physical fitness monitoring, similar to smart fitness devices.

Study of a recent, comprehensive survey reveals that though there are many possibilities such as chest, wrists, ankles etc., the selection of ideal deployment position for wearable sensing systems is highly debated (Attal et al. 2015). Feasibility and comfort are also factors to be considered, and these may be affected with prolonged use of the wearable systems, as in the case of systems worn on the chest, waist etc. A simpler and more comfortable solution is wrist-bands or anklets. As the activity of walking generally involves a sequence of swing-step-swing action with legs, and alternate swinging of hands, smart wrist-bands used for walk detection may be prone to error in case of only hand swing without actual walking. Similarly, smartphones when held in hand, often fail to differentiate between normal hand movement, periodic hand swing and an actual walk activity with hand swing. As such, we may consider human leg movement as a direct and better indicator of walking activity, as also observed by Cleland et al. (2013) in their work. However, this understanding may be proven wrong if a person decides to sit and imitate the leg movement of actual walk, as shown in Fig. 1.

The current work takes up this challenge of developing a wearable, leg-worn system in the form of a smart anklet that can be used to learn and differentiate between a normal walk (swing-step-swing with forward movement using both legs) and a mimicked walk activity which consists of repeated leg swing (swing-step-swing of legs in sitting posture). The system utilizes state of the art learning algorithms for enhancing the performance of activity recognition system, while reducing misclassification. This chapter is focused on the development of a smart wearable IMU-based system that can efficiently distinguish between normal walk and mimicked walk activities to eliminate the false detection of walk activity. It also contains detailed

Fig. 1 The two activities—a Normal Walk (towards the left) and a Mimicked Walk, or Leg-Swing in sitting state (towards the right)

discussions on the concepts and general techniques that will be fruitful for students and researchers working in the domain of Human Activity Recognition Systems (HARS).

The rest of the chapter is organized as follows: Sect. 2 gives a study of the recent research with respect to the wearable deployments and learning techniques used by state-of-the-art systems; Sect. 3 illustrates the variation in characteristics of data for the two activities, system setup, data acquisition, standardization and feature extraction processes, and preparation of final data corpus; Sect. 4 discusses the different learning techniques used; the evaluation metrics and experimental results are illustrated and analyzed in Sect. 5; lastly the concluding remarks and future scope are given in Sect. 6.

2 Related Works

The rapid progress in the field of Human Activity Recognition is a result of different challenges that researchers have overcome for many years. This has been aided by the rapid miniaturization and increasing availability of sensing devices. Another factor that plays a very important role is the adaption of learning algorithms that can be used to evaluate the performance of experimental setups. The systems that aim to recognize different human activities of daily living can be broadly aggregated as using wearable, or environment embedded devices. From a general understanding of their working principles, the wearable systems are able to identify activities at a more intrinsic level (drinking coffee etc.), while environment-embedded systems can be used for identifying person-level activities in a multi-person scenario (standing, sitting etc.). Also, the wearable systems have become easier to access and use, as even all smartphones these days come with a wide array of sensors and can act as wearable sensing devices too (Lane et al. 2010). The variation in types of activities detected is also huge in the case of the former, as revealed by a study of the state-of-the-art works.

The most common approach of physical activity recognition using wearable systems involves the use of Inertial Measurement Unit (IMU). The IMU sensors are essentially accelerometers, gyroscope and magnetometers, which have been heavily used in isolation and together for identification of different sets of physical activity. These activities consist of indoor (Attal et al. 2015; Tran and Phan 2016; Altun and Barshan 2010) and outdoor ones (Barshan and Yüksek 2014), upto locating and tracking of individuals (Hardegger et al. 2015; Zhang et al. 2017) and others. Health monitoring from physical factors (Avci et al. 2010), vitality (Cho et al. 2008) and social behaviour (Lane et al. 2011) have been used as indicators of well-being using off-the-shelf sensors and android based smartphones separately. The combination of smart-phones and wearable sensor has been used for monitoring academic performance of students and personality traits by Sano et al. (2015). Other than external activity, bodily disorders due to diseases can also be diagnosed by using such systems (Kumari et al. 2017). Habitual traits such as smoking can be recognized using smart-phone and smart-watches in the work by Shoaib et al. (2015). Physical activities of daily living such as walking, running, standing, sitting, fall detection., as well as abnormality of gait have been identified using body-worn acoustic (Cheng et al. 2011; Jang et al. 2007), foot-force (Li et al. 2018; Similä et al. 2017), IMU (Hegde and Sazonov 2015; Chandra et al. 2019; Yacchirema et al. 2018; Kumar et al. 2019), camera and pressure (van de Ven et al. 2015; Bounyong et al. 2016; Ozcan and Velipasalar 2015) and GPS (Zhang and Poslad 2014) sensing devices.

A proper recognition of activities is critical in determining the applicability of a Human Activity Recognition System (HARS) in a real-life scenario. The systems are modelled using standard classifiers and algorithms and tested to check their ability of identifying newer data patterns. These approaches can be broadly divided in two parts, namely machine learning and learning using neural networks. Some widely used machine learning algorithms based on mathematical models, have been used for designing efficient activity recognition systems such as Support Vector Machine (Barshan and Yüksek 2014; Ahmed et al. 2020), k-Nearest Neighbours (Altun and Barshan 2010; Jian and Chen 2015), Decision Tree (Fan et al. 2013; Jatoba et al. 2008), etc. On the other hand, Artificial Neural Networks (ANN) (Pires et al. 2020) are becoming increasingly popular due to their inherent ability of adapting to mutating input data via minimization of prediction loss. Networks with more hidden layers (deep neural networks) (Nweke et al. 2018), convolutional layers (CNN) (Zeng et al. 2014), and the deep CNNs (Yang et al. 2015; Panwar et al. 2017) are being used in state-of-the-art works for different challenges in activity recognition.

From the literature review, it is observed that a wide range of activity recognition problems can be solved with the help of smart wearable setups and adaptive learning algorithms. Among the various types of human activities identified, the problem of identifying walking activity is quite common and has been addressed in many works, such as in Barshan and Yüksek (2014), Kwapisz et al. (2011), and Banos et al. (2014). However, a notable technical gap remains in the fact that no such system has been designed that can distinguish between an actual walk activity and mimicked walk activity in sitting posture. So the problem of mis-identification of walk activity remains un-addressed amidst an ever increasing usage of smart wearable devices. Consequently, the challenge of designing a fool-proof wearable system that

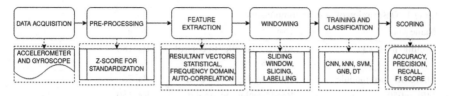

Fig. 2 Layout of the proposed system

can smartly differentiate between the leg-swing and actual walk activities, is yet to be met. Motivated by these observations, the proposed work contributes as follow:

- Our current work is, to the best of our knowledge, the only work where a IMU-based smart anklet is designed with the challenge of recognizing a real walk activity from leg swing (swing-step-swing of legs in sitting posture) as an alternative to wrist worn activity monitoring systems,
- A systematic, comparative study of the performance of classification algorithms such as deep convolutional neural networks and machine learning algorithms using real and mimicked walk data acquired from the proposed system.

3 System Details

This section illustrates the implementation and working of the overall system along with the computation framework used for proper identification of a real walk activity. Each sub-section in the following discussion provides detailed descriptions for all the steps followed in implementation of the proposed methodology namely, data acquisition, data standardization, feature extraction, data processing and labelling, followed by experimentation (training, testing) and performance evaluation. Figure 2 gives an overview of the implemented system model comprising the mentioned steps. This system framework illustration follows the general structure implemented in state-of-the-art wearable activity recognition systems (Pires et al. 2020; Avci et al. 2010; Chetty et al. 2015) etc.

3.1 Deployment Details

The system developed during the course of this work consists of a portable, battery powered, setup of inter-connected sensing devices controlled by a micro-controller circuit. The miniature size of all the components used in developing this setup has resulted in a light-weight, wearable module that can be worn as an anklet for activity monitoring. All the off-the-shelf, low-cost components used in this wearable module are mentioned in Table 1 along with their respective function and cost.

Table 1 Description of system components

Name	Use	Price
Arduino nano v3	Micro-controller to control sensors	Rs. 180
Micro SD card module	Data logging in SD card	Rs. 75
MPU-6050	Read accelerometer and gyroscope data	Rs. 150

3.2 Data Acquisition

The smart wearable anklet used in this work captures the variation in activity data retrieved from the 3-axis IMU module MPU-6050,[1] which consists of the respective vectors in x, y and z axis. The acquired data lies within ranges of +2g to −2g and +250 deg/s to −250 deg/s respectively, with corresponding default sensitivity scale factors in effect. These values are then logged in the SD card locally in CSV format. The experiments have been conducted in a supervised manner in indoors with 10 volunteers of varied physique. Every volunteer was asked to perform both normal walking, and leg swing (swing-step-swing with legs while sitting on a chair) activities, by wearing the smart anklet setup for specific durations upto 10 min. The ground truth information for all the acquired data was recorded for future reference. The sampling rate of the data is set 10 Hz, which gives us 10 data points every second. Every such data point in the raw data is a tuple of the following 7 components: $\{T_s, A_x, A_y, A_z, G_x, G_y, G_z\}$, where,

T_s denotes the current data and time instance, at which,

A_x gives the Linear Acceleration value along the x-axis,

A_y gives the Linear Acceleration value along the y-axis,

A_z gives the Linear Acceleration value along the z-axis,

G_x gives the Angular Velocity value along the x-axis,

G_y gives the Angular Velocity value along the y-axis,

G_z gives the Angular Velocity value along the z-axis,

Figures 3 and 4 show the variations in a sample corpus of raw 3-axis accelerometer and gyroscope data, for actual walk and leg swing activities respectively. While there is a distinct variation in the patterns of accelerometer data for the two activities as seen in Fig. 3, the gyroscope data for the two performed activities have more visual similarity, but lie in different range of values, as evident in Fig. 4. This temporal information, along with corresponding frequency domain features extracted in the later sections, is utilized by the learning algorithms in the experiments, for precise walk activity identification.

---

[1] https://robu.in/product/mpu6050gyrosensor2accelerometer.

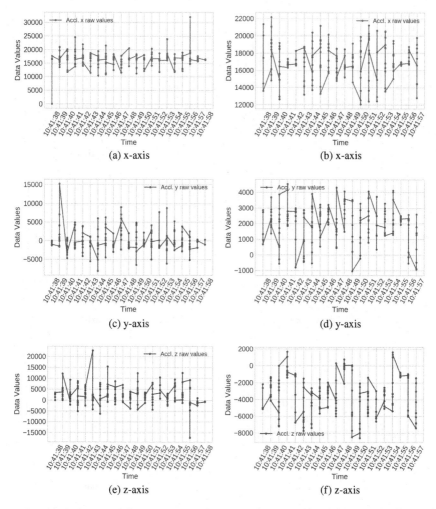

Fig. 3 Illustration of 3-axis Accelerometer data for the two activities: walk (on the left) and leg-swing (on the right)

3.3 Data Standardization and Feature Extraction

In this section, a sequence of steps have been followed to process the acquired data so that it can be used for the learning process. There has been extensive research regarding the feature extraction process, and in here some of these most common and relevant features are utilized, as discussed in the state-of-the-art review in Pires et al. (2020). The different intermediate forms of the processed data, along with the techniques used, are discussed below.

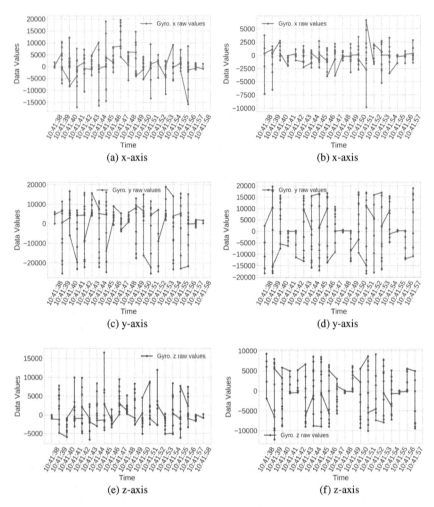

Fig. 4 Illustration of 3-axis Gyroscope data for the two activities: walk (on the left) and leg-swing (on the right)

3.3.1 Determining the Characteristics of Raw Data

In Fig. 5, a sample distribution of acquired data is shown for the x-axes of accelerometer (during actual walk activity) and gyroscope (for leg swing). From the histograms, it can be determined that the distribution of data is gaussian in nature. However, these normal distributions lie in different range of values for the two sensors, depending on the activity type. As a first step, such variations in the distribution of 3-axes data from both sensors is normalized.

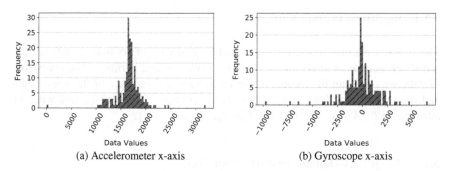

Fig. 5 Sample distribution of raw data for two activities

3.3.2 Standardisation of Data

The main motive behind standardisation is to minimise the variability of data and bring them to the same scale, where all the data points are re-scaled by shifting them with the mean of the set of values and dividing by their standard deviation. This widely used method is called *z-score* (Kreyszig 2009), and is calculated as:

$$z - score = \frac{x_i - \mu}{\sigma} \tag{1}$$

where, x_i denotes the current data element, μ denotes the mean of the set of elements and σ stands for the standard deviation. In this step, the complete corpus for both the accelerometer and gyroscope sensors is standardized in this manner, denoted by: $\{T_s, A'_x, A'_y, A'_z, G'_x, G'_y, G'_z\}$.

3.3.3 Calculation of Resultant Vectors

The standardised raw sensor data in a 7-tuple form is now processed, and from the values of the IMU data along the 3 axes, we initially determine the resultant acceleration (R_α) and angular velocity (R_ω) values using the equations:

$$R_\alpha = \sqrt{A'^2_x + A'^2_y + A'^2_z} \tag{2}$$

and

$$R_\omega = \sqrt{G'^2_x + G'^2_y + G'^2_z} \tag{3}$$

Now we have for every time instance T_s, the 6 acquired elements and 2 new features, given as: $\{T_s, A'_x, A'_y, A'_z, G'_x, G'_y, G'_z, R_\alpha, R_\omega\}$.

3.3.4 Statistical Feature Extraction

The simplest features that can be extracted from any set of continuous data are the statistical features, namely *mean, standard deviation* and *variance*. For every row of data, we calculate the mean, standard deviation and variance of the sensor data and their resultant values. After this step, the corpus consists of the following 11 attributes: $\{T_s, A'_x, A'_y, A'_z, G'_x, G'_y, G'_z, R_\alpha, R_\omega, \mu, \sigma^2, \sigma\}$.

3.3.5 Frequency Domain Feature Extraction

The temporal data acquired by our wearable module captures details of actual walk and leg swing activities, which consist of periodically repeated actions determined by the variation in patterns in data points. This periodicity can be better analyzed by using *Fourier transformation* (Bracewell and Bracewell 1986) on the signal We have implemented the widely used *Fast Fourier Transform* (FFT) algorithm (Cooley and Tukey 1965) for converting the acquired data from time-domain to frequency domain. Such transformation breaks down the whole spectrum as a distribution of periodic components. The generated FFT values are complex in nature, i.e. they contain a real part and an imaginary part which correspond to the magnitude and phase of the signal respectively, and both these values are used as features. Thus the *FFT co-efficients* for the 3-axis accelerometer and gyroscope data and their resultant values, generate 16 additional FFT features. For instance, the two FFT co-efficient features for Accelerometer x-axis data are: $\{A_{xFFT}', A_{yFFT}'\}$. The squares of absolute FFT co-efficient values, generate another feature, namely the *Power Spectral Density* (PSD) of the signal. It describes the frequency spectrum of the signal along with the corresponding power components therein. These PSD values are also added to our corpus as features. For instance, the two PSD features for Accelerometer x-axis data are: $\{A_{xPSD}', A_{yPSD}'\}$. Lastly, we calculate the corresponding *FFT Frequencies* for the PSD values, as another set of features for our corpus. So, in this step, our data set is enriched by another 32 features, along with every component in the element vector from previous section. Figures 6 and 7 give the plots of resultant accelerometer and gyroscope power features in terms of the inherent frequencies for the walk and leg swing activities respectively, and highlight the difference between the two activities in frequency domain.

3.3.6 Auto-Correlation

Another widely used feature in time series analysis, is auto-correlation (Bracewell 1965) which gives an idea about the linear relationship between the time-variant observations in the data set, and the model of non-random time series data. The auto-correlation function with lag k for a set of equally spaced (in terms of time) observations $y_1, y_2, \ldots, y_n$ corresponding to time instances $t_1, t_2, \ldots, t_n$ can be defined as:

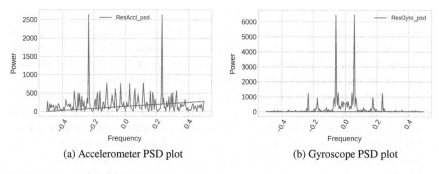

(a) Accelerometer PSD plot　　　　　　(b) Gyroscope PSD plot

Fig. 6 Power spectrum plots for walk activity

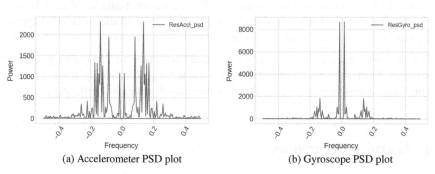

(a) Accelerometer PSD plot　　　　　　(b) Gyroscope PSD plot

Fig. 7 Power spectrum plots for leg swing activity

$$A_k = \frac{\sum_{i=1}^{n-k}(y_i - \mu)(y_{i+k} - \mu)}{\sum_{i=1}^{n}(y_i - \mu)^2} \tag{4}$$

The auto-correlation feature values for all the basic components in our data are computed to get 8 new corresponding features in our corpus from Sect. 3.3.5, namely: $\{A'_{xAutoCor}, A'_{yAutoCor}, A'_{zAutoCor}, G'_{xAutoCor}, G'_{yAutoCor}, G'_{zAutoCor}, R_{\alpha AutoCor}, R_{\omega AutoCor}\}$.

Thus, our final processed corpus consists of 3 statistical, 32 FFT and 8 auto-correlation component features along with the 3-axis standard values of acquired data and their resultant values for two sensors, to give a total of 51 components in every data point. This feature-rich corpus is now processed as described in the next section.

3.4 Data Processing

The data points, after adding features as discussed in Sect. 3.3, are now processed further before using the corpus for classification, as described below:

- *Sliding Window and slicing*: initially, we have used sliding windows such that subsequent element vectors have a 50% overlap between themselves. Given that labelled slices of data will be used during the learning process, this technique ensures that time dependent variations are appropriately represented in every instance of the processed data and in the trained model. Next, the corpus generated by the sliding window approach, is sliced into short, fixed sized windows of 2.5 s each (or 25 data points for given sampling rate 10 Hz). This fixed width windowing technique is widely used for sampling of activity data (Tran and Phan 2016; Anguita et al. 2012; Chetty et al. 2015). Each slice corresponds to one particular activity, and contains the extracted feature attributes from Sect. 3.3.
- *Labelling*: as the experiments are performed with supervised data acquisition in indoor environment, the corresponding labels of acquired data are already known from ground truth information. In this step every individual slice is labelled appropriately as either 0 (for leg swing) or 1 (actual walk activity).

The final processed corpus consists of element vectors with 2551 components in each (25 data points * 51 elements per data point * 2 (for 50% overlap) + 1 label). This comprehensive corpus is used for learning and classification in the next section.

4 Learning Techniques

During the experiments, the processed data is classified in two phases. Initially, a set of supervised machine learning algorithms have been used, to learn from sets of training data, which then predicted the class of test data. Next, the processed data is also fed as input to a deep convolutional learning network which attempts to identify the test data labels by learning features from training set. The individual classifiers are discussed below.

4.1 kNN Classifier

The kNN or k-Nearest Neighbours (Altman 1992) classification technique attempts to learn the nearness or closeness (in terms of distance) between a set of points, in order to determine their individual classes. This class is chosen by a simple voting mechanism, where the class with maximum number of votes in the defined neighbourhood is predicted as the class of the element vector.

4.2 SVM Classifier

The Support Vector Machine (SVM) algorithm (Cortes and Vapnik 1995) trains a model using labelled data to find an optimal plane of separation that can be used to classify the new test data. In our case of a binary problem, the objective is to identify a hyperplane which has maximum distance from data points of both the classes during training. In our work, we have used the SVM algorithm with a linear kernel as our corpus is feature-rich and the data is linearly separable.

4.3 GNB Classifier

The Gaussian Naive Bayes (GNB) classifier algorithm (Hand and Yu 2001) uses a concept of independent variables with the conditional probability model based on Bayes theorem. For a given continuous data-set with normal distribution, the conditional probability that a given event belongs to a certain class, is determined by the classification algorithm. This probability is calculated for all test instances given each class, repeated for all the classes. Finally the class with maximum probability is predicted by the classifier.

4.4 DT Classifier

The Decision Tree classifier is based on a cascading tree structure (Breiman et al. 1984), where the leaf nodes correspond to the class labels, and all internal nodes represent the tests on attributes. Beginning with the entire data at the root node, the decision rules are applied at each node, to predict the outcome and generate the next node. Here the Classification and Regression Tree (CART) algorithm (Breiman et al. 1984) based gini index cost function is utilized.

4.5 CNN Classifier

Apart from the aforementioned machine learning algorithms, we have also tested our data corpus with deep learning techniques, namely Convolutional Neural Network (CNN) (Fukushima 1980). Mostly used with 2-dimensional vectors such as images, the CNN learning mechanism can also be utilized in training a classifier to recognise human activity from sensor data. CNNs contain convolutional layers, pooling layers etc. along with fully connected layers in the end, which can extract features from the input data and map them to the respective classes/labels. The overall architecture of the CNN used in our work is described below. The same is also shown in Fig. 8.

- *First two layers*: The first two layers are *convolutional*, each of which uses the given kernel values to convolve the input (or previous layer's output stream). Also, in each layer the output is mapped by the Rectified Linear Unit (ReLu) activation function (Glorot et al. 2011), given by $relu(x) = max(0, x)$ to re-enforce non-linearity in the feature set after the convolution operations. For the first layer, length/input height of every feature vector is 2550. A set of 32 feature detectors (filters) with 3 kernels, learn the basic, local features from the data, and a output

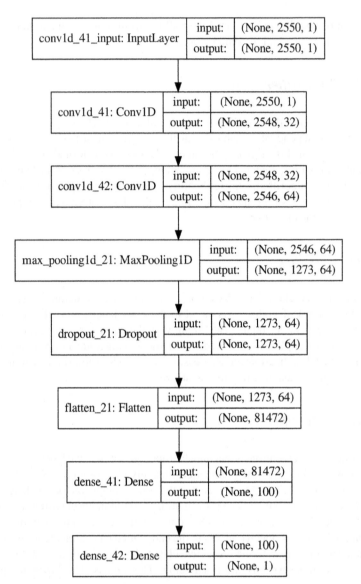

Fig. 8 The CNN architecture

neuron matrix of dimension 2548 × 32 is fed to the 2nd layer. This layer uses 64 new filters for learning other features to generate an output neuron of size 2546 × 64.

- *Third Layer*: this is a *max-pooling layer*, which eliminates the dependence on the feature position in the vector using sliding window of height 2 across data, for down-sampling and retaining only effective features (50% data discarded).
- *Fourth Layer*: The next layer is *dropout* (Hinton et al. 2016), which makes our CNN model ignore certain neurons randomly (at a rate of 40%) during training by assigning them zero weight. This ensures that there is a proper learning of the model, less sensitivity to minute variations in data or specific neuron weights, thus preventing overfitting.
- *Fifth Layer*: The *flatten* layer is used next to convert the previous output of dimension 1273 × 64, to a 1D vector such that it can be directly fed to the succeeding fully connected network. The output of this layer has a set of 81472 neurons.
- *Sixth Layer*: The final section consists of *two fully connected (dense) layers with an intermediate normalization layer*. The first dense layer takes the flattened feature vector as input and applies the ReLu activation function on this input, with a 100 neurons as output. The result is normalized by scaling it to a mean of zero and unit standard deviation in the intermediate layer. As ours is a two class problem, the final dense layer takes the normalized feature vector and applies the sigmoid activation function to predict the probability as output (between 0 and 1). These probabilistic values along with the true labels are then used by the cost function for model performance evaluation.

5 Results and Discussion

In this section, the different details pertaining to the experiments performed in this work, have been highlighted. This includes the experimental setup, system performance evaluation metrics and subsequent experimental results and observations.

5.1 Experimental Setup

The proposed work utilizes supervised learning, and the data instances are appropriately labelled based on volunteer and activity information gathered during data acquisition. The feature enriched data corpus has been classified using the supervised machine learning and deep learning algorithms described in Sect. 4. For evaluation purposes, we have taken 10 fold cross validation to ensure that the learning process involves proper representation of all data variations and to prevent holdout. The 10 random splits have been used for training and testing in a 70-30 ratio while maintaining a proper representation of both classes in every set. All the classifiers have been trained using each split's training set and evaluated using the corresponding test sets.

5.2 Evaluation Metrics

For evaluating the capability of the system, we have used some standard metrics, as discussed below.

- *Accuracy*—the proper recognition of each event, given by the formula:

$$Accuracy = \frac{TP + TN}{TP + TN + FP + FN} \tag{5}$$

- *Precision*—the correctly classified events out of all detected events, given by the formula:

$$Precision = \frac{TP}{TP + FP} \tag{6}$$

- *Recall*—the ratio of correctly detected events out of all true events, using the formula:

$$Recall = \frac{TP}{TP + FN} \tag{7}$$

- *F1 score*—harmonic mean of Precision and Recall, given as:

$$F1 - score = \frac{2 * Precision * Recall}{Precision + Recall} \tag{8}$$

where, *True Positive* (TP) means the model has correctly assigned the data to a class; *True Negative* (TN) means the model has correctly determined that the data does not belong to the class; *False Positive* (FP) means the model has wrongly assigned the data to a class; *False Negative* (FN) means the model has incorrectly determined that the data does not belong to the class.

5.3 Results of Classification

With the use of discussed metrics, classification performance for all the classifiers considering all the 10 random splits is shown in Figs. 9 and 10. Similarly, Figs. 11 and 12 illustrate the classifier-wise performance throughout the experiment. It is observed from the two figures that deep CNN is consistently efficient in classifying all the splits of actual walk and leg swing activity data accurately, with a maximum accuracy of 97%, followed by kNN algorithm with a maximum accuracy of 87% and DT classifier. The SVM classifier performs worse in terms of accuracy of identification, while GNB is worse in terms of F1 scores i.e. in terms of precision and recall.

Figure 13a shows the distribution of mean accuracy for all the learning techniques used for evaluation. Similarly, Fig. 13b shows the statistical distribution of F1 score for all the learning algorithms used. From these two figures, we see that CNN outperforms all other learning approaches by a good margin, and has minimum deviation among all. It is followed by kNN and DT. SVM and GNB classifiers rank lowest

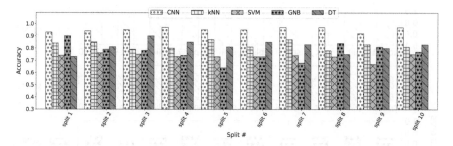

Fig. 9 Split-wise accuracy comparison for m-module

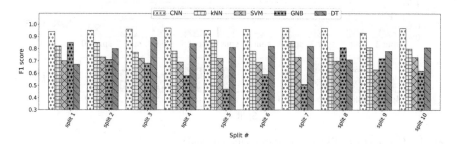

Fig. 10 Split-wise F1 score comparison for m-module

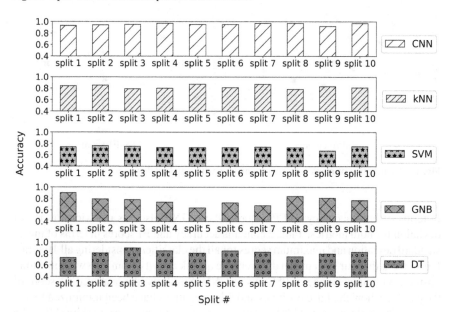

Fig. 11 Classifier-wise accuracy comparison for m-module

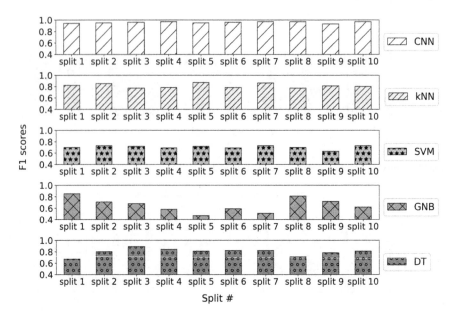

Fig. 12 Classifier-wise F1 score comparison for m-module

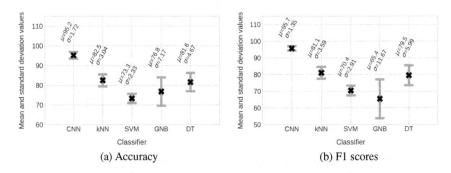

Fig. 13 Distribution of mean performance metrics with standard deviations

in terms of both accuracy and F1 score, where GNB also has the widest range of deviation for both metrics. Table 2 contains accuracy and F1 scores in terms of mean, standard deviation and maximum scores for all the classifiers considering all 10 splits. It is evident that our system can differentiate a walk activity from leg swing in sitting posture, with an accuracy of 95% using CNN. Similarly, the achieved F1 score of 97% reflects how the false positives and false negatives have been minimized by the system. The performance of kNN shows mean accuracy and F1 score above 80%. We also find that the GNB algorithm shows a highly varied performance, with maximum accuracy of 90% and lowest at 64%. A similar trend is seen in the F1 scores for GNB. SVM performs with moderate accuracy and F1 score below 75%.

Table 2 Performance statistics of classifiers

Classifier	Mean (Acc)	std (Acc)	Max (Acc)	Mean (F1)	std (F1)	Max (F1)
CNN	0.95	0.0172	0.97	0.96	0.0135	0.97
kNN	0.825	0.0304	0.87	0.81	0.036	0.87
SVM	0.733	0.0233	0.76	0.70	0.029	0.73
GNB	0.768	0.0716	0.90	0.654	0.117	0.85
DT	0.816	0.0467	0.90	0.795	0.0599	0.89

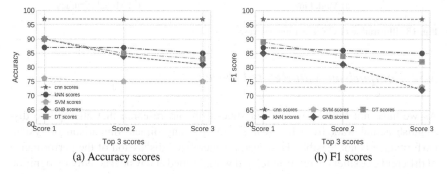

(a) Accuracy scores   (b) F1 scores

Fig. 14 Illustration of best three performances for each classifier

Considering all the splits of cross-validation, the three best identification instances for each classifier are illustrated in Fig. 14a and b. This gives a general idea of the system performance ad highlights that CNN is the ideal candidate as the proposed systems classifier of choice.

5.4 CNN Learning Curve

In the case of CNN, one complete learning process for a particular split of data is shown in Fig. 15a and b. In the first figure, we find that during training, the loss is minimized from 1.2 to almost 0, whereas in testing the loss could be minimized only upto 0.1, after which any training could not reduce the loss any further. Such situations have been used as a stopping point for the training process. Consequently, the second figure shows that the accuracy during testing had reached the maximum value, any training beyond which does not lead to better recognition accuracy.

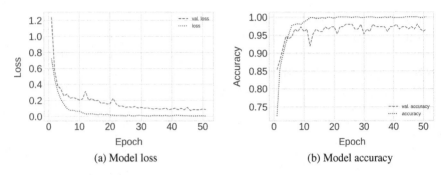

(a) Model loss   (b) Model accuracy

Fig. 15 Illustration of CNN learning curves

5.5   Comparison with Related Works

As we have not found any relevant state-of-the-art research that attempts to distinguish actual walk from leg swing activities, any direct comparison of system
performance is not feasible. However, a generalized discussion of the performance
of different classifiers in relevant related works aimed at physical activity recognition
is undertaken in this part of the discussion in the following points:

- Among the machine learning algorithms that have been used to classify the data,
  kNN has outperformed all others and is closely followed by DT. This is consistent
  with the relatively good performance of kNN (Jian and Chen 2015) and Decision
  Tree (Fan et al. 2013) in detecting different physical activities in comparison to
  SVM, NB, neural networks, as observed in Attal et al. (2015), Jatoba et al. (2008),
  Liang and Wang (2017), Kose et al. (2012), Pirttikangas et al. (2006) etc. The poor
  recognition accuracy of Naive Bayes is also seen in other works (Dernbach et al.
  2012).
- Learning algorithms in general are known to perform better in data corpus of
  limited size, and they generalize by learning from a fixed set of element vectors.
  Consequently, the classification performance for the conventional algorithms such
  as SVM, kNN, DT, GNB is lower than that of the CNN which benefits from
  repeated error minimization and learning from the extensive element vectors in
  the substantially large data corpus.
- Recent state-of-the-art works have used CNN to design models with a substantially higher success rate than conventional learning algorithms (Zeng et al. 2014;
  Yang et al. 2015; Panwar et al. 2017; Liang and Wang 2017), which resonate our
  observation in the current work.

Thus, the proposed system utilizes a novel corpus of IMU sensor signals using different classification techniques in a comparative study to determine system efficiency
in identifying a leg-swing activity from a real walk.

6 Conclusion and Future Scope

The inefficiency of wrist-worn or leg-worn smart wearable devices in monitoring of physical activities such as walking, has been addressed in this work. As an alternative, a low-cost wearable system with low sampling rate has been designed and deployed to gather novel physical activity data in two different situations. Initially, the actual walk data has been collected from volunteers, and then they have performed leg swing by mimicking the leg movement of real walks, in sitting posture. The data has been processed and classified using machine learning and deep convolutional networks to predict and classify the activities with a maximum accuracy and F1 score of 97% using the latter. This signifies the system's capability of reducing the false positive and false negatives and misclassification in general, which makes it useful in research and design of efficient fitness monitoring and healthcare systems that are capable of precisely identifying walks. In future, this work can be extended to include data gathered from other body locations for more variance in data patterns. Also, other sensing modalities and learning methods such as Recurrent Neural Networks etc. may be utilized for better recognition of actual walks from mimicked walk activity.

References

N. Ahmed, J.I. Rafiq, M.R. Islam, Enhanced human activity recognition based on smartphone sensor data using hybrid feature selection model. Sensors **20**(1), 317 (2020). https://doi.org/10.3390/s20010317

N.S. Altman, An introduction to kernel and nearest-neighbor nonparametric regression. Amer. Stat. **46**(3), 175–185 (1992)

K. Altun, B. Barshan, Human activity recognition using inertial/magnetic sensor units, in *International Workshop on Human Behavior Understanding* (Springer, 2010), pp. 38–51. https://doi.org/10.1007/978-3-642-14715-9_5

D. Anguita, A. Ghio, L. Oneto, X. Parra, J.L. Reyes-Ortiz, Human activity recognition on smartphones using a multiclass hardware-friendly support vector machine, in *International Workshop on Ambient Assisted Living* (Springer, 2012), pp. 216–223. https://doi.org/10.1007/978-3-642-35395-6_30

F. Attal, S. Mohammed, M. Dedabrishvili, F. Chamroukhi, L. Oukhellou, Y. Amirat, Physical human activity recognition using wearable sensors. Sensors **15**(12), 31314–31338 (2015). https://doi.org/10.3390/s151229858

A. Avci, S. Bosch, M. Marin-Perianu, R. Marin-Perianu, P. Havinga, Activity recognition using inertial sensing for healthcare, wellbeing and sports applications: a survey, in *23th International Conference on Architecture of Computing Systems 2010* (VDE, 2010), pp. 1–10

O. Banos, R. Garcia, J.A. Holgado-Terriza, M. Damas, H. Pomares, I. Rojas, A. Saez, C. Villa-longa, Mhealthdroid: a novel framework for agile development of mobile health applications, in *International Workshop on Ambient Assisted Living* (Springer, 2014), pp. 91–98. https://doi.org/10.1007/978-3-319-13105-4_14

B. Barshan, M.C. Yüksek, Recognizing daily and sports activities in two open source machine learning environments using body-worn sensor units. Comput. J. **57**(11), 1649–1667 (2014). https://doi.org/10.1093/comjnl/bxt075

S. Bounyong, S. Adachi, J. Ozawa, Y. Yamada, M. Kimura, Y. Watanabe, K. Yokoyama, Fall risk estimation based on co-contraction of lower limb during walking, in *2016 IEEE International Conference on Consumer Electronics (ICCE)* (IEEE, 2016), pp. 331–332. https://doi.org/10.1109/ICCE.2016.7430634

R. Bracewell, The autocorrelation function. The Fourier transform and its applications (1965), pp. 40–45

R.N. Bracewell, R.N. Bracewell, *The Fourier Transform and Its Applications*, vol. 31999 (McGraw-Hill, New York, 1986)

L. Breiman, J. Friedman, C. Stone, R. Olshen, *Classification and Regression Trees*. The Wadsworth and Brooks-Cole Statistics-Probability Series (Taylor & Francis, 1984), https://books.google.co.in/books?id=JwQx-WOmSyQC

I. Chandra, N. Sivakumar, C.B. Gokulnath, P. Parthasarathy, Iot based fall detection and ambient assisted system for the elderly. Clust. Comput. **22**(1), 2517–2525 (2019). https://doi.org/10.1007/s10586-018-2329-2

M.D. Chen, C.C. Kuo, C.A. Pellegrini, M.J. Hsu, Accuracy of wristband activity monitors during ambulation and activities. Med. & Sci. Sports & Exer. **48**(10), 1942–1949 (2016). https://doi.org/10.1249/mss.0000000000000984

R. Cheng, W. Heinzelman, M. Sturge-Apple, Z. Ignjatovic, A motion-tracking ultrasonic sensor array for behavioral monitoring. IEEE Sens. J. **12**(3), 707–712 (2011). https://doi.org/10.1109/JSEN.2011.2165942

G. Chetty, M. White, F. Akther, Smart phone based data mining for human activity recognition. Procedia Comput. Sci. **46**, 1181–1187 (2015). https://doi.org/10.1016/j.procs.2015.01.031

Y. Cho, Y. Nam, Y.J. Choi, W.D. Cho, Smartbuckle: human activity recognition using a 3-axis accelerometer and a wearable camera, in *Proceedings of the 2nd International Workshop on Systems and Networking Support for Health Care and Assisted Living Environments* (ACM, 2008), pp. 1–3. https://doi.org/10.1145/1515747.1515757

I. Cleland, B. Kikhia, C. Nugent, A. Boytsov, J. Hallberg, K. Synnes, S. McClean, D. Finlay, Optimal placement of accelerometers for the detection of everyday activities. Sensors **13**(7), 9183–9200 (2013). https://doi.org/10.3390/s130709183

J.W. Cooley, J.W. Tukey, An algorithm for the machine calculation of complex fourier series. Math. Comput. **19**(90), 297–301 (1965)

C. Cortes, V. Vapnik, Support-vector networks. Mach. Learn. **20**(3), 273–297 (1995)

S. Dernbach, B. Das, N.C. Krishnan, B.L. Thomas, D.J. Cook, Simple and complex activity recognition through smart phones, in *2012 Eighth International Conference on Intelligent Environments* (IEEE, 2012), pp. 214–221. https://doi.org/10.1109/IE.2012.39

L. Fan, Z. Wang, H. Wang, Human activity recognition model based on decision tree, in *2013 International Conference on Advanced Cloud and Big Data* (IEEE, 2013), pp. 64–68. https://doi.org/10.1109/CBD.2013.19

K. Fukushima, Neocognitron: a self-organizing neural network model for a mechanism of pattern recognition unaffected by shift in position. Biol. Cybern. **36**(4), 193–202 (1980). https://doi.org/10.1007/978-3-642-46466-9_18

X. Glorot, A. Bordes, Y. Bengio, Deep sparse rectifier neural networks, in *Proceedings of the Fourteenth International Conference on Artificial Intelligence and Statistics* (PMLR, 2011), pp. 315–323

D.J. Hand, K. Yu, Idiot's bayes-not so stupid after all? Int. Stat. Rev. **69**(3), 385–398 (2001)

M. Hardegger, D. Roggen, G. Tröster, 3d actionslam: wearable person tracking in multi-floor environments. Pers. Ubiquit. Comput. **19**(1), 123–141 (2015). https://doi.org/10.1007/s00779-014-0815-y

N. Hegde, E.S. Sazonov, Smartstep 2.0-a completely wireless, versatile insole monitoring system, in *2015 IEEE International Conference on Bioinformatics and Biomedicine (BIBM)* (IEEE, 2015), pp. 746–749. https://doi.org/10.1109/BIBM.2015.7359779

G.E. Hinton, A. Krizhevsky, I. Sutskever, N. Srivastva, System and method for addressing overfitting in a neural network. US Patent 9,406,017 (2016)

Y. Jang, S. Shin, J.W. Lee, S. Kim, A preliminary study for portable walking distance measurement system using ultrasonic sensors, in *2007 29th Annual International Conference of the IEEE Engineering in Medicine and Biology Society* (IEEE, 2007), pp. 5290–5293. https://doi.org/10.1109/IEMBS.2007.4353535

L.C. Jatoba, U. Grossmann, C. Kunze, J. Ottenbacher, W. Stork, Context-aware mobile health monitoring: evaluation of different pattern recognition methods for classification of physical activity, in *2008 30th Annual International Conference of the IEEE Engineering in Medicine and Biology Society* (IEEE, 2008), pp. 5250–5253. https://doi.org/10.1109/IEMBS.2008.4650398

H. Jian, H. Chen, A portable fall detection and alerting system based on k-nn algorithm and remote medicine. China Commun. **12**(4), 23–31 (2015). https://doi.org/10.1109/CC.2015.7114066

M. Kose, O.D. Incel, C. Ersoy, Online human activity recognition on smart phones, in *Workshop on Mobile Sensing: From Smartphones and Wearables to Big Data*, vol. 16 (2012), pp. 11–15

E. Kreyszig, *Advanced Engineering Mathematics*, 10th edn. (Wiley, New York, 2009)

V.S. Kumar, K.G. Acharya, B. Sandeep, T. Jayavignesh, A. Chaturvedi, Wearable sensor-based human fall detection wireless system, in *Wireless Communication Networks and Internet of Things* (Springer, 2019), pp. 217–234. https://doi.org/10.1007/978-981-10-8663-2_23

P. Kumari, L. Mathew, P. Syal, Increasing trend of wearables and multimodal interface for human activity monitoring: a review. Biosens. Bioelectron. **90**, 298–307 (2017). https://doi.org/10.1016/j.bios.2016.12.001

J.R. Kwapisz, G.M. Weiss, S.A. Moore, Activity recognition using cell phone accelerometers. ACM SIGKDD Explor. Newsl. **12**(2), 74–82 (2011). https://doi.org/10.1145/1964897.1964918

N.D. Lane, M. Mohammod, M. Lin, X. Yang, H. Lu, S. Ali, A. Doryab, E. Berke, T. Choudhury, A. Campbell, Bewell: a smartphone application to monitor, model and promote wellbeing, in *5th International ICST Conference on Pervasive Computing Technologies for Healthcare* (2011), pp. 23–26

N.D. Lane, E. Miluzzo, H. Lu, D. Peebles, T. Choudhury, A.T. Campbell, A survey of mobile phone sensing. IEEE Commun. Mag. **48**(9), 140–150 (2010). https://doi.org/10.1109/MCOM.2010.5560598

G. Li, T. Liu, J. Yi, Wearable sensor system for detecting gait parameters of abnormal gaits: a feasibility study. IEEE Sens. J. **18**(10), 4234–4241 (2018). https://doi.org/10.1109/JSEN.2018.2814994

X. Liang, G. Wang, A convolutional neural network for transportation mode detection based on smartphone platform, in *2017 IEEE 14th International Conference on Mobile Ad Hoc and Sensor Systems (MASS)*, (IEEE, 2017), pp. 338–342. https://doi.org/10.1109/MASS.2017.81

H.F. Nweke, Y.W. Teh, M.A. Al-Garadi, U.R. Alo, Deep learning algorithms for human activity recognition using mobile and wearable sensor networks: state of the art and research challenges. Expert Syst. Appl. **105**, 233–261 (2018). https://doi.org/10.1016/j.eswa.2018.03.056

K. Ozcan, S. Velipasalar, Wearable camera-and accelerometer-based fall detection on portable devices. IEEE Embed. Syst. Lett. **8**(1), 6–9 (2015). https://doi.org/10.1109/LES.2015.2487241

M. Panwar, S.R. Dyuthi, K.C. Prakash, D. Biswas, A. Acharyya, K. Maharatna, A. Gautam, G.R. Naik, Cnn based approach for activity recognition using a wrist-worn accelerometer, in *2017 39th Annual International Conference of the IEEE Engineering in Medicine and Biology Society (EMBC)* (IEEE, 2017), pp. 2438–2441. https://doi.org/10.1109/EMBC.2017.8037349

I.M. Pires, G. Marques, N.M. Garcia, F. Flórez-Revuelta, M. Canavarro Teixeira, E. Zdravevski, S. Spinsante, M. Coimbra, Pattern recognition techniques for the identification of activities of daily living using a mobile device accelerometer. Electronics **9**(3), 509 (2020). https://doi.org/10.3390/electronics9030509

S. Pirttikangas, K. Fujinami, T. Nakajima, Feature selection and activity recognition from wearable sensors, in *International Symposium on Ubiquitious Computing Systems* (Springer, 2006), pp. 516–527. https://doi.org/10.1007/11890348_39

A. Sano, A.J. Phillips, Z.Y. Amy, A.W. McHill, S. Taylor, N. Jaques, C.A. Czeisler, E.B. Klerman, R.W. Picard, Recognizing academic performance, sleep quality, stress level, and mental health using personality traits, wearable sensors and mobile phones, in *2015 IEEE 12th International*

Conference on Wearable and Implantable Body Sensor Networks (BSN) (IEEE, 2015), pp. 1–6. https://doi.org/10.1109/BSN.2015.7299420

M. Shoaib, S. Bosch, H. Scholten, P.J. Havinga, O.D. Incel, Towards detection of bad habits by fusing smartphone and smartwatch sensors, in *2015 IEEE International Conference on Pervasive Computing and Communication Workshops (PerCom Workshops)* (IEEE, 2015), pp. 591–596. https://doi.org/10.1109/PERCOMW.2015.7134104

H. Similä, M. Immonen, M. Ermes, Accelerometry-based assessment and detection of early signs of balance deficits. Comput. Biol. Med. **85**, 25–32 (2017). https://doi.org/10.1016/j.compbiomed. 2017.04.009

Statista, Forecast wearables unit shipments worldwide from 2014 to 2023, https://www.statista. com/statistics/437871/wearables-worldwide-shipments/. Accessed 25 Jan 2020

D.N. Tran, D.D. Phan, Human activities recognition in android smartphone using support vector machine, in *2016 7th International Conference on Intelligent Systems, Modelling and Simulation (ISMS)* (IEEE, 2016), pp. 64–68. https://doi.org/10.1109/ISMS.2016.51

P. van de Ven, H. O'Brien, J. Nelson, A. Clifford, Unobtrusive monitoring and identification of fall accidents. Med. Eng. & Phys. **37**(5), 499–504 (2015). https://doi.org/10.1016/j.medengphy. 2015.02.009

D. Yacchirema, J.S. de Puga, C. Palau, M. Esteve, Fall detection system for elderly people using iot and big data. Procedia Comput. Sci. **130**, 603–610 (2018). https://doi.org/10.1016/j.procs.2018. 04.110

J. Yang, M.N. Nguyen, P.P. San, X.L. Li, S. Krishnaswamy, Deep convolutional neural networks on multichannel time series for human activity recognition, in *Twenty-Fourth International Joint Conference on Artificial Intelligence* (AAAI Press, 2015)

M. Zeng, L.T. Nguyen, B. Yu, O.J. Mengshoel, J. Zhu, P. Wu, J. Zhang, Convolutional neural networks for human activity recognition using mobile sensors, in *6th International Conference on Mobile Computing, Applications and Services* (IEEE, 2014), pp. 197–205. https://doi.org/10. 4108/icst.mobicase.2014.257786

Z. Zhang, S. Poslad, Improved use of foot force sensors and mobile phone gps for mobility activity recognition. IEEE Sens. J. **14**(12), 4340–4347 (2014). https://doi.org/10.1109/JSEN.2014. 2331463

P. Zhang, X. Chen, X. Ma, Y. Wu, H. Jiang, D. Fang, Z. Tang, Y. Ma, Smartmtra: robust indoor trajectory tracing using smartphones. IEEE Sens. J. **17**(12), 3613–3624 (2017). https://doi.org/ 10.1109/JSEN.2017.2692263

Facial Recognition Application with Hyperparameter Optimisation

Hannah M. Claus, Cornelia Grab, Piotr Woroszyllo, and Patryk Rybarczyk

Abstract This chapter explores optimisation techniques for a facial recognition application using an Artificial Neural Network (ANN). Facial recognition applications utilise a certain set of hyperparameters to achieve the best recognition accuracy depending on the data set. There are several approaches to find the best set of these hyperparameters. They can be put into two categories: Manual optimisation and automated optimisation. By experimenting with both approaches in order to find the best, certain patterns could be observed. Manual optimisation achieved the highest recognition accuracy 96.6% with a specific set of parameters used in the ANN. Using automated optimisation algorithms, the accuracy could be improved. The experiments were run in Google Colab using certain classifiers, such as Grid Search and Random Search. The optimisation algorithms that were used demonstrated an increase of the accuracy of almost 2% and a decrease of the prediction error. Based on the findings it can be concluded that the automated optimisation strategies are more efficient and provide a better set of parameters to use within an ANN model, so that an improved prediction accuracy can be achieved. Using the automated optimisation combined with manual approximation to find better values turned out to be the best solution.

Keywords Hyperparameter optimisation · Random search · Grid search

1 Introduction

Facial recognition is a term that is commonly used in areas within Big Data. Despite its natural meaning of recognising a human's face, it is connected to far more complex processes. What appears self-evident to humans requires considerable training and testing time for an algorithm. Moreover, it is not only important to recognise a face, but to link it to a unique identity. A study by Sheehan and Nachman (2014) found that human faces have evolved over time to become more individual. Hence,

H. M. Claus (✉) · C. Grab · P. Woroszyllo · P. Rybarczyk
University of Bedfordshire, School of Computer Science and Technology, Luton, UK, LU1 3JU, Luton, England

© The Author(s), under exclusive license to Springer Nature Switzerland AG 2022   141
Y. Baddi et al. (eds.), *Big Data Intelligence for Smart Applications*,
Studies in Computational Intelligence 994,
https://doi.org/10.1007/978-3-030-87954-9_6

facial recognition has become more important in the past few years because "facial recognition technology treats the face as an index of identity" (Gates 2011). This can be observed in the recent development of implementing more closed circuit television (CCTV) systems to identify criminals faster or to prevent attacks detected through conspicuous behavioural patterns. Nevertheless, in order to implement such a system, it has to be trained and tested before its application. The facial recognition application uses an Artificial Neural Network (ANN) which takes its origin from the human neural networks by having the "ability to learn by interacting with its environment or with an information source" (Hassoun 1995). ANNs can be used for several purposes which are not limited to image inputs. However this chapter will focus on the training and testing of an ANN for a facial recognition application.

Biometric authentication is a widely known procedure to safely identify a person. Like other biometric authentication techniques using fingerprints, the iris, voices, and other unique traits (Bhatia et al. 2013), the face has a great significance. While a person would have to be close enough to a camera, microphone or a similar authenticity device in order to be identified, facial recognition has the advantage of not needing the physical or near contact with a person. Cameras can be attached to public buildings to process non-digital situations, or digital images can be evaluated by an algorithm. Hence, the overall security can be enhanced significantly. This technology makes it possible to identify several people at the same time, given the algorithm in use has a high accuracy with a low error ratio. Public places, airports, governmental buildings, hospitals, and every other place can be observed 24 h per day without breaks. Thus, governments all around the world strive to enhance their security systems by implementing good facial recognition algorithms (Mann and Smith 2017). Despite the fact, that the prevention of criminal behaviour has issues beyond the authentication process because the legal and ethical aspects are still not defined yet (Hornung et al. 2010), it is important to create an intelligent application that has the best accuracy and the lowest error ratio that is possible. With such a great tool comes a great responsibility to ensure that the identities detected are indeed correct. For this reason, the ANN needs to be implemented in a way that guarantees the best outcomes. This can be achieved by finding the best values for the hyperparameters that control the training process of the ANN. There are several approaches to finding these values and improving the recognition accuracies. This chapter will experiment with two of these approaches and find the most efficient one, namely manual and automated optimisation. The ANN will be optimised by tuning the hyperparameters and training the ANN with each set of parameters to compare the accuracies.

This optimisation procedure is of great importance because it can improve the ANN's recognition accuracy just by the slightest change of values and therefore, this can influence whether a criminal can be found or not. Thus, effective optimisation of the hyperparameters and using a diverse set of training input is necessary in order to safely implement a facial recognition application.

This chapter will document the optimisation of a developed Machine Learning (ML) solution that will achieve the highest recognition accuracy with the minimal recognition error by finding an optimal set of hyperparameters. Additionally, the two methods of finding an optimal set of parameters, manually or automated, will be compared. Furthermore, this documentation will also focus on the state-of-the-art research on facial recognition and discuss certain considerations and issues.

2   Problem Statement

An ANN is to be designed and trained that implements the optimal set of hyperparameters to recognise the faces of 30 different people from a database. The search for the set of parameters is conducted using two different methods: 1. trial and error also known as manual optimisation and 2. automated optimisation, such as Grid Search and Random Search. The first part of this chapter discusses the manual optimisation in detail. The second part continues with the utilisation of Grid Search and Random Search. These methods will be compared and evaluated. The result is a set of hyperparameters which achieves the highest recognition accuracy with the lowest error.

3   Theoretical Framework

In order to be able to follow the upcoming experiments fully, the theoretical framework needs to be established. Here, the most important concepts will be explained and discussed.

3.1   Artificial Neural Network

In the following, the basic concepts of ANNs will be explained. An ANN has a clear structure with a foundation of an input layer of neurons, a number of hidden layers of neurons and a final layer of output neurons (Wang 2003). Furthermore, there are four main elements of an ANN model: First, there are the connections between the layers created by synapses, each link is characterised with a weight that will be multiplied by the incoming signal. Second, the adder sums up all the products of the inputs and the synapses' weights. Third, there are several activation functions that can be used to limit the outputs. Fourth, the external bias can decrease or increase the input of the activation function (Csaji and Ten Eikelder 2001). Putting this information into a context, it can be said that the inputs of each layer are the outputs of the preceding layer only. Thus, the activation function is applied to the weighted sum of the layer in order to obtain a new input for the next layer. This kind of network is called a feed-forward ANN because the inputs are fed into one direction only, namely forward.

Additionally, there are two important methods that need to be mentioned here: K-fold cross-validation (KCV) and Back-propagation (BP). They both have a significant impact on ANNs and will be discussed in the experiments conducted for this documentation.

3.2 K-Fold Cross-Validation

Starting with the KCV, it is an approach for "model selection and error estimation of classifiers" (Anguita 2012). Furthermore, the dataset will be split into a specific number of subsets which is determined by k. A number of subsets will be used to train the ANN model, the rest will be the dataset for the testing phase in order to evaluate the performance. An example can be seen in a script provided by Dr. Benham Kia, where $k = 4$ and $3/4$ are defined as the training set, while $1/4$ defines the testing set (Kia 2018). This separation of datasets is recommended in order to achieve a high recognition accuracy that stays when new datasets are entered after the training phase. The goal is to "maximize [the] predictive accuracy on the new data points" (Dietterich 1995), without risking to train the model too good, so that it fits the noise in the data and studies the different peculiarities of the training sets instead of finding a general predictive rule (Dietterich 1995). This would result in an overfitted model with too many parameters and, therefore, a high variance. The opposite case would mean that the ANN model is too complex but does not have enough parameters to train with, hence, the bias would be too high (Jabbar and Khan 2014), which is called an underfitted ANN model. However, using KCV is said to prevent an ANN model from overfitting (Lin et al. 2008) which is why it was utilised within the experiments for this chapter.

3.3 Back-Propagation

Continuing with BP, this learning rule can be "employed for training nonlinear networks of arbitrary connectivity" (Rumelhart et al. 1995). It plays an important role regarding real-world applications. The algorithm looks for the minimal values of the error function using gradient descent (Rojas 1996). The BP "repeatedly adjusts the weights of the connections in the network so as to minimize a measure of the difference between the actual output vector of the [network] and the desired output vector" (Rumelhart et al. 1986). To put it differently, after each training phase, the BP goes back into the model and adjusts the ANN's weights and biases in order to minimise the cost function and have a better prediction in the next iteration.

Under these circumstances, using a ML algorithm seems like a justified approach when creating a solution for face recognition. Nevertheless, a ML algorithm can only work well if the parameters that are being used are the best ones. There is a great number of different parameters that can change the outcome of a training phase for each classifier. The library MLPClassifier, which was used within conducting the experiments, includes 23 parameters in total (scikit learn.org 2020). Each parameter has at least two different values as an option. With atleast $2^2 3$ options the dimensions take on a great size. Using just trial and error as a method to find the optimal set of parameters is time consuming. Hence, this process can be optimised automatically. For this reason there are several optimisation algorithms that can be used. The two algorithms that are in focus are Grid Search (GS) and Random Search (RS).

3.4 Grid Search

Starting with Grid Search, it is a simple approach that tries every possible combination of parameters within a defined grid. First, the dimensions have to be set with specific hyperparameters that want to be observed. Each dimension has a range of possible values. As already mentioned, there are 23 parameters with different ranges, so the search was limited to only a few parameters (Bergstra and Y. Bengio 2012).

3.5 Random Search

However, Random Search chooses the values randomly from the configured space. This makes it possible to use more values for each variable, because the chance of choosing the same value twice is low, whereas GS can only work with the predefined values (Bergstra and Y. Bengio 2012).

Moreover, face recognition has a complex training procedure that needs to be observed attentively, in order to achieve a good accuracy that has a minimal error ratio. Indeed, face recognition has many advantages, but they can be threatened by superfluous training of the model and, therefore, risking overfitted or underfitted ANN models. By tuning the hyperparameters efficiently, the accuracies can be improved and a safe algorithm can be established.

4 Related Work

Improving facial recognition accuracies is a goal not only defined by academic researchers but also by governments and security agencies. A difference of 0.1% in the accuracy can determine whether a person might be the wanted criminal or just someone who looks similar. Hence, there are several approaches to finding and choosing the best set of hyperparameters. The ANNs are as complex as the strategies to optimise them. There is a great number of different algorithms and again a great number of optimisation strategies. Claesen and De Moor state that automation is inevitable when optimising ANNs but there has yet to be found a self-configuring learning strategy that evades overfitting or underfitting intuitively (Claesen and De Moor 2015). Nevertheless, the discussion can be held in a wide spectrum of possible solutions when only talking about optimising machine learning algorithms in general. As soon as the application of the algorithm and the input data is defined, the problem is more specific. There is no ultimate guide on how to perfectly optimise a learning algorithm to get 100% recognition accuracy. There is bias in the data, there is bias in society, hence, the algorithm can only mirror the data that it receives (Buolamwini et al. 2018). But finding the optimal way to tune hyperparameters needs several experiments. The aim is to find regularised solutions that can be applied in all scenarios (Sra et al. 2012). A benchmark with according variables should be declared to be applicable in automated searches. But until this solution is found, the optimisation process is still a process that includes trials and errors.

A recent paper by Bergstra and Bengio (2012) demonstrated the differences between Grid Search and Random Search. There it is stated that Random Search explores the space of hyperparameters more widely because it chooses the values of the parameters randomly. In this case, a significant difference can be found between both search algorithms. Grid Search might promise to find the optimal set eventually, but only in the defined grid and limited through time. Random Search on the other hand does not promise to find the one perfect set of parameters but a wide range of different options that are chosen randomly. Thus, a broader spectrum of sets can be tested and an approximation can be made. For this reason, Grid Search and Random Search are the two automated optimisation algorithms that will be further discussed in this chapter.

5 Methodology

This chapter is the documentation and evaluation of optimising the search of a specific set of parameters used in a ML solution for a biometric recognition task. Therefore, an ANN needed to be used. The programming platform Google Colab was used intensely to conduct the experiments for this task. Moreover, a folder was provided that included in total 1500 images of 30 different people, 50 images each in different lighting conditions (Belhumeur and Kriegman 1997). This defined the testing and training datasets for the ANN. There were two additional files provided, namely: "process_yale_images" (Schetinin 2021) and "classify_yale" (Schetinin 2021). They included the project scripts that created the basis of the ANN.

There are a few questions that need to be considered, in order to start with the optimisation process:

- How much time can be put into the search and testing process?
- What level of complexity should the ANN apply? (Claesen and De Moor 2015)
- Which learning algorithm will be used?
- How many parameters should be defined through the search?

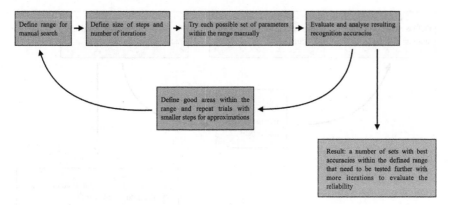

Fig. 1 Model for the manual optimisation process

- What dataset can be used as an input source?

After defining the answers to the above questions, the experiments have to be split up into four categories: 1. manual optimisation, 2. grid search, 3. random search, 4. final approach. The experiments will deepen their complexity with each approach, starting with the easiest one.

The concept for manual optimisation is straight-forward. It follows the simple trial-and-error method. It was known that there would not be enough time to experiment with all parameters that the MLPClassifier offers. Hence, only the number of hidden neurons (NOHN) and the number of principal components (PCs) were explored because of their significant influence within the ANN. Furthermore, the size of the steps to the next value and the number of iterations had to be defined. When all these values were established, the search began by testing each possible set of parameters manually. All results were documented in an Excel sheet to compare the results and evaluate them. When a good range could be identified, the values were optimised again and further experiments were conducted until a list of the best sets could be found in the defined time frame (Fig. 1).

Following the manual optimisation, the automated search algorithms were used. For both Grid Search and Random Search, only the range or the grid had to be defined and the search could start. Depending on the dimensions, the time to get results could vary. Nevertheless, the results showed the best combination of the values for the parameters and after several iterations of this process, the best set of parameters in the defined range could be found. The results of both automated optimisation strategies can be combined with manual optimisation as the final approach. This way the efficiency of the automation can determine a good range of parameters and the best set can be found by manually approaching the narrow range (Figs. 2, 3).

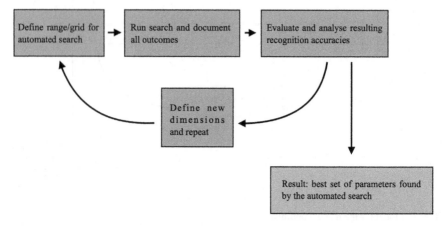

Fig. 2 Model for the automated optimisation process

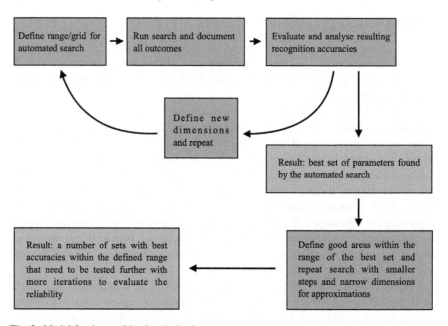

Fig. 3 Model for the combined optimisation process

After experimenting with these different strategies, specific patterns will show and determine which strategy can be defined as the best one for this task. The following discussions will go deeper into the process of experimenting with manual and automated optimisation and evaluate the results.

6 Part 1

Part 1 discusses the manual search of a set of parameters that meets the requirements.

6.1 Designing a Solution

First, Google Drive had to be connected with the "process_yale_images" file, in order to unzip the 1500 images in the folder. They were then prepared for the ANN and put into a matrix with $n = 1500$ rows and $h * w$ columns with n being the number of images and $h = 77$ and $w = 68$ being the $height(h)$ and $width(w)$ of the images in pixels (Schetinin 2021). This is how the size of the input layer could be determined with $h * w$, $77 * 68 = 5,236$. The 50 images of each person differ in quality of lighting, where the first picture is the brightest, the last picture the darkest (Figs. 4, 5).

Consequently, functions from the ML sklearn library were imported (scikit learn.org 2020). Using KCV, the dataset was split into k folds, $k = 5$ as a common recommendation (Anguita 2012). Hence, 20% of the dataset was kept for the testing phase. Furthermore, the principal component analysis (PCA) was applied using the variable $nof\_prin\_components$, which determines the number of principal components (PCs). PCA is a technique "to reduce multidimensional data to lower dimensions while retaining most of the information" (Karamizadeh et al. 2013). Hereby, the eigenface method is used to extract the characteristic features of a face and lower the dimensions of the images. Thus, eigenvectors are calculated, which are features that characterise the variation between face images (Turk and Pentland 1991). By comparing the Euclidian distance between the eigenvectors in the dataset, the smallest distance must be found in order to identify the correct person. This means, that the multidimensional images are reduced to a linear combination to improve the effi-

Fig. 4 Lightest shade of female face

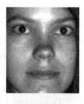

Fig. 5 Darkest shade of female face

ciency of the algorithm. As a result, the PCA variable determines how much variance there will be in a set of images (Karamizadeh et al. 2013).

After choosing a specific number of principal components, the number of hidden neurons, NOHN, had to be chosen. Unlike the KCV, the procedure of finding the optimal size of hidden neurons is not defined yet. However, there is a common understanding, that an excessive use of hidden neurons will cause overfitting, as this will create a more complex model than needed (Liu et al. 2007). To simplify the process, one could argue that the number of hidden neurons should be between the number of total input and output nodes and the number of hidden neurons should be approximately 2/3 of the size of the input layer, plus the size of the output layer. For this reason, this parameter can only be determined in combination with the KCV in order to see at which point the performance of the algorithm stops improving. This indicates that the training has started to "fit the noise in the training data" (Liu et al. 2007), hence, the model is overfitted. This procedure is a trial-and-error method with the aim to find the best possible values. With this in mind, the number of hidden neurons should be in the range between $30 - 5,236$, where $5,236$ is the number of pixels in one image and 30 is the number of different people, hence the size of the output layer. 2/3 of $5,236$ results in approximately $3,508$, adding 30, shows the limit of $3,538$ hidden neurons. For simplifying the experiments, $3,500$ NOHN are chosen.

The last lines of the script will be explained in the following to understand the training process of the ANN. This is the script for the first experiments when searching for the best parameters manually.

```
clf = MLPClassifier(hidden_layer_sizes=(nohn,), solver='sgd',
activation='tanh', batch_size=256, verbose=True,
early_stopping=True).fit(X_train_pca, y_train)
```

6.2 MLPClassifier

MLPClassifier is the function of the ML sklearn library (scikit learn.org 2020) that trains the model iteratively. MLP is short for multi-layer perceptron which describes the ANN, because it has multiple layers.

The first parameter *hidden_layer_sizes* passes the value that is consciously choosen for the number of hidden neurons.

The second parameter *solver* passes the solver for weight optimisation, thus, in which way the weights of the ANN should be adjusted. As already introduced, BP uses gradient descent, and therefore, *sgd* stands for stochastic gradient descent. Thus, this ANN uses BP.

The third parameter *activation* determines which activation function will be used in the hidden layers. As this ANN should be a complex model, a non-linear activation function is needed. In order to make the optimisation process more efficient, the hyperbolic tangent *tanh* is chosen.

The fourth parameter $batch\_size$ determines the size of mini batches for the stochastic optimisation of the ANN. Mini batches are used to stabilise the training process and to reduce the variance of the gradient estimate (Wang et al. 2019).

The fifth parameter $verbose$ decides whether progress messages should be printed in the standard output. Here, it is set to true.

The sixth parameter $early\_stopping$ is set to true in order to stop the training of the model when the validation score is not improving, for example to signalise a potential overfitted model. In this case, it will automatically take 10% of the training data as validation and terminate the training once there is no improvement after a specific amount of iterations. In this case it is set to $tol = 0.000100$ for 10 consecutive epochs.

Then, the function $fit(X, y)$ is called to fit the model to the data matrix and target. Here, X_train_pca depicts the data where the principal components were already applied. y_train describes the target data.

The training of the model starts by running this script. Once the training is completed, the testing dataset can be applied through the function $predict(X)$. Here, the testing data matrix with the applied principal components will be used together with the testing target data.

For these reasons, the goal of the experiments is to find the right combination of both parameters, $nof\_prin\_components$ and $nohn$, and to create a good ANN for this specific task.

A variety of experiments with different parameters was conducted to use the trial-and-error method in order to find a good set of parameters that fit the requirements. In the following, the experiments will be described in detail.

6.3 Experiments

In this part of the report, the different experiments conducted will be described and discussed in detail. In total, there were four stages of experiments. Each set of parameters is iterated 20 times, so that the resulting values could be representative for the experiments. In general, it can be said, that the two wanted parameters, the number of iterations (NOI), the testing accuracy (TA), the validation score (VS), and the loss played the most important roles. Due to the $early\_stopping$ parameter, each run through of the script had a limited amount of iterations within the ANN and stopped once the VS did not improve for 10 consecutive epochs. The maximum amount of iterations within the model is 200, which is reached rarely. In order to find a good solution, only sets of parameters were considered that achieved equal to or higher than 90% TAs. The sets of parameters will be shown like this: $(nof\_prin\_components, nohn)$.

It is important to mention that the conditions for the tables are not always completely the same. The ranges differ for each table. Hence, the overall average can be biased because even the outer ranges with medium recognition accuracies count.

This is just representing the overall results of a particular table in order to find a good set of parameters in a great variety of experiments.

6.3.1 Stage I

In the first stage, there are four tables dedicated to a specific *nohn* value (NOHN) respectively. The different NOHNs are 100, 150, 200, and 250, choosing to start with a low number of hidden neurons. Simultaneously, the value for $nof\_prin\_components$ increased by 10, initially starting at 70 principal components (PCs) and stopping at 450. The range is manually adjusted for a few tables, because after a certain amount of PCs the iterations stopped automatically and overfitting had started. At the beginning, only the number of iterations and the testing accuracy are documented. In the following, each individual table will be analysed separately.

The table Nohn100 had 100 NOHNs and the range goes from 100 PCs to 250 PCs. For each set of parameters, the script is run 20 times consecutively. The maximum average testing accuracy (MATA) is at ca. 84.5%, the minimum at 79.9%, with an overall average of 82.6%. Only one single iteration with 110 PCs had a 90% TA, which is not enough for the wanted result. The set of parameters with the best average accuracy is (200, 100).

For this reason, the NOHN is increased by 50 in the table Nohn150, using the range from 70 PCs to 310 PCs. 12 sets of parameters showed 90% accuracy in at least one iteration. Nevertheless, the MATA is at ca. 86.3%, which is only 1.8% better than the MATA from the last table. Surprisingly, the overall average is at 84.3%, which is also 1.8% better than the previous one. Hence, the set (170, 150) already achieved a better result.

Increasing the NOHN even further to 200, the table Nohn200 was created. Here, the range of the PCs starts at 70 PCs and ends after 450 PCs. Although the MATA has reached 89.2% which already has an additional 2.9% accuracy compared to the previous MATA, the overall average almost stays the same at 84.4%. This might be caused by the wider range of PCs, even though 34 sets of parameters have 90% accuracy in at least one iteration. The best performing set is: (200, 200) with 89.2% TA.

The last table from this stage is called Nohn250 with 250 NOHN and a range of PC from 100 to 400. In this table, out of 31 sets, 28 sets achieved an accuracy of 90% or more in at least one iteration. However, the MATA is at 88.3% and the overall average of all accuracies within the table are at 87.4%. In this case, the MATA is smaller than the MATA from the previous table, but the overall average is higher. The difference in the overall average could be the result of different PC ranges. Nevertheless, the set with the best performance has the values (250, 250). Figure 6 shows a summary of the best performances from this stage.

Table	nof_prin_components	nohn	Average testing accuracy (after 20 iterations)
Nohn100	200	100	84.5%
Nohn150	170	150	86.3%
Nohn200	200	200	89.2%
Nohn250	250	250	88.3%

Fig. 6 The best set of parameters from each table from stage I

In conclusion, four sets of parameters could be chosen that achieved the best performance in their respective tables. The best testing accuracies were in a range between 84% and 90%.

6.3.2 Stage II

The second stage introduced a radical approach to finding a good NOHN. The first stage had shown that the values might be too small for the model to achieve a better performance. In order to keep the trial-and-error method, this stage created new fixed NOHN with varying PCs. This time, the NOHNs are 500, 1500, and 2500. Hence, instead of increasing by 50, the NOHN increased by 1000. The tables concentrated on increasing the PCs by 100 and finding good sets of parameters. Notably, at this stage the VS (validation score) additional to the TA (testing accuracy) and the NOI (number of iterations) was documented. In this case, the VS is initiated by the parameter *early_stopping* from the script, as already explained before. This means, that 10% of the 80% training data are used to tune the parameters and to find the optimal NOHN to determine when to stop the algorithm in order to apply BP. Here, the cross-validation already compares the performance of the training algorithm and chooses the best results. Hence, the VS will indicate how good the performance of a specific set of parameters is after it has been trained with the algorithm and the training data.

Starting with the table PCA I, the NOHN is set to 500, whereas the PCs started at 100 and ended at 900. Again, each set of parameters is run 20 times in order to get a statistically viable representation of the trained algorithm outcomes. In this table, the MATA is at 89.8% with a connected average VS of 86.7% for the set of parameters (300, 500). This accuracy is already better than all accuracies from the tables before. Hence, the new approach shows progress. The maximum average validation score (MAVS) is at 89.8%, but the connected average TA is at 89.2% for the set (200, 500), which is a slightly lower result. Thereupon, it is important to note, that a high TA outranks a high VS because the model is designed to have a good accuracy with new data, hence, a good TA is what is looked for. Nevertheless, the second set would also be the second best performing set in general, thus, these two sets can be noted as the best performing sets of this table. Adding to this, it is interesting to see that after the MATA the accuracies of the next sets start decreasing with a local minimum at 62.7% with the set (900, 500) and an average VS of 28.4%. This indicates that this set of parameters makes the model too complex and therefore, it creates unreliable results that should not be used in the final algorithm.

Continuing with the table PCA II, the NOHN is defined as 1500 with PCs in a range from 100 to 1200. The MATA is at 94.8% with an associated average VS of 91.3% with the set (300, 1500). Surprisingly, by keeping the same amount of PCs but adding another 1000 hidden neurons, it keeps showing the best performance. Now, the MAVS is at 94.5% with an associated average TA of 94.2% through the set (200, 1500), again it is in the same style as in the table before. In order to find a closer solution, a sub-experiment was conducted with the middle value between 200 and 300 PCs, which resulted in the set of (250, 1500). This experiment did not achieve a better TA, but it is close enough to the MATA with 94.7% (important to note that the MATA is rounded to 94.8% in order to decrease the decimal places, but the table shows 94.75% which means that the difference between the two TAs is only 0.05% instead of 0.1%). However, the average VS is at 93.2%, which is less than the MAVS but better than the average VS for the MATA. Hence, this sub-experiment showed an approximation to an optimal set of parameters with the second best average TA and the second best VS. At the same time, the set (1200, 1500) portrayed what happens, when the NOHN is too high, because the majority of iterations stop at 12, which means that the algorithm stopped learning after 2 iterations, which resulted in an average VS of 2.6% and an average TA of 3.5%. Thus, the average results have a deflection and can not be taken into the calculation. This portrays an example of a potentially overfitted algorithm.

The last table for this stage, PCA III, had 2500 hidden neurons and started with 100 PCs increasing to 1100 PCs. Furthermore, the MATA is at 94.6%, which is lower than the one from the last table, but this MATA appeared twice with two different associated average VS. The set (300, 2500) with the MATA had an average VS of 93.2%, the set (400, 2500) had an average VS of 91.3%. Hence, the first set of parameters had a better overall performance. In fact, it had the same number of PCs as the two previous MATAs, namely 300. The MAVS of 96% is again achieved by the set (200, 2500) with an associated average TA of 93.8%. Important to mention is not only the fact, that the MAVS is again achieved by a set with 200 PCs, but also the fact that every iteration of said set achieved a VS of 93% or higher, which seems like a more reliable set of parameters than the set (400, 2500), where only half of the iterations achieved a TA of 93% or higher. But repeating the behaviour of the previous experiments shows that the higher the number of PCs, the lower the TAs and VSs after reaching the MATA.

As can be seen, a pattern started to show. The sets with 300 PCs achieved the highest average testing accuracy, whereas the sets with 200 PCs achieved the highest average validation score.

Figure 7 shows final results taken out of all tables of the second stage. The table PCA III depicts that increasing the NOHN unnecessarily does not improve the TA, hence, the best performance is achieved with 1500 and 2500 hidden neurons. It is important to mention that the number of iterations indicate when the cross-validation

Table	nof_prin_compone nts	nohn	Average testing accuracy (after 20 iterations)	Average validation score (after 20 iterations)
PCA I	200	500	89.8%	89.2%
PCA I	300	500	89.8%	86.7%
PCA II	200	1500	94.5%	94.2%
PCA II	250	1500	94.7%	93.2%
PCA II	300	1500	94.8%	91.3%
PCA III	200	2500	96 %	93.8%
PCA III	300	2500	94.6%	93.2%

Fig. 7 The best set of parameters from each table from stage II regarding the MATA and the MAVS with the result of another sub-experiment from PCA II

stopped the training process in order to prevent overfitting and rather apply BP to improve the algorithm even more. The tables also show that with increasing the number of PCs, the average number of iterations decrease, which only supports the fact that the sets with high PCs only achieve the minimum TA and VS and, therefore, might lead to overfitting. All in all, it can be said that the sets (250, 1500), (300, 1500), and (200, 2500) show the best performances.

6.3.3   Stage III

In the previous stage it is shown that sets with 300 PCs achieved the best TA and the sets with 200 PCs achieved the best VS. Hence, the middle (250) proved to be a good approximation to achieve a good TA in combination with a good VS. For this reason, the first approach in this stage is to see how the TAs improve when the number of 250 PCs stays the same but the NOHN increases instead. The second approach is another radical experiment with a higher number of PCs (550) and again increasing NOHNs. This stage documented the same values as the previous stage II.

The first table, Neuron I, has 250 PCs and starts with 50 hidden neurons with increases following logarithmic scale increments due to previous findings up to 10,000 hidden neurons. As already stated, exceeding the overall number of 3500 is not the goal, but it is done nevertheless, because of experimental reasons. The MATA here is achieved twice at 96.8% with an average VS of 96.1% for the set (250, 5000) and an average VS of 96% for the set (250, 10000). It must be said that the average NOI is at 51.2 for the first set and 43.2 for the second set, hence, the learning process is faster than with lower NOHNs, i.e. at least 100 iterations for 50 to 500 hidden neurons. It is also interesting to see, that the experiment with the set (250, 1500) is now conducted for a second time (first time is in the table PCA II). Comparing both experiments, the following insights can be obtained: The TA for

Table	nof_prin_compone nts	nohn	Average testing accuracy (after 20 iterations)	Average validation score (after 20 iterations)
Neuron I	250	1500	95.1%	92.9%
Neuron I	250	5000	96.8%	96.1%
Neuron I	250	10000	96.8%	96 %
Neuron II	550	5000	96 %	85.5%
Neuron II	550	10000	95.9%	87.7%

Fig. 8 The best set of parameters from each table from stage III regarding the MATA and the MAVS with the result of another experiment from Neuron I to compare with the result from PCA II

the first experiment is at 94.7% with a VS of 93.2% and an average NOI of 68.3. The TA for the second experiment is at 95.1% with a VS of 92.9% and an average NOI of 72.6. It is likely that the average values would not be exactly the same, but it is clear that the difference is between 0.4% for the TA and 0.3% for the VS, which is low. Thus, it can be said, that this set seems to be the most reliable one within the acceptable range of NOHNs. The difference between the TAs of the set with the MATA and the chosen set is at 1.7%. Now, one would have to choose whether to risk overfitting but achieving a good accuracy or staying in the acceptable range and still achieving 94–95% accuracy.

Looking at the table Neuron II, the number of PCs stays consistent with 550 and the NOHN increases from 50 to 10,000, again in logarithmic scale increments. The MATA is at 96% with an associated average VS of 85.5% through the set (550, 5000). Surprisingly, the MAVS is only at 87.7% with an average TA of 95.9% through the set (550, 10000). These sets have similar results, also compared to the results with 250 PCs from the previous table. Moreover, looking at the set (550, 1500), it can be seen that the average TA of 92.1% is already decreasing, when compared to the table PCA II, where it is shown that all accuracies after the set (300, 1500) were decreasing. Hence, it is only natural, that the NOHN would have to be increased in order to balance the constant decrease. But more hidden neurons would then exceed the maximum. It can be stated, that the previous experiment in table Neuron I achieved a better MATA.

Noting that the maximum NOHN is consciously exceeded, the following table shows the best performances (Fig. 8).

In conclusion, increasing the NOHN radically changes the TA by 1.7%. Moreover, the good accuracies were only achieved because the NOHN exceeded the maximum of 3,500 by a great factor which made the model more complex than it has to be, hence, it risks being overfitted. Furthermore, it could be seen that the results of the set (250, 1500) only had a difference of 0.4/0.3% compared to the results of the first experiment with the same set.

6.3.4   Stage IV

Stage IV of the experiments is the last stage. It is not a necessary experiment given the high number of hidden neurons it applies, it is rather an excursus on the limitations of Machine Learning algorithms in general. Adding to this, the loss provided by the loss function is now noted in order to see which parameters were able to minimise the losses and still keep a good accuracy.

The table Top values I conducted experiments with two different NOHNs, 5000 and 10000, and with the number of PCs increasing from 100 to 600 by 100. Here, the MATA is achieved twice at 95.3% with an associated average VS of 95.8% and an average loss of 0.28 for the set (300, 5000) and with an average VS of 96.2% and an average loss of 0.34 for the set (300, 10000). The MAVS is at 97.4% with an associated average TA of 94.9% and an average loss of 0.49 for the set (200, 10000). Unfortunately, the stated TAs do not create a new maximum value, they rather showed the losses when using complex models. Nevertheless, a new overall maximum value for the validation score is achieved which is 1.3% higher than the MAVS from table Neuron I.

Continuing with the table Top values II, the experiments depicted an approximation for the number of PCs in combination with 5000 hidden neurons. Here, a total of 30 iterations were undertaken for each set of parameters. As already shown in previous experiments, the range between 200 and 300 PCs seemed to have the best performances without regarding the NOHN. In order to find a better set of parameters with the best performing amount of hidden neurons (5000), several experiments were conducted. Still, the MATA only achieved 96.3% with an average VS of 96.2% and an average loss of 0.34 for the set (280, 5000), which is 0.5% less than the results of the sets (250, 5000) and (250, 10000) from Neuron I. The MAVS is at 96.6% with an average TA of 96.1% and an average loss 0.33 for the set (230, 5000).

Figure 9 summarises well that the best performances are achieved if the number of PCs is between 200 and 300. Because of the automatic stop function in the script, overfitting can not happen, it is only a risk for further experiments that do not include said function.

Table	nof_prin_comp onents	nohn	Average testing accuracy (after 20 iterations)	Average validation score (after 20 iterations)	Average loss
Top values I	200	10000	94.9%	97.4%	0.49
Top values I	300	5000	95.3%	95.8%	0.28
Top values I	300	10000	95.3%	96.2%	0.34
Top values II	230	5000	96.1%	96.6%	0.33
Top values II	280	5000	96.3%	96.2%	0.34

Fig. 9 The best sets of parameters from each table from stage IV regarding the MATA and the MAVS

6.3.5   Conclusion

All things considered, the experiments not only showed the development of finding good sets of parameters, they also showed the growth of understanding within the limitations of an ANN. With each experiment certain values could be excluded because the algorithms became either too complex or were not complex enough. Fortunately, the function *early_stopping* prevented the algorithm from overfitting and used 10% of the training set for cross-validation. After a few experiments a pattern could be seen, which showed that the range of 200–300 PCs achieved the best recognition accuracies. Now, only the specific amount of hidden neurons has to be chosen.

6.4   Results

The experiments that were conducted, showed a great variety of results. The task was to find the best set of parameters with the lowest error ratio and a good recognition accuracy. The following table shows the best results from all stages. The loss depicts the error ratio, hence, the lower the loss value, the lower the error ratio (Fig. 10).

Table	nof_prin_comp onents	nohn	Average testing accuracy (after 20 iterations)	Average validation score (after 20 iterations)	Average loss (after 20/30/30 iterations)
Nohn100	200	100	84.5%	-	-
Nohn150	170	150	86.3%	-	-
Nohn200	200	200	89.2%	-	-
Nohn250	250	250	88.3%	-	-
PCA I	200	500	89.8%	89.2%	-
PCA II	300	1500	94.8%	91.3%	-
PCA III	200	2500	96 %	93.8%	-
Neuron I	250	1500	95.1%	92.9%	-
Neuron I	250	5000	96.8%	96.1%	-
Neuron II	550	5000	96 %	85.5%	-
Top values I	300	5000	95.3%	95.8%	0.28
Top values II	230	5000	96.1%	96.6%	0.33
Top values II	280	5000	96.3%	96.2%	0.34

Fig. 10 The the best sets of parameters from each table from each stage with their average TA, VS, and loss

There are three sets of parameters that need to be compared: 1. (200, 2500), 2. (250, 5000), and 3. (300, 5000). The first set is still in the range of acceptable NOHNs, where 3500 is the maximum. If 3500 is the official limit of hidden neurons, then the first set would be the best one. If the number of hidden neurons is not as important, then sets with higher NOHNs can be considered. In this case, the second set has the overall best recognition accuracy of 96.8%. As stated several times, a function prevents the algorithm from overfitting, hence, the risk is not as high. However, the error ratio needs to be considered as well. According to the error ratio, the third set would show the best performance.

Now, choosing the safer path with 1500 NOHN, an approximation is made, which is shown in the table Approximation. Here, the pattern of 250–300 PCs showing the best performances is taken and tested again. The NOHN stayed the same at 1500, while the number of PCs were 250, 275, and 300. Adding to that, the loss is now documented, as well, in order to be able to compare the results to the sets of parameters from stage IV. Moreover, the experiments were conducted with two different testing data sizes. The initial experiments had 20% of the overall data as testing data, which is again the case for the first experiment in the table Approximation. The second one has 33% of the overall data as testing data, which means that $k = 3$ in the KCV process. This will indicate how well the parameters perform with a smaller training data size, which fits the use of real-world applications more. Thus, lower accuracies are expected because there is less data to use for training purposes (Fig. 11).

It is important to mention that the set (300, 1500) with 20% testing data is now shown twice, one time in the table PCA II with the average TA of 94.8% and average VS of 91.3%, and the second time now in the table Approximation with the average TA of 96.4% and the average VS of 91.7%. The average values of these both experiments result in 95.6% for the TA and 91.5% for the VS. Nevertheless, the set with 275 PCs shows the best recognition accuracy and is only 0.2% lower than the overall MATA from the set (250, 5000). The average losses of 0.31 are too high, given the fact that this value is the exact average value between 0.34 and 0.28 average losses. The VS of 92.6% is lower compared to the other sets, but as already stated, the actual testing accuracy outmatches the validation score, hence, the TA counts more than the VS.

Looking at Fig. 12, the experiment with 33% testing data shows lower results, given the fact that there is less data to train with. The testing accuracies still reach

nohn	nof_prin_compone nts	Average testing accuracy (after 20 iterations)	Average validation score (after 20 iterations)	Average loss (after 20 iterations)
1500	250	96.2%	93.4%	0.34
1500	275	96.6%	92.6%	0.31
1500	300	96.4%	91.7%	0.30

Fig. 11 The results of the first experiment from the table Approximation with $test\_size = 0.2$

nohn	nof_prin_compone nts	Average testing accuracy (after 20 iterations)	Average validation score (after 20 iterations)	Average loss (after 20 iterations)
1500	250	92.8%	91.2%	0.37
1500	275	92.7%	90.3%	0.37
1500	300	92.7%	89.3%	0.35

Fig. 12 The results of the first experiment from the table Approximation with $test\_size = 0.33$

at least 92.7% with only 0.04-0.06 more losses than the average of the previous experiment. Surprisingly, the TAs are very close with only 0.1% difference. But as previous experiments have shown, more iterations can change the overall results and potentially increase or decrease them. Adding to that, the lower accuracies could be explained when putting the number of PCs and NOHNs into relation with the training data size. If the training data size decreases from 80% to 66%, then the values for the parameters would need to be adjusted in the same manner, in order to contain the relative accuracies. For experimental purposes, the absolute values were compared to show that the chosen values would still perform well enough with a larger testing data size.

In conclusion, it can be said that the sets with values 250–300 PCs and 1500 hidden neurons show the best and most reliable performances and are still in the limit of the acceptable complexity within the model. The exact accuracies can vary with each new iteration, hence, it is hard to pick only one exact value. But the set (275, 1500) seems to be a good approximation and leads to an average between 250 and 300 PCs.

6.5 Analysis

After finding a good approximation towards a set with a recognition accuracy over 90% and an error ratio of approximately 0.3, it needs to be shown why the model behaved the way it did. The following Fig. 13 shows a graph designed in MATLAB with the testing accuracy as a function of the number of PCs and the NOHN.

This figure shows that the function has the highest accuracy values between 200 and 300 PCs, increasing with the NOHN, as is proven by the tables in the spreadsheet. Furthermore, Fig. 14 and Fig. 15 show the explicit maxima that can be seen in Fig. 13.

These figures prove that the pattern that is detected during the experiments indicates that the number of PCs between 200 and 300 show the best performances. Hence, the ANN should at least have 250 principal components. Figure 13 also displays that all accuracies between the PC values 100 and 800 achieve at least 90% through 5000 hidden neurons.

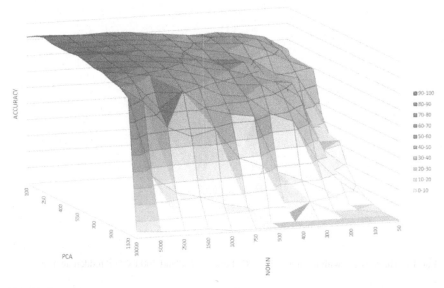

Fig. 13 The behaviour of the ANN with different values for PCs and hidden neurons in relation to the testing accuracy

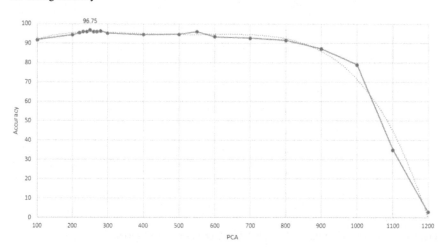

Fig. 14 The maxima with the number of PCs between 200 and 300 for 5000 hidden neuron

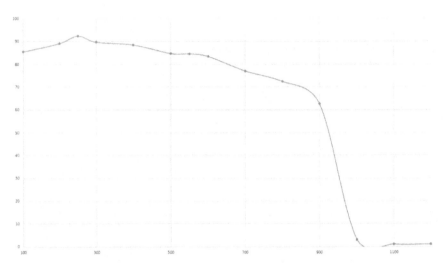

Fig. 15 The maxima with the number of PCs between 200 and 300 for 500 hidden neurons

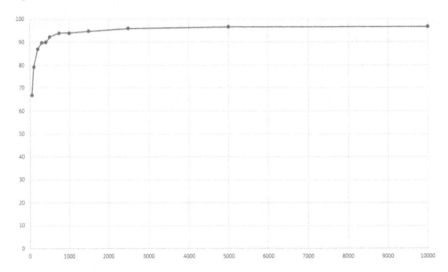

Fig. 16 The behaviour of the model with 250 PCs and increasing NOHNs

Figure 16 depicts the development of the accuracies by increasing the NOHN at 250 PCs. This behaviour is easily explained. If the NOHN is too low, then the model is not complex enough, hence, the accuracies are not high. Conventionally, the accuracy would be too good with increasing NOHNs because the ANN model would start to fit the noise and therefore, start overfitting. But because there is a function applied in the script that stops the learning process before it starts overfitting, the accuracy seems to become consistent with growing NOHNs. Unfortunately, that does not stop the model from becoming too complex for the task. That is why the usual number

nohn	nof_prin_compone nts	Average testing accuracy (after 20 iterations)	Average validation score (after 20 iterations)	Average loss (after 20 iterations)
1500	209	93.7%	91.5%	0.45
1500	230	94.6%	92.2%	0.38
1500	251	94.5%	90.9%	0.38

Fig. 17 The results of the third experiment from the table Approximation with test_size = 0.33 and PC put into relation with test_size

of hidden neurons should be between the general amount of input and output nodes, thus, 3500 in this case.

Moreover, one of the reasons why the accuracies with the bigger testing data size were lower, besides the fact that there is a smaller amount of data to train with, is the fact that the absolute values of the parameters were trained and tested with. For the testing data size of 20%, 1200 images will be trained with. Here, it can be assumed, that by choosing 33% to be testing data, the values would need to be adjusted relatively. Now, only 1000 images will be used for training. This might also be the reason, why the set (250, 1500) from the table Approximation achieved the lowest value in the first experiment with 20% testing data and the best performance with 33% testing data. However, this hypothesis needs to be proven in another experiment.

Nevertheless, out of curiosity, a short excursus was undertaken, in order to see how the relative parameters would perform. Two more experiments were added to the table Approximation where the number of PCs was adjusted, so it was always in the same relation to the training data as with the testing data size of 20%. The Fig. 17 shows the values for $test\_size = 0.33$, hence 33% of the general dataset.

It can be seen that the TAs from Fig. 17 show at least 1% better accuracies than from Fig. 12. By adjusting the number of PCs, the general accuracies improve, but the losses become higher because there is less data to train with. Now the chance of only training with, for instance, the lighter images or with the Caucasian looking people, hence, leaving the darker images or the south-east Asian and African people for the testing, are higher, which means that the algorithm could struggle with data that it has not been trained with. Changing the variable $test\_size$ to 0.1, thus, having only 10% as testing data, changes the outcomes again, because now the number of PCs is adjusted and the sets are trained and tested with new values. In this case, there is more data to train with, so higher accuracies are expected, which can be seen in Fig. 18.

Here, the number of PCs has to be increased because the amount of training data increased. When there are 300 PCs for 1200 training images (25%), then there must be 251 PCs for 1005 training images (still 25%) and 338 PCs for 1350 training images (also 25%). This was done with all PCs here to find the relative values. As can be seen, the TAs were higher than for 33% testing data size, but lower than for 20%. Hence, this only shows that all experiments from the several stages were conducted with the optimal value for k.

nohn	nof_prin_compone nts	Average testing accuracy (after 20 iterations)	Average validation score (after 20 iterations)	Average loss (after 20 iterations)
1500	281	95.9%	93.3%	0.29
1500	309	96.4%	95 %	0.27
1500	338	94.1%	92.3%	0.29

Fig. 18 The results of the fourth experiment from the table Approximation with test_size = 0.1 and PC put into relation with test_size

In the final analysis, the best performance can be achieved with the number of principal components between 200 and 300. After reaching 1500 hidden neurons, the model seems to slow down with the learning process and the accuracies are all over 90% with increasing NOHNs.

7   Part 2

Part 2 discusses the automated search of a set of parameters by implementing optimisation algorithms such as Grid Search and Random Search.

7.1   Grid Search

Optimising is a key aspect when using an ANN. In order to find a more efficient approach for using the optimal set of hyperparameters, the process can be automated. For Grid Search, a specific grid space had to be defined. It was recommended to use a small number of dimensions, to prevent the grid from becoming too complex. The chosen dimensions for this experiment were: 1. Number of hidden neurons (NOHN), 2. Activation function, 3. Solver, and 4. Batch size. The experiments conducted in Part 1 had predefined values for the dimensions, where only the NOHN and PCs were changed. Optimisation algorithms make it possible to broaden the dimensions, hence, instead of using the same values as in Part 1, new values were added to compare the outcomes. The values of the activation function included $tanh$, which is the hyperbolic tan function, and $relu$, which implements the rectified linear unit function (scikit learn.org 2020a). The solver parameter has three possible values of which two specific ones were used: sgd, the stochastic gradient descent, and $adam$, a stochastic gradient-based optimiser (scikit learn.org 2020). The batch size was provided values of 2^n starting with 2 and closing with 512. The rest of the parameters used their default values. Notably, the activation function's default value is $relu$, and the solver's default value is $adam$. This means, that in this experiment

nof_prin_componen ts	nohn	batch_size	solver	activation	Average testing accuracy (after 20 iterations)	Average validation score (after 20 iterations)	Average loss (after 20 iterations)
200	100	8	adam	tanh	96.6%	97.0 %	0.01
250	100	4	adam	tanh	96.9%	96.7%	0.01
300	200	4	adam	tanh	96.5%	95.6%	0.01

Fig. 19 The best results through Grid Search

the default values are compared with other matching values, that were used in Part 1.

Grid Search was used on the MLPClassifier with the library GridSearchCV (scikit learn.org 2020a). The challenge was to set a good range of values for the NOHN. The fact that each added value made the process more complex, contributed to the decision to start with small ranges and conduct several smaller experiments.

The results from Fig. 19 show the three best results with the best accuracies for different numbers of PCs in the range that was established in Part 1. It is noticeable that *adam* as a solver and *tanh* as an activation produce the best outcome. The batch size was reduced from 256 from Part 1 to 8 and 4 which contributes to the NOHN being so low. The very low loss is a benefit and already shows an improvement when compared to Part 1 where the average loss was mostly at 0.3. Unfortunately, the accuracy did not improve much when compared to the results of the manual optimisation. From previous experiments, it could be seen that the optimal number of PCs is between 200 and 300. This is why this was tested in Grid Search as well. Figure 19 shows that 250 PCs achieved the best accuracy which is also the middle of the recommended range. With this in mind, further experiments were conducted by using extra hidden layers. The number of hidden layers so far has been 1. By adding a second or even a third hidden layer the ANN becomes more complex. Many experiments were started, but after 12 h of running time, the programme was dropped due to Google Colab platform restrictions. Unfortunately, this meant that the effort in order to find a better set of parameters through Grid Search became too time consuming. By adding more values to the dimensions, the programme ran longer and 10 h became the average time and the resulting accuracies were not satisfying. At this point, the automated optimisation was not more efficient than the manual optimisation. As a result, another optimisation algorithm was used.

7.2 Random Search

Random Search chooses its values randomly and can therefore explore a wider range of possible combinations with a lower chance of using the same values twice. Using 250 PCs to simplify the search, there were several parameters with different values (scikit learn.org 2020b). These parameters with their possible values were defined:

- Nohn: the range from 100 to 6,000 in natural numbers
- Solver: $adam, sgd, lbfgs$
- Learning rate: $constant, adaptive$
- Batch size: values starting with 4 and closing with 256
- Activation: $identity, logistic, tanh, relu$
- Initial learning rate: 0.0001–0.01
- Momentum: 0–1

NOHN, solver, batch size and activation have already been introduced. The new solver value $lbfgs$ is an optimiser that originates from the family of quasi-Newton methods. Two new values were added to the activation, namely $identity$, which is a no-op activation that returns $f(x) = x$, and $logistic$, which defines the logistic sigmoid function. Moreover, three additional parameters were included: 1. Learning rate, 2. Initial learning rate, and 3. Momentum. The learning rate schedules the updates for the weights. Therefore, $constant$ means that there is a constant learning rate provided by the parameter which defines the initial learning rate, whereas $adaptive$ implies that it keeps the initial learning rate until the training loss stops decreasing, then the learning rate in that moment is divided by 5. Hereby the initial learning rate controls the step-size regarding the update of the individual weights. The values for momentum also influence the update during the gradient descent. It is important to mention that both the parameters $momentum$ and $learning\_rate$ are only used when the solver is set to sgd. Therefore, the initial learning rate can also only apply for sgd or $adam$ (scikit learn.org 2020). These experiments also took a lot of time to process and run, but the running time was still below the one needed for Grid Search. The maximum time was 7 h with an average of approximately 5 h.

The results that can be seen in Fig. 20 show that Random Search was able to find specific values for the parameters that improved the accuracies by 1–2% compared to the previous experiments. A pattern can be identified, which is indicated by the fact that the best accuracies are achieved when using a constant learning rate, the

nof_pri n_comp onents	nohn	batch _size	solver	activation	mome ntum	learning_ rate_init	learning_rate	Average testing score (after 20 iterations)	Average loss (after 20 iterations)
250	4920	64	lbfgs	relu	0.3	0.001	constant	97.8 %	0.01
250	3151	4	lbfgs	relu	0.65	0.001	constant	97.5%	0.01
250	3149	8	lbfgs	identity	0.2	0.0009	adaptive	97.9%	0.01
250	2741	4	lbfgs	tanh	0.2	0.0002	constant	98.1%	0.01
250	672	8	adam	logistic	0.3	0.001	constant	98.1%	0.01
250	2659	64	lbfgs	tanh	0.2	0.0013	constant	98.0%	0.01

Fig. 20 The best results through random search

lbfgs solver, and *tanh* as an activation function. The fifth set of parameters from the table seems like an outlier because of the low NOHN but still achieves a high accuracy.

7.3   Further Optimisation

Grid Search and Random Search provided various sets of parameters. Notably, Random Search only showed results that were chosen randomly, so it is not proven that the provided values for the parameters are indeed the best ones. Thus, the experience from Part 1 shows that changing just one value of a parameter manually can have a significant impact. By choosing the two best sets of parameters from Random Search that share the most values, new experiments were conducted manually. Because the fourth and sixth set of parameters from Fig. 20 are similar and only differ significantly in the size of mini batches, the search was extended. By trying out different batch sizes it could be seen that 256 achieved the highest accuracy.

By training the ANN and testing the prediction accuracy, new values for the NOHN could be identified (Fig. 21).

Although the NOHN ranges from 1750 to 2741 the difference in the accuracy is very small. All accuracies that were achieved during this optimisation had an average recognition accuracy between 98.1 and 98.2% and an average loss of 0.01. The NOHN is still within the accepted range between 30 and 3,538. In order to choose one set of parameters only, there are more iterations of the training process needed. 20 iterations for each set is enough to use it in this comparison but to distinguish and identify one particular set that can be chosen for later uses as well, therefore, at least several thousands of iterations are needed.

nof_prin_com ponents	nohn	batch_size	solver	activation	Average testing score (after 20 iterations)	Average loss (after 20 iterations)
250	2741	256	lbfgs	tanh	98.10 %	0.01
250	2659	256	lbfgs	tanh	98.17 %	0.01
250	2250	256	lbfgs	tanh	98.12 %	0.01
250	2000	256	lbfgs	tanh	98.19 %	0.01
250	1750	256	lbfgs	tanh	98.17 %	0.01

Fig. 21 The best results through manually optimising random search results

8   Results

All methods that were used provided resulting sets of parameters that achieved at least 96% accuracy. Important to mention is the fact that the aspect of time limited the search for the optimal set of parameters for this face recognition task. Nevertheless, the result of the optimisation of the outcomes from Random Search showed a variety of possible NOHN that could be used within the ANN. 98.2% prediction accuracy is a good outcome that can be used in several applications.

9   Analysis

Analysing the search for an optimal set of parameters showed that it matters which optimisation method is used. A lot of time ran into the capture of the outcomes of each iteration in Part 1. In total, 3,240 iterations were undertaken and documented with an average of three values per iteration. Assuming that the average time of one iteration and its documentation took 0.5 min, this results in 1,620 min or 27 full hours. The best accuracy achieved by manually looking for the parameters was 96.6%. Comparing this to the time that Grid Search needed, an average of 10 h to get one good set of parameters where the best one so far was 96.9%, the total amount of time needed is similar. For three good sets of parameters through Grid Search, it takes 30 h. There the argument of being more efficient only works when stating that the time that the programme needs, can be used working on other tasks. But in order to work on this chapter the actual results are needed, hence, Grid Search did not turn out to be of benefit. Nevertheless, Random Search posed a new approach by displaying different combinations of values of parameters with an average running time of 5 h. By choosing the best results from Random Search and extending the optimisation by adding manual approximations, a good set of parameters was found. Given that the approximation took exactly one hour and finding six good sets of parameters, each taking 5 h, the total time needed equals to 31 h. This is still the longest time needed, but it also provided the best recognition accuracy and embedded two optimisation methods. Provided that more time is available to conduct deeper experiments, an even better recognition accuracy could be achieved.

Figure 22 shows the range of values for the hyperparameters that could be identified through these experiments for achieving the best recognition accuracy and the lowest error ratio.

Grid Search might be able to provide the best values for the parameters eventually, but the cost is too high to still be called an efficient optimisation. However, Random Search does not guarantee to find the one best set of values, but there is a wider range of better results in less iterations. And combining this with manually tuning the already optimised search a good range can be defined as shown in Fig. 22.

nof_prin _compo nents	nohn	batch _size	solver	activati on	mome ntum	learning_ rate_init	learning_rate	Average testing score (after 20 iterations)	Average loss (after 20 iterations)
250	1750- 2741	256	lbfgs	tanh	0.2	0.0002	constant	98.1% - 98.2%	0.01

Fig. 22 The best range of values for the hyperparameters

10 Common Issues in Face Recognition

Recent studies have found that algorithms used for machine learning show biases towards classes like race and gender (Buolamwini et al. 2018). If the training dataset does not show a balanced representation of a variety of people, the use of that algorithm can discriminate against minorities and reinforce disadvantages towards these groups. After looking closely at the provided dataset (Belhumeur and Kriegman 1997), it proved the missing diversity of people. The dataset has a total of 30 different people, of which only approximately 30% represent women, people of colour, and people with south-east Asian descents together. Since 1997, when the dataset was created, the demographics of society have changed, therefore, the dataset needs to be changed, as well. It would be interesting to see, how the performance of the algorithm would change, if the testing data was more diverse than the training data. The performance would probably decrease, based on the fact, that the algorithm was not trained well enough to identify a wider range of ethnicities and genders. Hence, by choosing a training dataset, a greater representation of the world's population must be considered. Although a perfect training dataset does not exist, there are ways to achieve better performances by using different databases, for instance the database "Labeled Faces in the Wild" created by the University of Massachusetts (University of Massachusetts 2021). Equally important is the representation of different age groups. A child or a young adult has different facial features than a middle-aged or elderly person. The provided dataset showed people in a narrow range of ages, the majority of which seemed middle-aged. Another experiment could also document the behaviour of the algorithm when trained with datasets of people who have birthmarks, facial hair, or injuries. For instance, the algorithm could have focused its recognition on people's eyebrows, thus, people who do not have eyebrows due to medical conditions would not be identified at all. Identically, the algorithm could have focused its recognition on the fact that every person has a set of two eyes, which could potentially lead to a chimpanzee being identified as well. Moreover, another experiment could analyse the performance of recognition if people wear facial accessories, such as glasses, piercings or even makeup. As a last proposal, the database "Labeled Faces in the Wild" (University of Massachusetts 2021) offers pictures of people with various facial expressions and with different backgrounds. A potential question would be: If a person closes their eyes, would the algorithm still be able to identify them? What happens, if people show their teeth? In order to answer these questions, new experiments would have to be conducted. As a result, these issues can only be noticed through a complex training and testing phase with a diverse dataset.

11 Conclusion

All things considered, the development and optimisation of a solution for a Machine Learning task was a detailed process. In the same fashion as natural neural networks learn, artificial neural networks learn by finding familiar patterns. The goal was to find a set of parameters for the model to achieve the best recognition accuracy with the lowest error ratio. Several experiments were conducted to display a wide variety of solutions. The risk of overfitting the model could be ignored because of the convenient script that stopped the training process before it could fit to the peculiarities of the images. The first pattern that was noticed was the fact that sets of parameters with numbers of principal components between 200 and 300 achieved the best performances, once the model had achieved a level of complexity. By increasing the complexity of the ANN, hence, by increasing the number of hidden neurons, the testing accuracies were improved. Two different methods of tuning the values of the hyperparameters were experimented with. The first part of this chapter dealt with the manual optimisation of the values in favour of the recognition accuracy. The highest accuracy achieved was 96.6% with an average loss of 0.3 and spending 27 h on it. In the second part, two different optimisation algorithms were utilised, namely Grid Search and Random Search. Grid Search provided a highest recognition accuracy of 96.9% and an average error ratio of 0.01 and 30 h of work. Random Search used the same amount of time but provided an accuracy of 98.1% with a loss of 0.01. By manually optimising the values from Random Search, an accuracy of 98.2% could be achieved. In other words, the combination of both automated optimisation followed by manual optimisation for an approximation of a better accuracy turned out to be the best solution.

For further use of the shown sets of parameters the risk of overfitting/underfitting the model needs to be considered. Adding to this, the impact of biased datasets was discussed with new proposals for experiments and databases. The experiments showed that the accuracies can change with each iteration, hence, replicating the experiments will never show the exact same values but they will likely be in the same range. It would also be interesting to see how the performances would change with bigger or smaller datasets and whether the relative values would show a similar behaviour with different data sizes.

All in all, the empirical investigation of finding an acceptable set of parameters to use in a specific Machine Learning algorithm proved to be an interesting but time consuming challenge which provided a satisfying result. The requirements of the task were met by creating a wide range of experiments and analysing the performances of the algorithm precisely. This chapter proposed ideas for new experiments to understand the functions of this particular algorithm in more detail and to improve further accuracies.

Acknowledgements The authors are grateful to Dr Schetinin from the University of Bedfordshire for useful comments, guidance and support.

References

M.J. Sheehan, M.W. Nachman, Morphological and population genomic evidence that human faces have evolved to signal individual identity, vol. 5 (2014)

K. Gates, *Our Biometric Future: Facial Recognition Technology and the Culture of Surveillance* (NYU Press, 2011)

M.H. Hassoun, *Foundamentals of Artificial Neural Networks* (MIT Press, Cambridge, Massachusetts, 1995)

R. Bhatia et al., Biometrics and face recognition techniques. Int. J. Adv. Res. Comput. Sci. Softw. Eng. **3**, 93–99 (2013)

M. Mann, M. Smith, Automated facial recognition technology: Recent developments and approaches to oversight (2017)

G. Hornung, M. Desoi, M. Pocs, Biometric systems in future preventive scenarios—legal issues and challenges, in *BIOSIG 2010: Biometrics and Electronic Signatures, Proceedings of the Special Interest Group on Biometrics and Electronic Signatures* (Bonn, Germany, 2010), pp. 83–94, Gesellschaft fuer Informatik e.V

S.-C. Wang, Artificial neural network, in *Interdisciplinary Computing in Java Programming*. The Springer International Series in Engineering and Computer Science, vol. 743 (2003), pp. 81–100

B. Csaji, H. Ten Eikelder, Approximation with artificial neural networks (2001)

D. Anguita, L. Ghelardoni, A. Ghio, L. Oneto, S Ridella, The "k" in k-fold cross validation, in *20th European Symposium on Artificial Neural Networks, Computational Intelligence and Machine Learning (ESANN)* (2012), pp. 441–446

B. Kia, Session 14: K-fold cross validation (2018)

T. Dietterich, Overfitting and under computing in machine learning. ACM Comput. Surv. **27**, 326–327 (1995)

H. Jabbar, R. Khan, Methods to avoid over-fitting and under-fitting in supervised machine learning (comparative study), in *Computer Science, Communication & Instrumentation Devices* (2014)

S.-W. Lin, K.-C. Ying, S.-C. Chen, Z.-J. Lee, Particle swarm optimization for parameter determination and feature selection of support vector machines. Expert. Syst. Appl. **35**, 1817–1824 (2008)

D.E. Rumelhart, R. Durbin, R. Golden, Y. Chauvin, *Backpropagation: Theory, Architectures, and Applications* (Lawrence Erlbaum, Hillsdale, New Jersey, 1995)

R. Rojas, The backpropagation algorithm, in *Neural Networks*. (Springer, Berlin, 1996)

D.E. Rumelhart, G.E. Hinton, R.J. Williams, Learning representations by back-propagating errors, vol. 323 (1986), pp. 533–536

scikit learn.org, Mlpclassifier—scikit-learn 0.24.1 documentation (2020)

J. Bergstra, Y. Bengio, Random search for hyper-parameter optimization. J. Mach. Learn. Res. **13**, 281–305 (2012)

M. Claesen, B. De Moor, Hyperparameter search in machine learning, in *The XI Metaheuristics International Conference* (2015)

J. Buolamwini, T. Gebru, S. Friedler, C. Wilson, Gender shades: intersectional accuracy disparities in commercial gender classification, vol. 81 (2018), pp. 1–15

S. Sra, S. Nowozin, S. Wright, *Optimization for Machine Learning* (MIT Press, Cambridge, Massachusetts, 2012)

P. Belhumeur, D. Kriegman, The yale face database (1997)

V. Schetinin, Process_yale_images (2021)

V. Schetinin, Classify_yale (2021)

S. Karamizadeh, S.M. Abdullah, A.A. Manaf, M. Zamani, A. Hooman, An overview of principal component analysis. J. Signal Inf. Process. **4**, 173–175 (2013)

M. Turk, A. Pentland, Eigenfaces for recognition (1991)

Y. Liu, J. Starzyk, Z. Zhu, Optimizing number of hidden neurons in neural networks (2007), pp. 121–126

W. Wang, N. Srebro, A. Garivier, S. Kale, Stochastic nonconvex optimization with large minibatches. Proc. Mach. Learn. Res. **98**, 1–26 (2019)

scikit learn.org, Gridsearchcv—scikit-learn 0.24.2 documentation (2020)

scikit learn.org, Randomizedsearchcv—scikit-learn 0.24.2 documentation (2020)

University of Massachusetts, Labeled faces in the wild (2021)

Internet-Assisted Data Intelligence for Pandemic Prediction: An Intelligent Framework

H. M. K. K. M. B. Herath ⓘ

Abstract The latest advances in the field of information and communication technology (ICT) are enabling organizations to evolve and expand in the age of "Big Data". The growth of big data and the development of Internet of Things (IoT) technology have aided the viability of modern smart city initiatives. The governments and industries will use these technological advancements, as well as the widespread use of ubiquitous computing, to address healthcare requirements in a variety of ways. The novel COVID-19 and other major pandemic events are known for being unexpected, unpredictable, and dangerous. With the rapid increase in coronavirus cases, big data has the potential to facilitate the prediction of outbreaks. As we witnessed, COVID-19 has caused tremendous harm to humanity all over the globe. Owing to the knowledge gained from the novel COVID-19, early pandemic prediction and responses are crucial. Various approaches for predicting pandemics have been proposed, but none have yet been developed based on people's everyday behaviors and environmental changes. The aim of this chapter is to develop a framework for pandemic prediction in a smart city by utilizing big data intelligence provided by people and environmental changes. The framework was tested using data from the novel COVID-19 virus, which was spread across Sri Lanka in 2020. Based on the experimental findings, the proposed framework has the potential to predict pandemics in the notion of the smart city.

Keywords Big data intelligence · Cloud computing · Cyber physical systems · Internet of Things (IoT) · SEIR model

1 Introduction

A pandemic is an infectious disease outbreak that has spread around a vast geographic region, such as several continents or the entire world and affecting a large number of people. Three major global pandemic outbreaks have occurred in the last few

H. M. K. K. M. B. Herath (✉)
Faculty of Engineering Technology, The Open University of Sri Lanka, Nugegoda, Sri Lanka

© The Author(s), under exclusive license to Springer Nature Switzerland AG 2022   173
Y. Baddi et al. (eds.), *Big Data Intelligence for Smart Applications*,
Studies in Computational Intelligence 994,
https://doi.org/10.1007/978-3-030-87954-9_7

decades: 1. Severe Acute Respiratory Syndrome (SARS), 2. Middle East Respiratory Syndrome (MERS), and 3. Ebola Virus (EVD) (Drosten et al. 2003). SARS was initially identified in Southern China in November 2002, and was caused by the SARS-associated coronavirus (SARS-CoV), which spreads rapidly via droplets. In comparison to SARS, new MERS cases have been found in Saudi Arabia. By the end of January 2020, a total of 2519 laboratory-confirmed MERS cases and 866 associated deaths had been recorded worldwide since it was first detected in 2012. Table 1 depicts the comparison between the different viruses and their characteristics.

The novel COVID-19 or also known as SARS-CoV-2 infection is increasingly growing at alarming rates around the world. As of April 2021, more than 139 million people have been infected, with around 2.99 million people dead as a result of the disease (Herath et al. 2021a). At the time of this writing, the United States has more COVID-19 infection cases, while Australia is in control of the pandemic.

Despite their economic growth and position as hubs for cultural trade and technological advancement, megacities, defined by the United Nations (UN) as cities with populations of more than 10 million people, face growing environmental and population health threats (Lai et al. 2020). By 2030, the United Nations predicts 43 megacities, the majority of which will be in developed countries. By 2050, the world's population would have swelled to almost 10 billion people, with 68% of them residing in cities (United Nations 2018). Megacities have a significant impact on various aspects (Borraz and Galès 2010) of human life, including transportation, healthcare, energy, education. Climate hazards, pandemics, civic demonstrations, and organized crime have all made cities the "locus of risks". Most megacities are especially vulnerable to infectious diseases due to their population density and accessibility, as shown by MERS, EVD, SARS, and COVID-19 epidemics. Nowadays, the novel COVID-19 outbreak has emphasized the importance of technology in everyday life (Herath 2021).

Table 1 Comparison between Coronavirus disease 2019 (COVID-19), Severe Acute Respiratory Syndrome (SARS), Middle East Respiratory Syndrome (MERS), and Ebola Virus Disease (EVD)

Description	COVID-19	SARS	MERS	EVD
Incubation period (days)	2–14	2–10	2–14	2–21
Contagious during incubation	Yes	Yes	Yes	No
Vaccine as of 2020	No	No	No	Yes
Infected body part	Blood vessels	Respiratory system		All muscles
Main transmission	Respiratory droplet secretions			Body fluids
Shape and size in nm	Spherical 80–120	Spherical 80–90	Spherical 90–125	Filament 14,000 × 80 nm
Mortality rate %	2.65	14–15	34	Up to 90
Pronounced R-naught	2–2.5	3.1–4.2	< 1	1.5–1.9

Table 2 Platforms to enable smart city technologies

Smart city platforms	Enabled services for smart applications
Internet of Things (IoTs) (Santana et al. 2017)	Sensors and actuators, middleware, data collection
Big data (Santana et al. 2017)	Data processing, data storing, data analysis, data visualization
Cyber-physical systems (Santana et al. 2017)	Computation in physical systems, city actuation
Cloud computing (Santana et al. 2017)	Hosting services, hosting storage and computation, elasticity and scalability

Smart city prospects are very promising, and various edge computing manufacturers, such as IBM and Intel, are launching various projects to consolidate their guidance in this field. Smart health, smart transportation, smart security, smart government, smart buildings, smart grid, smart tourism, smart lifestyle, smart home, and smart climate are known to be ten critical fields that will play key roles in creating a smart community (Caragliu and Bo 2019). Big data in healthcare is problematic not only because of its sheer scale but also because of the heterogeneous quality of the data and the pace at which it must be handled. The integration of big data into smart healthcare systems has ushered in a new age in electronic and mobile healthcare that is bearing results in terms of cost-effectiveness and productivity gains. Rapid advancements in ICT, the Internet of Things (Herath et al. 2021b), cloud computing, and smart applications have allowed sharing of real-time data. The large volume of various urban data allows for a more comprehensive and holistic view of urban environments and real-time events. Cyber-physical systems, the Internet of Things, big data, and cloud computing are the four primary tools used by software networks for smart cities, according to Santana et al. (2017). Table 2 describes the different smart city platforms and their available technologies for smart applications.

1.1 The Role of Big Data in Smart Cities

Big data refers to the massive evolution of structured, unstructured, and semi-structured data. Through careful interpretation, businesses can gain greater perspectives and make smarter decisions by processing them optimally. Big data analytics can derive benefit by developing a modern wave of technologies and architecture that can achieve high velocity, discover, and analyze large volumes and varieties of data (Beyer and Laney 2012; Laney 2001; Ward and Barker 2013). The growth of big data and the development of Internet of Things (IoT) technology have aided the viability of smart city initiatives (Hashem et al. 2016). The IoT facilitates the convergence of sensors, Radio-frequency identification (RFID), and Bluetooth in the real-world setting using highly networked services, and big data offers the ability for cities to gain useful information from a vast volume of data gathered from multiple sources.

Table 3 A comparisons of various smart applications and their uses

Smart application	Specific use(s)	IoT technologies	Communication technologies
Smart healthcare (Hashem et al. 2016)	Health-related data collection, pandemic related data collection, and monitoring	Sensors, wearable devices, cameras, RFID cards, smartphones	Bluetooth, ZigBee
Smart transportation (Hashem et al. 2016)	Monitoring and management of route, traffic, and parking spaces	Cameras, RFID cards, smartphones	RFID, 3G, 4G
Smart governance (Hashem et al. 2016)	To develop smart policies with the aim of better managing citizens	Smartphones, cameras, sensors	Wi-Fi, LTE, LTE-A, WiMax, Bluetooth, LoRaWAN
Smart grid (Hashem et al. 2016)	Manage the power supply	Smart meters, smart readers	Wi-Fi, Zigbee, Z-Wave

A list of the various smart city applications is described in Table 3. In this chapter, the author has mainly focused on smart healthcare and smart transportation technologies in the IoT.

The author has recently identified a number of variables linked to the early identification of the COVID-19 pandemic (air pollution, hospital admission rate, death cases, parking density). The latest pandemic is also having an effect on government and medical facilities. As a result, there is also a need to identify potential pandemics before they occur. One of the most critical technologies for the early prediction of a pandemic scenario is big data. The aim of this chapter is to develop a framework by enabling big data intelligence for future pandemic predictions in smart cities. The rest of this chapter is categorized into four sections. In Sect. 2, a literature review is carried out. Section 3 discussed the proposed framework architecture. Results and observations are discussed in Sect. 4. Finally, Sect. 5 brings the proposed system to a conclusion.

2 Literature Review

This section addresses related works in the fields of IoT, smart healthcare, and big data-driven technologies in a brief manner. Google Scholar was used to reviewing academic works from 2015 to 2020. The author has considered the factors that impact the novel COVID-19 since this chapter primarily focuses on the big data-enabled framework for pandemic prediction.

Remote sensing and the Hadoop ecosystem have recently provided effective and low-cost solutions in a variety of domains, especially in healthcare. On the one hand, sensing devices enable doctors and nurses to track patients remotely, whether in the hospital or at home, and to diagnose medical emergencies in real-time. Hadoop-based tools, on the other hand, provide hospitals with cost-effective and fast data storage and processing platforms, especially in extreme situations like virus contamination or global pandemic situations. Noh and Lee (2015) have introduced an integrated big data platform model which focuses on open-source S/W configuration.

Harb et al. (2020) have proposed an efficient big data analytics tool for real-time patient tracking and decision-making, which would benefit both hospital and medical personnel. They have proposed a framework that is made up of four layers such as 1: real-time patient surveillance, 2: real-time evaluation and data storage, 3: patient classification and disease diagnosis, and 4: data retrieval and visualization based on big data analytics and data processing techniques. The efficacy of their proposed platform in terms of patient classification and disease diagnosis in healthcare applications has been demonstrated in experimental findings.

Prospects of the modern healthcare industry have accelerated by combining big data and smart systems. Pramanik et al. (2017) have addressed a systematic assessment of various big data and smart system developments, as well as a critical examination of current advanced healthcare systems. They've also suggested a big data-enabled smart healthcare system architecture, which provides theoretical representations of intra and inter-organizational business models in the healthcare sector.

Assist from sensors and IoT devices are important for monitoring the health conditions of a person. Syed et al. (2019) have discussed healthcare services to diseased and healthy populations through remote surveillance using intelligent algorithms, software, and techniques, as well as fast analysis and expert intervention for effective treatment decisions. They have also presented an outline of various data analysis algorithms and big data platforms for smart healthcare applications.

Jia et al. (2020) have presented a detailed study on characteristics of novel COVID-19, as well as key big data capabilities, such as online public opinion analysis, virus-host analysis, pandemic visualization, COVID-19 close contact inspection, and pandemic prediction assessment.

The adoption of big data strategies to increase the quality of healthcare delivery has become unavoidable due to the growing volume of data in the healthcare industry. Aboudi and Benhlima (2018) have proposed an extensible big data model which focused on both stream and batch computing to improve the efficiency of healthcare services even further by providing real-time warnings and allowing precise forecasts on patient health status.

The implementation and incorporation of smart city technologies can be supported by smart city software platforms. Santana et al. (2017) have reviewed the state-of-the-art of smart city software platforms and they have studied 23 proposals and classified them into four groups based on the most often used supporting technology, as well as practical and non-functional requirements: Big Data, Internet of Things, Cyber-Physical Systems, and Cloud Computing. The author has studied existing smart city system domains that have the ability to detect pandemic scenarios in a smart city by studying Santana's article. Table 4 shows the different domains of smart city systems which are related to pandemic scenario prediction.

Table 4 Domains of smart city systems (related to the pandemic prediction)

Systems	City sensing	Traffic control	Air pollution	Healthcare	Disaster prevention
GAMBAS	–	√	–	–	–
SmartSantander	√	√	√	–	–
Padova Smart City	√	–	–	–	–
OpenIoT	√	–	–	–	–
WindyGrid	√	√	√	√	–
ClouT	√	√	–	√	√

A comprehensive study and framework of the use of big data in the battle against major public health outbreaks such as COVID-19 are also needed, which will serve as a valuable guide for governments and medical institutions. The author has acquired a clear understanding of existing big data-enabled technologies after completing the literature review. Based on the knowledge and experience gained from the COVID-19 pandemic, the author has proposed a single framework to predict future pandemics in a smart city. Section 3 of this chapter describes the methodology of the proposed framework.

3 Development of the Framework

This section discusses the development of the proposed pandemic prediction framework, which is based on big data intelligence and IoT technologies. After data was collected from the big data sources, it was used to predict the spread of pandemics and alert the relevant institution for future actions.

Big data obtained from various domains, such as vehicle parking density, air quality standards, and healthcare-related data, was widely used in the proposed design framework. The author has extracted a few domains that could be useful in predicting a pandemic. These criteria were chosen based on the knowledge and experience of several pandemics that have occurred in recent years. The big data repositories used in the development of the proposed framework are described in Table 5.

Figure 1 shows the program flowchart of the proposed framework. The big data collection procedure, as shown in the figure, separates data into three categories (Healthcare data, $PM_{2.5}$ index data, and parking data). From the A&P model, the system was collected the dependent variables related to air quality and parking density measures. The SEIR model was generated using information from hospital data. Input variables D_H, D_A, and D_T were sent to the Machine learning (ML) algorithm. The ML process was used to predict the current state of the pandemic after the data was collected from the SEIR and A&P models.

Table 5 Domains and data repositories used for the proposed framework

Domain	Data source/s	Data uses
Healthcare	Johns Hopkins University Center for Systems Science and Engineering data repository (Dong et al. 2020)	COVID-19 confirmed, death, and recovered cases
Air pollution	Colombo US Embassy's air quality monitor data repository (Embassy and Air Pollution: Real-time Air Quality Index (AQI). 2021)	Air quality index data
Parking space	Data.World data repositories (Data.World 2021)	Parking station ID, year/month, parking usage, and percentage of usage

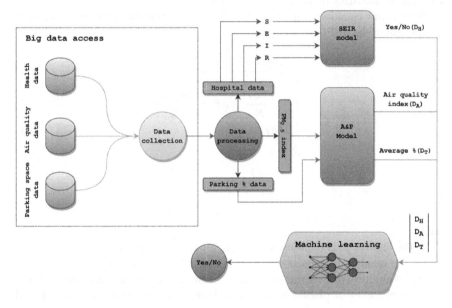

Fig. 1 Program flow chart of the proposed pandemic detection framework

Data was collected from the various sources mentioned in Table 5 which were stored in big data repositories. The author has mostly centered on the data providers relating to parking management, air quality, and healthcare-related data. According to the latest pandemic experience, as a pandemic strikes, road traffic density has decreased (available parking spaces have reduced). As a result, air pollution has reduced. These criteria were chosen to identify a pandemic scenario. The SEIR model, which is used to simulate pandemics, was developed using healthcare-related data sources. The variable $S(t)$, $E(t)$, $I(t)$, $D(t)$, and $R(t)$ were calculated from COVID-19 data repositories which are available at (Johns Hopkins University Center for Systems Science and Engineering data repository (Dong et al. 2020)).

3.1 Development of the SEIR Algorithm

The SIR model is well-known in the field of epidemiology, and it has yielded many significant results. The susceptible, infected, and recovered people are represented by the letters S, I, and R in the model (Smith and Moore 2004; McCluskey 2010; Zhu and Ying 2014). The coronavirus disease, on the other hand, has a latent period in which the individual is infected but not yet symptomatic. In the SEIR (Susceptible-Exposed-Infected-Recovered) model, a new part E is added to define the symptomless but infected process (Li et al. 2021). In this section, we proposed an SEIR model for predicting pandemics based on the number of susceptible, exposed, infected, and recovered individuals collected from big data repositories. Relationships between each part of the SEIR model are shown in Fig. 2.

Differential equations represented in Eq. 1 describe the relationship of each component of the SEIR model.

$$
\begin{aligned}
\dot{S} &= \frac{-rbSI}{N} - \frac{r_2 b_2 SE}{N} \\
\dot{E} &= \frac{rbSI}{N} - aE + \frac{r_2 b_2 SE}{N} \\
\dot{I} &= aE - yI - uI \\
\dot{R} &= yI \\
\dot{N} &= -uI \\
\dot{D} &= uI
\end{aligned}
\tag{1}
$$

were,

r: The number of contaminated contacts who are susceptible,

r_2: The number of contacts who have been exposed to the virus,

b: Probability of disease transmission between a susceptible and infectious person,

b_2: Probability of disease transmission between a susceptible and an exposed person,

a: The rate at which an infected person becomes infected after being exposed,

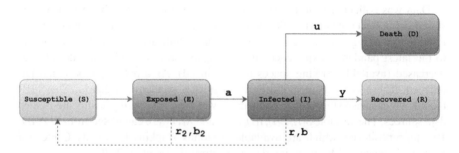

Fig. 2 Relationship between each component of the SEIR model

y: The recovery rate,
u: The mortality rate.

$$N(t) = S(t) + E(t) + I(t) + R(t) - D(t) \tag{2}$$

The total population ($N(t)$) is described using Eq. 2. As shown in the equation, the total population is a relationship between susceptible population ($S(t)$), exposed population ($E(t)$), infected population ($I(t)$), recovered population ($R(t)$), and the number of death cases ($D(t)$). Figure 3 illustrates the working principle of the proposed SEIR algorithm.

After the acquisition of healthcare-related data from the big data repositories, the "*big data processing*" function was performed to extract the model data for the "*parameter estimation*" process. The SEIR model was generated using the parameters extracted from the "*parameter estimation*" process. The model also used cloud computing technologies to display real-time data and graphs related to the pandemic.

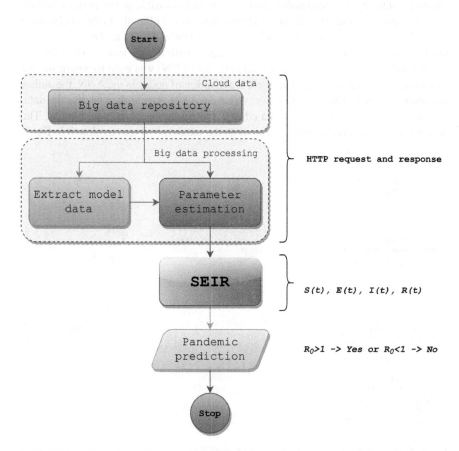

Fig. 3 Program flow chart of the proposed SEIR algorithm

3.2 Development of the A&P Algorithm

The development of an air quality and parking space measuring (A&P) algorithm to predict possible pandemics in a smart city is described in this section. The author has used big data repositories to derive dependent variables including the air quality index and the data related to the parking space available in a parking station. The algorithm was developed using the Python development environment. Following the extraction of data, the system evaluated the value to determine the time span during which the pandemic would have an effect on the city. We know from experience of the latest pandemic that as people are exposed to the virus, the number of hospital admissions rises. Therefore, People are placed in isolation at home or in a quarantine area. As a result, traffic in the area reduced, and air quality improved. The author has used the same phenomena to develop the algorithm. Figure 4 depicts the program flow chart of the proposed A&P algorithm.

In the pre-processing phase, data normalization was performed and features were identified. During the pandemics, parking spaces accessible at the parking station were occupied at a higher percentile, according to public officials. In this study, more than 80% of parking usages were selected as a potential danger in the area. It's also used as a reference for air quality measurements, with a 0–50 index. Following the selection of features, an artificial neural network (ANN) was used to determine the city's pandemic status $(P(V_i, A_i))$. For the activation of nodes in the ANN, the author has used the ReLU activation function. After the ANN algorithm generates the data, the pandemic level (P_i) was estimated and the pandemic curve was plotted. The author has developed a formula to calculate the pandemic stages as shown in Eq. 3.

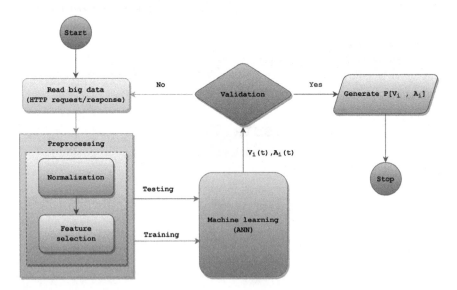

Fig. 4 Program flow chart of the proposed A&P algorithm

Table 6 Pandemic levels (P_i) and required action

Pandemic level (P_i)	Status	Responses
1.3	Danger	A possible pandemic strike on the area, with a high mortality rate, necessitates urgent lock-down or protective precautions
1.0	Highly threat	The area could be struck by a pandemic, necessitating decisive intervention
0.8	Threat	Possible threat, action required
0.4	Normal	Normal city conditions, do not necessitate consideration of the situation
− 0.3 to 0.0	Do not concern	There is no disease in the city, so no intervention is required

$$P_i = 0.78 - 0.63 \sum \frac{V_i}{n} - 2.32 \sum \frac{A_i}{n} + 0.51 R_0$$

Where,

$P_i = pandemic\ level\ at\ ith\ day$

$V_i = rate\ of\ parking\ usage\ at\ ith\ day$ \hspace{2cm} (3)

$A_i = rate\ of\ air\ quality\ at\ ith\ day$

$R_0 = basic\ reproduction\ number$

$i = day\ number$

$n = no\ of\ days$

Table 6 depicts the pandemic levels and the actions that must be taken to prevent the spreading of the disease or virus. The author has identified three potential regions of the developed pandemic curve in a smart city, as seen in the table. Where the pandemic level is in the 0.8–1.3 range, it is considered a threat to the city, and relevant institutions or government officials must act immediately. If P_i is in the range of −0.3 to 0.0, there is no disease in the city, and no action is needed.

4 Experiment Results

The aim of this study is to develop a framework for predicting pandemics in smart cities using big data from healthcare, air quality, and parking spaces. The author used data obtained from the Johns Hopkins University Center for Systems Science and Engineering (CSSE) data repositories, the US Embassy's Colombo Air Quality Monitor data repository, and parking station database accessible at Data.World to evaluate the efficiency of the proposed framework. For the experimenting on the proposed framework, the Python development environment was used.

4.1 Results of the Pandemic Prediction SEIR Algorithm

For the extraction of confirmed, death, and recovered COVID-19 cases, the author has used the COVID-19 upstream repository which is maintained by the Johns Hopkins University Center for Systems Science and Engineering (CSSE). The COVID-19 status was generated after the data was collected. Figure 5 depicts the extracted data from the repository by using the filtering keywords (LK) in the Python environment.

In Sri Lanka, Fig. 6 depicts the behavior of confirmed, death, and recovered COVID-19 cases. The graph was generated in the Python environment using variables extracted from the COVID-19 data repository.

The classic epidemic curve was obtained by plotting the values of the S, E, I, and R compartments at each time step are shown in Fig. 7. After the data was collected from the big data, it was processed for the S, E, I, and R models. The SEIR model was generated based on the COVID-19 cases for the first 60 days in Sri Lanka. Initially, the system was set to $y = 0.12$, and $\beta = 0.3$ for the total population (N) of 415 cases. In the diagram, the population is completely vulnerable at the outset, with just a few infected individuals. As the virus progresses, the number of infected people increases

```
Pandemic prediction SEIR model [Developed by: H.M.K.K.M.B. Herath]
Collecting data from https://datahub.io/core/covid-19/r/time-series-19-covid-combined.csv
-----------------------------------------------------------------------------------------
      Day        Date Country/Region  Confirmed  Recovered  Deaths
0       1   1/22/2020      Sri Lanka          0          0       0
1       2   1/23/2020      Sri Lanka          0          0       0
2       3   1/24/2020      Sri Lanka          0          0       0
3       4   1/25/2020      Sri Lanka          0          0       0
4       5   1/26/2020      Sri Lanka          0          0       0
..     ...        ...            ...        ...        ...     ...
295   296  11/12/2020      Sri Lanka      15723      10653      48
296   297  11/13/2020      Sri Lanka      16191      11031      53
297   298  11/14/2020      Sri Lanka      16583      11324      53
298   299  11/15/2020      Sri Lanka      17287      11495      58
299   300  11/16/2020      Sri Lanka      17674      11806      61

[300 rows x 6 columns]
```

Fig. 5 Result window of the processed data from big data repositories

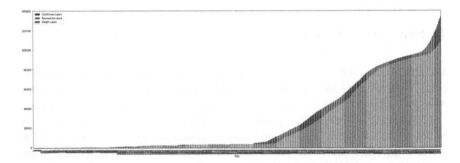

Fig. 6 Result window of the acquired COVID-19 cases through big data repository

S E I R model [Developed by: H.M.K.K.M.B. Herath]

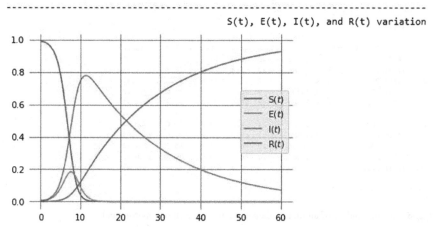

Fig. 7 Result window of the pandemic prediction S, E, I, and R model

until it reaches a peak, after which it starts to decrease as there are fewer and fewer people to infect. When the R_0 (basic reproduction number) value exceeds one ($R_0 >$ 1), the infection rate exceeds the recovery rate, and the infection spreads across the population. $R_0 = 2.5$ ($R_0 > 1$) was calculated using the SEIR model, indicating that the pandemic is spreading across the community.

4.2 Results of the Air Quality Based Pandemic Prediction Algorithm

The results and observations of the air quality-based pandemic prediction model are discussed in this section. The proposed algorithm was developed in the Python development environment. The author collected major data on air pollution from the data repository of the Colombo US Embassy's air quality monitor. Traffic congestion on the roads reduces as people become isolated due to the pandemic. This would lead to improved air quality than on a typical day. This phenomenon was used by the author as one factor in predicting a pandemic in a city. Fine particulate matter ($PM_{2.5}$) is an atmospheric pollutant that poses a health risk when concentrations are high. When $PM_{2.5}$ levels are high, the air becomes hazy and visibility is reduced. Data on fine particulate matter ($PM_{2.5}$) was used to design the algorithm. Figure 8 shows the result window of data acquisition from data repositories in the air quality-based pandemic prediction model.

Figure 9 depicts the data gathered during the last 17 months in Colombo. The lower air quality index is shown in green, while the highest air quality index is shown in red. During the COVID-19 period, Colombo, Sri Lanka, had lower air pollution due to

```
Air quality detection model [Developed by: H.M.K.K.M.B. Herath]
HTTP Requesting: [US Embassy's Colombo air quality monitor data repository]
Connection Established
HTTP Response:  200
---------------------------------------------------------------------------------------------
            Date Country     City  count   min    max  median  variance
0      4/26/2021      LK  Colombo     17  23.0   26.0    25.0      7.35
1     12/31/2020      LK  Colombo     22  22.0   25.0    23.0      6.34
2      1/14/2021      LK  Colombo     12  23.0   25.0    24.0      4.47
3      1/31/2021      LK  Colombo      9  22.0   25.0    24.0     10.28
4       3/2/2021      LK  Colombo     18  20.0   24.0    22.0     14.77
..           ...     ...      ...    ...   ...    ...     ...       ...
295    4/15/2021      LK  Colombo     21  66.0   88.0    74.0    680.62
296    4/26/2021      LK  Colombo     23  66.0   94.0    74.0    876.92
297     5/6/2021      LK  Colombo     22  70.0   94.0    88.0    754.80
298     5/9/2021      LK  Colombo     21  66.0   94.0    88.0   1240.90
299   12/31/2020      LK  Colombo     23  70.0  100.0    88.0    609.05

[300 rows x 8 columns]
```

Fig. 8 Result window of air quality data extraction on Python environment

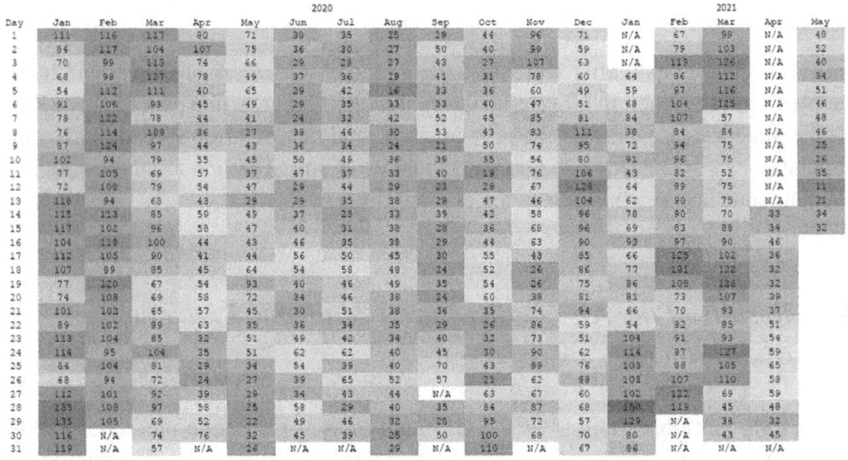

	2020												2021				
Day	Jan	Feb	Mar	Apr	May	Jun	Jul	Aug	Sep	Oct	Nov	Dec	Jan	Feb	Mar	Apr	May
1	111	116	117	80	71	30	35	25	29	44	96	71	N/A	67	99	N/A	48
2	84	117	104	107	75	36	30	27	50	40	99	59	N/A	79	103	N/A	52
3	70	99	113	74	66	29	28	27	43	27	107	63	N/A	119	126	N/A	40
4	68	98	127	78	49	37	36	28	41	31	78	60	64	86	112	N/A	34
5	54	112	111	40	65	29	42	16	33	36	60	49	59	97	116	N/A	51
6	91	106	93	45	49	29	35	33	33	40	47	51	68	104	125	N/A	46
7	78	122	78	44	41	24	32	42	52	45	85	81	84	107	57	N/A	48
8	76	114	108	36	27	38	46	30	53	43	83	111	38	84	84	N/A	46
9	87	124	97	44	43	36	34	24	21	50	74	95	72	94	75	N/A	25
10	102	94	79	55	45	50	49	36	39	35	56	80	91	96	75	N/A	26
11	77	105	69	57	37	47	37	33	40	19	76	106	43	82	52	N/A	35
12	72	108	79	54	47	29	44	29	23	28	67	128	64	89	75	N/A	11
13	118	94	68	43	29	29	35	38	28	47	46	104	62	90	75	N/A	21
14	115	113	85	59	49	37	28	33	39	42	58	96	78	90	70	33	34
15	117	102	96	58	47	40	31	38	28	36	68	96	69	83	88	34	32
16	104	118	100	44	43	46	35	38	29	44	63	90	93	97	90	46	
17	112	105	90	41	44	56	50	45	30	55	43	85	66	125	102	36	
18	107	89	85	45	64	54	58	48	24	52	26	86	77	131	122	32	
19	77	120	67	54	93	40	46	49	35	54	26	75	86	109	128	32	
20	74	108	69	58	72	34	46	38	24	60	38	81	81	73	107	39	
21	101	102	65	57	45	30	51	38	36	35	74	94	66	70	93	37	
22	89	102	89	63	35	36	34	35	29	26	86	59	54	82	85	51	
23	113	104	85	32	51	49	42	34	40	32	73	51	104	91	93	54	
24	114	95	104	35	51	62	62	40	45	30	90	62	114	87	127	59	
25	84	104	81	29	34	54	39	40	70	43	89	76	103	88	105	65	
26	68	94	72	24	27	39	65	52	57	21	62	88	108	107	110	58	
27	112	101	92	39	29	34	43	44	N/A	63	67	60	102	122	69	59	
28	135	108	97	58	25	58	29	40	35	84	87	68	150	119	45	48	
29	135	105	69	52	22	49	46	32	28	95	72	57	129	N/A	34	32	
30	116	N/A	74	76	32	45	39	25	50	100	68	70	80	N/A	43	45	
31	119	N/A	57	N/A	28	N/A	N/A	29	N/A	110	N/A	67	86	N/A	N/A	N/A	

Fig. 9 Result window for last 17 months air quality index of the Colombo city

lower traffic effects, as seen in the illustration. From May to October 2020, people in Sri Lanka were contaminated with COVID-19, and they were forced to stay at home as a precaution. As a result of this occurrence, the city's traffic density decreased, and air pollution was decreased.

Figure 10 depicts the processed and analyzed data for the last 17 months in Colombo city of Sri Lanka. Lower air quality indexes were also identified during the COVID-19 pandemic, according to the data. The second potential pandemic threat was observed in April 2021, as seen in the figure.

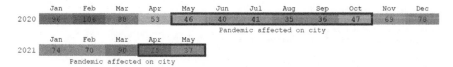

	Jan	Feb	Mar	Apr	May	Jun	Jul	Aug	Sep	Oct	Nov	Dec
2020	96	106	88	53	46	40	41	35	36	47	69	78

Pandemic affected on city

	Jan	Feb	Mar	Apr	May
2021	74	70	90	25	37

Pandemic affected on city

Fig. 10 Analyzed air quality index for past 17 month

4.3 Results of Parking Data Based Pandemic Prediction Algorithm

This section outlines the findings and observations of vehicle parking availability during the pandemic period. Figure 11 depicts the impact of parking capacity over the past 11 months in 2020, based on the data extracted from the data repositories. The results were produced using the Python programming language. According to the test results, the highest percentile was registered from May to October, when COVID-19 was at its peak in the city.

Figure 12 depicts the analysis of the experiment findings based on the data collected. According to the test findings, the yellow color field is the threat zone. The monthly average vehicle parking rate was found to be 38.31% ($N_{mean} = 552$). According to previous data, the author has observed that 80% of parking spaces were not occupied on a typical day. Therefore, 80% was chosen as the reference value of the model. As a result, the period from June to September 2020 was described as a pandemic-affected period in the city. During the pandemic, the average percentage of parking spaces was occupied by 88.75% ($N_{mean} = 1282$). The city's pandemic period (Jun–Sep) was identified in the highlighted region (red) of the graph.

Figure 13 shows the pandemic curve (described in Eq. 3) generated by the framework. According to the pandemic curve, the pandemic has arisen in the city when the pandemic level reaches greater than or equivalent to one ($P_i \geq 1$). In the illustration, the danger zone is represented by red nodes.

```
Pandemic prediction using parking spaces [Developed by: H.M.K.K.M.B. Herath]
HTTP Requesting: [https://data.world]
Connection Established
HTTP Response:  200
--------------------------------------------------------------------------------
    Station ID Date/Year  No of Vehicles  Percentage(%)
0            1    20-Feb          576.0           40.0
23           1    20-Mar          492.0           34.0
46           1    20-Apr          502.0           35.0
69           1    20-May          837.0           58.0
92           1    20-Jun         1264.0           88.0
115          1    20-Jul         1398.0           97.0
138          1    20-Aug         1260.0           87.0
161          1    20-Sep         1198.0           83.0
184          1    20-Oct         1063.0           74.0
207          1    20-Nov          558.0           39.0
230          1    20-Dec          523.0           36.0

Parking details for STATION ID: 1
```

Fig. 11 Result window for parking data acquisition in Python

Parking details for STATION ID: 1

Monthly average for past 12 months:552 (38.31%)
Possible pandemic detection:Jun-Sep (Avergae (88.75%))

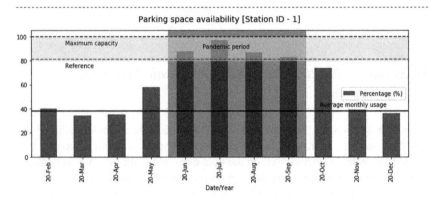

Fig. 12 Result window of pandemic detection through parking data

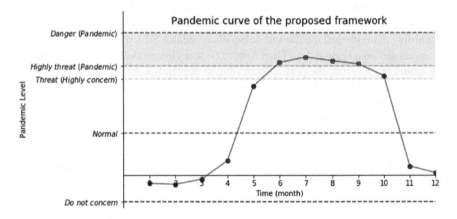

Fig. 13 Result window of the pandemic curve generated in proposed framework

The framework was tested using data obtained from the recent COVID-19 incident at Colombo, Sri Lanka. The proposed framework has the potential to predict epidemics using big data, according to the test results. The mean absolute error (MAE) of the proposed system was found to be 0.16. As a result, the system operates at a higher degree of efficiency.

5 Conclusion

Researchers, information technicians, and programmers from all over the world have been experimenting with a variety of technological methods and developments in order to improve pandemic detection and response using modern IoT-based tools. With the rapid increase of infectious diseases, big data has the potential to facilitate the prediction of outbreaks. The author of this chapter suggested the development of a framework focused on big data related to healthcare, air quality, and the availability of parking spaces in the city. Data from a recent COVID-19 incident in Colombo, Sri Lanka, was used to test the framework. For the COVID-19 pandemic, each domain has tested and confirmed the results. The mean absolute error (MAE) of the proposed system was found to be 0.16. According to the findings of the experiments, the proposed system has the potential to predict pandemics. The author's next step is to develop an algorithm that considers all geographical regions and compares the efficacy of various smart city models.

References

M.A. Beyer, D. Laney, *The Importance of 'Big Data': A Definition*, Gartner, Stamford, CT, pp. 2014–2018 (2012)

O. Borraz, P. Le Galès, Urban governance in Europe: the government of what? Pôle Sud **1**, 137–151 (2010)

L. Benhlima, Big data management for healthcare systems: architecture, requirements, and implementation. Adv. Bioinf. (2018)

A. Caragliu, C.F. Del Bo, Smart innovative cities: the impact of Smart City policies on urban innovation. Technol. Forecast. Soc. Chang. **142**, 373–383 (2019)

Colombo US Embassy Air Pollution: Real-time Air Quality Index (AQI) (2021). https://aqicn.org/city/sri-lanka/colombo/us-embassy. Accessed 1 May 2021

E. Dong, H. Du, L. Gardner, An interactive web-based dashboard to track COVID-19 in real time. Lancet. Infect. Dis **20**(5), 533–534 (2020)

Data.World. (2021). https://data.world/home/. Accessed 1 May 2021

C. Drosten, S. Günther, W. Preiser, S. Van Der Werf, H.R. Brodt, S. Becker, H. Rabenau, M. Panning, L. Kolesnikova, R.A. Fouchier, A. Berger, Identification of a novel coronavirus in patients with severe acute respiratory syndrome. N. Engl. J. Med. **348**(20), 1967–1976 (2003)

H.M.K.K.M.B. Herath, G.M.K.B. Karunasena, S.V.A.S.H. Ariyathunge, H.D.N.S. Priyankara, B.G.D.A. Madhusanka, H.M.W.T. Herath, U.D.C. Nimanthi, *Deep Learning Approach to Recognition of Novel COVID-19 Using CT Scans and Digital Image Processing* (2021)

H.M.K.K.M.B. Herath, G.M.K.B. Karunasena, H.M.W.T. Herath, Development of an IoT Based Systems to Mitigate the Impact of COVID-19 Pandemic in Smart Cities, in *Machine Intelligence and Data Analytics for Sustainable Future Smart Cities* (Springer, Cham, 2021), pp. 287–309

H.M.K.K.M.B. Herath, Internet of things (IoT) enable designs for identify and control the COVID-19 pandemic, in *Artificial Intelligence for COVID-19* (Springer, Cham, 2021)

I.A.T. Hashem, V. Chang, N.B. Anuar, K. Adewole, I. Yaqoob, A. Gani, H. Chiroma, The role of big data in smart city. Int. J. Inf. Manage. **36**(5), 748–758 (2016)

H. Harb, H. Mroue, A. Mansour, A. Nasser, E. Motta Cruz, A hadoop-based platform for patient classification and disease diagnosis in healthcare applications. Sensors **20**(7), 1931 (2020)

Q. Jia, Y. Guo, G. Wang, S.J. Barnes, Big data analytics in the fight against major public health incidents (Including COVID-19): a conceptual framework. Int. J. Environ. Res. Public Health **17**(17), 6161 (2020)

Lai et al. 2020 Apr 14Y. Lai, W. Yeung, L.A. Celi, Urban intelligence for pandemic response: viewpoint. JMIR Publ. Health Surveill **6** (2), e18873 (2020). https://doi.org/10.2196/18873

D. Laney, 3D data management: Controlling data volume, velocity and variety. META Group Res. Note **6**(70), 1 (2001)

C. Li, J. Huang, Y.H. Chen, H. Zhao, A fuzzy susceptible-exposed-infected-recovered model based on the confidence index. Int. J. Fuzzy Syst. 1–11 (2021)

C.C. McCluskey, Complete global stability for an SIR epidemic model with delay–distributed or discrete. Nonlinear Anal. **11**(1), 55–59 (2010)

K.S. Noh, D.S. Lee, Bigdata platform design and implementation model. Indian J. Sci. Technol. **8**(18), 1 (2015)

M.I. Pramanik, R.Y. Lau, H. Demirkan, M.A.K. Azad, Smart health: big data enabled health paradigm within smart cities. Expert Syst. Appl. **87**, 370–383 (2017)

E.F.Z. Santana, A.P. Chaves, M.A. Gerosa, F. Kon, D.S. Milojicic, Software platforms for smart cities: concepts, requirements, challenges, and a unified reference architecture. ACM Comput. Surv. (CSUR) **50**(6), 1–37 (2017)

L. Syed, S. Jabeen, S. Manimala, H.A. Elsayed, Data science algorithms and techniques for smart healthcare using IoT and big data analytics, in *Smart Techniques for a Smarter Planet* (Springer, Cham, 2019), pp. 211–241

D. Smith, L. Moore, The SIR model for spread of disease-the differential equation model. Convergence (2004)

United Nations. 68% of the world population projected to live in urban areas by 2050, says UN. 2018 Revision of World Urbanisation Prospects (2018)

J.S. Ward, A. Barker, A. Undefined by data: a survey of big data definitions (2013). arXiv:1309. 5821.

K. Zhu, L. Ying, Information source detection in the SIR model: a sample-path-based approach. IEEE/ACM Trans. Netw. **24**(1), 408–421 (2014)

NHS Big Data Intelligence on Blockchain Applications

Xiaohua Feng, Marc Conrad, and Khalid Hussein

Abstract With the exponentially increasing number of AI (Artificial intelligence) applications on Big Data being developed, AI cyber security defense becomes ever required. Blockchain technology invented in 2008 with BitCoin could be benefited alongside the customer of Big Data and so on. Following a rapid progress in its advance, this subject has recently become a hot discussion topic in the ICT (Information and communications technology) world. In this chapter, Big Data security is discussed from the beginning to the impact which could benefit IT engineers, ICT students and CS academic researchers. As a case study, because medical record is personally identifiable privacy information, it needs strictly access control security. Blockchain technology has the features of trustworthy cyber security, anti-fake, anti-alteration, integrity, immutability and transaction accounting transparency reputation, these made it a good candidate for being applied to NHS medical records. Currently Blockchain technology, as one of the most important smart technologies, had been very widely used in smart applications that could influence the world. An analysis on such topic is provided in this chapter.

Keywords Big data · Blockchain · Immutability · Consensus · Distributed ledger · PBFT (Practical Byzantine Fault Tolerance) algorithm · Consistency · Cyber security · NHS (National Health Service) · Medical record · Cryptography · AI · Smart application

X. Feng (✉) · M. Conrad · K. Hussein
University of Bedfordshire, University Square, Luton 1 3JU, UK
e-mail: xiaohua.feng@beds.ac.uk

M. Conrad
e-mail: marc.conrad@beds.ac.uk

K. Hussein
e-mail: khalidhussein@study.beds.ac.uk

© The Author(s), under exclusive license to Springer Nature Switzerland AG 2022   191
Y. Baddi et al. (eds.), *Big Data Intelligence for Smart Applications*,
Studies in Computational Intelligence 994,
https://doi.org/10.1007/978-3-030-87954-9_8

1 Introduction

An investigation of feasibility to Blockchain technology application on NHS medical records was carried out recently. The main aim of the research was to develop secure authentication system using Blockchain technology to allow users, GPs (General practitioner), heath organizations to access to the medical and demographic records location independently. The idea was that the Blockchain record would have only smart contract which would allow access to the entity to the required record, which meant the Blockchain would not withhold any records or medical information. All the medical record will be kept safely in network transparently and securely. The objectives of this chapter were carried out in terms of developing an intricate system that would help in the building of a more secure and stable health system, along with the use of AI (Artificial Intelligence) and Information and Communications Technology ML(machine learning)/DL (deep learning) technology and so on, which would provide the integrity, availability and authenticity. Furthermore, to develop an intricate system through using the Blockchain technology; to reduce the cost of PII (Personally identifiable information) data management. In order to addressing and resolving cyber security issues, decentralized framework was used, tackling the challenges that arising from the Blockchain technology, such as stigma from its association with crypto in general. The predicament of Blockchain still had not fully met speed requirement, time-consuming, and laborious. A case study of NHS Big Data Infrastructure security deploying Blockchain technology would be introduced in the project. Taking the longest chain was the chosen one principle into account, Blockchain guaranteed anti-alteration and the digital signature algorithm guaranteed anti-fake integrity. A shared, replicated, permissioned ledger with consensus, provenance, immutability and finality smart application was introduced to NHS, UK. The physically NHS Blockchain experiments could lead the trends to solving some of the outstanding NHS cyber security issues. Nowadays in healthcare field, with Blockchain developed rapidly, Blockchain technology become one of healthcare revolution. Because Blockchain can change how doctor access patients' data, how clinical research would take place and other aspects of the healthcare system, Blockchain technology played one of the modernization solutions in NHS.

2 Background of Blockchain

2.1 About Blockchain

2.1.1 Blockchain Definition

Blockchain was defined as a system used to make a digital record of all the occasions a cryptocurrency (a digital currency such as bitcoin) was bought or sold, and that was constantly growing as more blocks were added Cambridge (2021). In other words,

Blockchain was a system in which a record of transactions made in BTC (bitcoin) or another cryptocurrency were maintained across several computers that were linked in a peer-to-peer network. A Blockchain was essentially a digital ledger of transactions that is duplicated and distributed across the entire network of computer systems on the Blockchain. The kind of technologies adopted Blockchain advanced features was what Blockchain technology about.

2.1.2 Blockchain Technology

Blockchain technology had been widely applied in many areas. Blockchain technology was most simply defined as a decentralized, distributed ledger technology that records the provenance of a digital asset (Builtin 2018). Due to the advantages of Blockchain feature, applications were quite broad in the growth. Blockchain had a nearly endless number of applications across almost every industry. For instance, the ledger technology could be applied to track fraud in finance, securely sharing patient medical records between healthcare professionals and even acted as a better way to track intellectual property in business and music rights for artists. Metropolitan police could also make use of Blockchain technology to secure cyber-bulling and cyber-stalking forensics evidence (Feng and Short 2017). There were some of examples of applications of this technology being discussed in Sect. 2.1.3. A combination of technologies of cryptography public and private key technology, decentralized network to share distributed accounting technology and motivate mechanism of record, store and integrity security service for network transactions formed Blockchain technology.

A schematic diagram of how Blockchain working being demonstrated in Fig. 1. It demonstrated how Blockchain worked in principle. Basically, Blockchain was a distributed digital transactions ledger. As soon as the transactions were recorded and

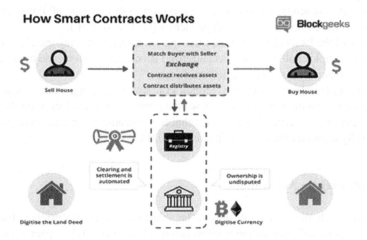

Fig. 1 The flow of transactions in the Blockchain model (Steemit 2017)

verified, it was almost impossible that anybody could alternate it. Everybody in the Blockchain had the transactions record. That meant people could not change every transactions records in the world, which made Blockchain immutability integration characteristics (Zhao et al. 2021).

2.1.3 How Blockchain Worked

The principle of work was, whenever each transaction happened, this accounting detail was sent to everyone in this Blockchain network nodes. Therefore, each nodes kept a record of this transaction. After verified, the records were kept in this Blockchain. When the transaction accounts (transaction bookkeeping) up to nearly one megabyte, these transactions were packed into a block. A packed block contained about 4000 transaction records. This block was chained to the previous Blockchain and expect the next block.

Blockchain offered permanent and transparent transaction records. The whole network had a consistent account record. Everybody was able to check and trace the record. In Blockchain distributed database, account was not centralized control by any one node, but controlled by all the nodes, joint maintenance and joint bookkeeping by all.

Any single node could not assume it was a main part to alternate any record. In order to change a record, which require to control the majority (more than 51%) of the nodes. While Blockchain had unlimited nodes and increasing nodes all the time, that made the attempt of making any change impossible.

The essence of Blockchain technology was an open accounting/recording system which verify each other. This system record all the accounts happened all the transactions or activities. Every transactions changes were recorded in the network summary/total account. Every attendee all had a complete network summary. The transparency characteristics guaranteed everybody could statistic all the attendee's records and the current situation independently (D'Aliessi 2016).

Due to all the Blockchain open and transparent to everyone, anyone could check the source. People could trust this decentralized Blockchain system, and not to worry if there were any hiding conspiracies inside there.

To summarize, since Blockchain technology allowed checking, tracing and storing all types of information in its chain, these characteristics made Blockchain the most security system on earth. Blockchain technology became more and more popular and broadly used on many areas (Zhao et al. 2021).

2.1.4 Blockchain Applications

Blockchain technology was not only to be able to create cryptocurrency, but also to support personally identifiable information (Feng et al. 2021), peer review, e-voting and other democratic decision and audit tracing trading, financial transaction, culture, medical, science, business, logistics, messaging services and smart contract, legal and justice system; as well as public health. For instance, NHS could store patients' medical records in Blockchain to make medical records location independently accessed. Pharmaceutical industry could use Blockchain to verify medicine, in order to prevent false, anti-fake and so on.

Zhao et al. (2021) had analyzed some details, Blockchain had four innovative features if compare with conventional technologies: "peer-to-peer network, cryptography, chaining-based data structure and PoW (proof of work)".

If a Blockchain operated on a peer to-peer network where there was decentralized control. This characteristic removed likelihood of any single point of fault in the system to affect other nodes. That reduced any negative influence on compromise of individual points, which improved the weakness of designed systems.

Cryptography is the foundation of any security system. Especially, cryptographic hash offered one-way cryptographic hash encryption, was normally used broadly. Furthermore, public key infrastructure cryptography algorithm were used to generate and verify digital signature, that was ensuring accountability and non-repudiation for all of digital transactions integration, for authentication of users.

Blockchain data structure, in a conventional Blockchain, was a single chain of blocks, where every block had exactly one parent block, apart from the beginning block in the chain, (which was referred to as genesis or parent block). This data structure was the reason of Blockchain technology was referred. One block consisted of multiple transactions. Typically, the block was one Megabyte per block maximum size, (about 4000 transactions). This parameter would directly affect the throughput of the Blockchain platform. By chaining the blocks together, any alteration of any block, could be easily detected. Furthermore, due to big data redundancy existed in the Blockchain system, it was almost impossible for any adversary to change any transaction in the record on the Blockchain. This Blockchain was an implementation of the secure distributed ledger. Each block had a predefined header structure. The block header consists of several vital fields. One of them was the hash value of the parent block that connected the new block to an existing Blockchain (D'Aliessi 2016).

PoW (proof of work) was the most important characteristic. Normally, traditional consensus algorithms, which resorted to the use of election to control the current membership of the blocks (including who should be the primary) and multiple rounds of message exchanges to reach a consensus when a predefined condition was satisfied. PoW converted the consensus issue into a competitive puzzle solving game and whoever solves the puzzle first, gained to determine the next block and win a reward as if the one had successfully mined a pot of treasure. This was the main reason that participants who entered the proof of work competition were often referred to as miners. A miner's task was to assemble the next block as quickly as possible and try out different random numbers that would make the hash value of the block header smaller than the difficulty target value. Due to the nature of the cryptographic hash, one cannot predict which random number would solve the question. Hence, the only way to find the solution was to try many times until one was found. This puzzle design is later formalized as a non-interactive zero-knowledge proof issue.

These included memory security, thread security and process security, output and input data security and AI security and big data management (Zhao et al. 2021).

In addition, Zhao et al. (2020) had applied Blockchain technology on Hierarchical Processing and Logging of Sensing Data and IoT Event. Besides, IBM also had mainframe application on hybrid Blockchain technology. Another company that using the unique technology was Ford, IBM had working together with Ford, plan to track its raw materials transactions (IBM 2020).

2.2 Blockchain for NHS

2.2.1 Cyber Security Issue on NHS

The current healthcare system had not fully modernized. It relied heavily on the interaction between patient and doctors and works on limited data. The limitation aspect of the healthcare resulted in a healthcare system that kept taking advantage of the data efficiently. Also, the current process of getting healthcare is long and tedious at its best. To improve healthcare system effectively handling of the patient is necessary (Feng et al. 2021).

2.2.2 Blockchain Characteristics

A Blockchain is a decentralized, peer to peer, anonymous, electronic accounting ledger, public transaction transparency with digital signature and cryptographic supported system. It possessed with anti-fake, anti-alteration record immunity characteristics. The more secure authentication and stable health system using public Blockchain technologies were introduced to NHS, which would provide better integrity, availability and authenticity, to allow medicine doctors and heath staff related accessing to the NHS spine project, using Blockchain technology, working on NHS physical Blockchain (Feng et al. 2021).

In public Blockchain, a new block creation principle was only a single member could create a new block. Furthermore, new block verification and propagation rule made the new block could only be verified, by the population; that is, the length of the chain (although the new block creator will obtain a commission and package reward). The valid new blocks were propagated its popularity throughout the entire network in the world. In that way, ensured the longest chain principle of Blockchain. This principle guaranteed anti-alteration incidence feature. The medical authority was in the position which granted the access control to medical record. The system should allow all transactions to the record safer. To adding amending or accessing or updating the record to database via a mechanism of verification through Blockchain, after the verification confirmed then the transaction would be authorized and performed (Yang et al. 2017).

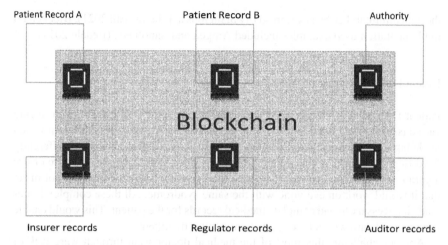

Fig. 2 Decentralized solution using Blockchain (Weisenthal 2019)

Blockchain based health network held the complete medical history for each patient, could be more efficient and secure solution (IBM 2020) as shown in next section Fig. 2.

The Fig. 1 shown the flow of transactions in the Blockchain technology. Some Blockchain application examples could be discussed next (Steemit 2017).

2.2.3 Blockchain Application Examples

Clinical Trial Application

NHS clinical trial was a simple method of testing new drugs like Covid19 Coronavirus Vaccine and its effectiveness in a controlled environment. Not only that the pharmaceutical industries needed to invest heavily in clinical trials. There might appear counterfeit/frauds happened in clinical trials. Clinical trials produce many big data. It includes statistics, reports, surveys, medical imagery, blood tests and so on. Although anyone review the clinical reports could ensure that the results of the trial were true. But that was not always the case.

Since Blockchain could act as the medium to facilitate clinical trials. As Blockchain provided data integrity, it could act as a proof when the authenticity of the documents needed to be verified. It required other nodes for the verifications process. Overall, the distributed networks ensure that the data integrity should be maintained and made no data could be modified without authorized access. The simple idea of data integrity could change how clinical trials took place. Blockchain would be a system with no loophole and Blockchain improve NHS healthcare positively. Blockchain used SHA-256 hash, one of cryptography algorithms to ensure that the data immutability by any third-party malicious act (Feng et al. 2021). Blockchain implementation used examples included Amgen and Sanofi etc. (Iredale 2019).

Patient Data Management Application

Patient Data Management had been an outstanding issue. NHS health were facing varied issues and health concerns. This meant that each patient was different which made big data management of patients not easy. As each disease worked differently for different patients, it was not possible to produce a clear logical structure or use a common treatment strategy. If treatment worked on one patient, it did not mean that it would work on everyone with the same syndrome. All these complex issues made it necessary to had complete medical records for the patient. This would enable personalized care with a focus on patient-centric treatment.

Another challenge that most of the medical doctor went through were lack of information availability. This could result in patient treatment adding more cost to overall healthcare. The patient big data were not safe as which should be. The systems on which the data were stored were not completely secure. Doctors also use insecure way to share information. For example, we could easily see doctors sharing information through social media. This could easily lead to the patient PII data leak. Moreover, the patients were never in charge of their data. Data ownership were also an issue where the organizations mainly posed the patient's big data without the proper permission of the patient.

The solution would be, Blockchain provided a better trade-off to handle patient data management. It provides a structured way to store data which could be accessed by the right professionals. Patients' data could be stored on the Blockchain and could only be kept accessible to the patient, which the doctor who is handling the case. Access could be revoked anytime which ensuring that the patient had full authority on whose medical reports. That did not mean that not all PII data were inaccessible. Insensitive data could be available publicly which then could be accessed by healthcare organization or stakeholders through API. This ensured proper collaboration across different systems. Doctors could also request data whenever possible and location independently (Feng et al. 2021). Patients could also share their data without revealing their identity. This was especially favorable to the healthcare institutes as they could use the anonymous data to improve their healthcare research and systems. Patients always had control of their data, and they decided who had authentication to access the data.

When combined with IoT big data, researches could constantly monitor the patient's condition such as heartbeat and other vital body functions. This would help improve patients health care were taken care of. IoT with the help of Blockchain could enable faster decision making and saving patients' life in critical life-threatening conditions and get the forensics evidence (Alexakos et al. 2020) for the multi devices.

To implement use of Blockchain as their solutions like digital health company DokChain, the charitable foundation Patientory and so on (Iredale 2019).

2.3 PBFT Algorithm in Application

Blockchain Used PBFT (Practical Byzantine Fault Tolerance) Algorithm to keep consistency. In order to keep accuracy and consistency at the real world, the decentralized characteristics of Blockchain operation quality depended on the distributed nodes. However, it was not possible that all the Blockchain network notes are always working in a good order. If a node out of order, the algorithm of PBFT (Practical Byzantine Fault Tolerance) could be applied to the Blockchain. Byzantine fault is a condition of a computer system, particularly decentralized computing systems, where node may be faulty and there was imperfect information on whether a node had failed. PBFT algorithm was a very good consensus algorithm, introduced by Leslie B. Lamport, the 2013 Turing Award winner, an American computer scientist. He was best known for his seminal work in decentralized systems introduced the Byzantine Generals Problem to solve cybercriminals maliciously attacked node and so on. Practical Byzantine fault tolerance (PBFT) is the property of a system that could resist the class of failures derived from the Byzantine Generals' Problem. De Angelis (2018) from Southampton university, discussed the algorithm in their paper "PBFT versus proof-of-authority: applying the CAP theorem to permissioned Blockchain".

Although there were faulty and malicious nodes exist in a decentralized computing system, such as where nodes may fail and there was imperfect information on whether a node had failed, node crash down due to kernel panic, or hacker the cybercriminal perpetrators' threatening node. To reaching consensus, correctness; validity still could be maintain through collaboration. It was a question that how to keep consistency in such a condition though. The Byzantine Generals Problem solution was, if the majority reaching consensus of a consistency, which could achieve a fault tolerance, that is a correctness solution. Although there were faulty node and malicious node exist in a decentralized Blockchain, because the majority nodes could still maintain consistency and correctness, the overall Blockchain result could still keep a consistency and correctness. An agreement or treaty could still reach consistency and correctness.

If malicious node $f > 1$, total number of nodes

$$n > 3f \tag{2.1}$$

where n is the total number of nodes in the Blockchain, and f is the number of questionable nodes in the Blockchain.

A consistency and correctness could be achieved under the equation of (2.1) condition. In fact, if more than 51% nodes in the Blockchain (the majority) were good nodes, that Blockchain could perform its function well well as desired. Using recurrence recursive algorithm, nested block took the largest vectors, consistency and correctness were still achieved by PBFT (Practical Byzantine Fault Tolerance) kept consistency and correctness. If the condition could not be satisfied, it would be impossible to keep a consistency and correctness.

For example, let f was the number of questionable nodes, and n was the total computer nodes number in the system.

When $n > 3f$ being satisfied, the integrity could be kept. There were consistency and correctness achieved in that Blockchain.

For instance, let $f = 2$, $n > 7$, consistency and correctness would be kept. PBFT recursion algorithm could make Blockchain network consistency and correctness (Feng et al. 2021). DEA (Data Envelopment Analysis) was an optimization method for measuring the efficiency and productivity of decision-making units with multiple inputs/outputs (Ali et al. 2010) could be used on NHS data.

3 Blockchain Security Applied Solution

3.1 Blockchain Algorithms

Blockchain was designed about 10 min to produce a block in the world. There were 21 million bitcoin in the world in total, which could be used as accounting reward for mining blocks. As nobody could possibly to predict which member would solve the question for the next new block, this unpredictability was essential for the integrity of the network because if it is known a member could be the one, who decides on the next new block, then this member could be corrupted by bribes or being threatened by others to include questionable transactions that maybe led to some kind of challenges, or to exclude legible transactions that might cause DDOS (distributed deny of service) cyber-attack.

The algorithm mining principle would be:

a (character) string = Front block header + transactions + Time Stamp + random number.

Hash = SHA256(SHA256(string)) = 000......00,000......; (a 256 digit string).

Then, this 256-digit hash was the newest block's header.

Based on the n calculation.

The probability of success in PoW was, in binary, the probability of to get a 0, was $\frac{1}{2}$.

$\frac{1}{2} \times \frac{1}{2} \times \frac{1}{2} \times \ldots \ldots \times \frac{1}{2} = (\frac{1}{2})^n$.

In order to adjust the value of number n could realize every 10 min to create a new block (which contained several thousand transactions).

The complexity: if there were 10'000 mining machines in the world, each mining machine calculation ability is 14 T per second.

$1\,T = 10^{12}$.

That is, they could calculate 14 T (14×10^{12}) Hash digest value computation in every second $= 1.4 \times 10^{13}$ per second.

Therefore, in 10 min, the every mining machine computation could be

$= 10 \times 60 \times 1.4 \times 10^{13}$.

$= 600\,s \times 1.4 \times 10^{13}$ at each mining machine.

For the $10'000$ mining machine, the computation $= 600 \times 1000 \times 1.4 \times 10^{13}$; which is about 8×10^{19};

that meant every 10 min, you could calculate $8 \times 1.4 \times 10^{19}$ times, in terms of produce a new block.

If $n = 66$, the calculation probability is $(2^{-1})^{66}$, that is an average computation $= (2)^{66}$ times, which is about 8×10^{19}.

So the n was 66.

The first person who calculated the front 66 "0" n, could produce the new block. That meant he was the successful miner.

Due to Blockchain was a decentralized technology, its application on AI, process data for health system to realizing cyber security would be worthwhile. Decentralized solution provided a more secure system for accessing NHS big data, and provided the availability and integrity globally, because nowadays many people travel, working abroad, and they needed access to their medical record for treatment, or for insurance purposes. Location independently could satisfy their requirement. Blockchain system would cut the cost for exchanging data and processes and at the same time improve the authentication process. The research would take the NHS Spine as a model and improve the authenticity and confidentiality by deploying Blockchain and decentralized technology. The Project started long ago and since then the technology and the demands advanced very much (Feng et al. 2020). Adding the introducing of Internet of things (IoT) and Machine learning (ML), which created more potential for vulnerabilities (Feng et al. 2021; Weisenthal 2019) "Authenticity trust issues, regulatory and all the requirements that enable safety and confidentiality, and security issues are barrier to information sharing, the health care industry needs a more efficient and secure system for merging medical records."

Electronic medical records are currently maintained in data Centers, and access is limited hospital and care provider network. Centralization of such information make it vulnerable to security breaches and can be expensive. A decentralized, Health Insurance Portability organization compliant Blockchain based health network holds the complete medical history for each patient, can be more and efficient and secure solution" as shown in Fig. 2. The Fig. 2 demonstrated the IBM solution schematic diagram to date, it is still an on-going development. Whenever a further technical progress on identity and authentication technical support of the big data infrastructure security deployment of Blockchain technologies had been achieved, an updated figure would be shown (nd). Currently, SHA 256 Hash Algorithm was introduced to NHS project (nd). SHA256 algorithm is a cryptographic algorithm that used by Blockchain to secure transactions.

There are five main requirements for the hash Algorithms:

1. One-Way: That mean you cannot reverse the hash back to the original document.
2. Repeatable/Deterministic: That mean if I take the same document and apply the hash algorithm. I should get the same value.
3. Quick computation speed: No need to take long time to execute.

4. The Avalanche Effect: This is a very important characteristic of the Blockchain. Which mean that if there is a small change in the document, this will produce completely different hash vale.
5. Must Withstand Collision: Collision should not be possible that means, you cannot alter a file in a way that it should have the same hash.

Mining is SHA 256 algorithm hashing. Although SHA 256 algorithm was not the best solution, it was a good start to be introduced to NHS project. Blockchain are cryptographically linked together this is where actually the chain formed together (nd).

3.2   Anti-False Transaction

By trace back to the whole Blockchain record to check if that transaction was reasonable, that broadcast was accepted and confirmed/verified by this Blockchain network then packed to a new block in that Blockchain. But, if which transaction were not reasonable, that broadcast would be refused and not be verified by this Blockchain networkintegrity. Once the block where the transaction was embedded in is deep enough in the chain), the record becomes immutable, that is, it is impossible to change the record without being detected.

3.2.1   Verification

Blockchain verification: used digital signature, random generalize a string.

Private key string (secret), to generalize a public-key string and an address (for others to pay you). In Blockchain, PKI cryptography was used. The private key could encrypt a string, while public-key could decrypt the string.

For a transaction, a record was produced, then hashing that transaction record, obtain a SHA256 digest. Use the private key to encrypt that SHA256 digest to get a cipher. Broadcasting this transaction and produced its cipher and the public key corresponded to the Blockchain network.

In order to verify the truthiness, the others to hash this broadcasted transaction, to get a digest 2, then using the broadcasted public-key decrypt the broadcasted cipher, to achieve a digest 1. Then comparing the digest 2 with digest 1, if they were the same, that verified the cipher was true from the original broadcasting, because you have the only unique private key. But, if the digest 2 is different with the digest 1, that meant this transaction was false, all the user in this whole Blockchain network would refuse this transaction, which use digital/electronic signature to realize an anti-fake function in the Blockchain network to make immutability characteristics.

3.2.2 Anti-Fake

In Blockchain technology, an anti-false transaction method was by trace back to the whole Blockchain record, to check whether that particular newly created transaction was reasonable. If the condition were satisfied, that broadcast was accepted and verified by the Blockchain network then packed to a new block in that Blockchain. But, if that transaction was not reasonable, that broadcast would be refused and could not get verified by the Blockchain network, that meant the fake block was failed. In this way, Blockchain technology executed anti-false transaction block could be maintained (Feng et al. 2021).

3.3 Anti-Alteration Integrity

The longest chain was the chosen one principle of Blockchain. This principle ensured any modification attempt not feasible. When people in the network received a new block, wait for the next new block. If a perpetrator tried to delete a transaction, change this record, the anti-alteration method was, although everybody took the first received new block into account, to carry on their work, until a newer block connected to the Blockchain, that made the newer chain was longer than the others. So when the newer one broadcasting to the network, everybody in this network known this was longest chain. The others would stand in that line again, then continued the mining work. The one did not have the longest chain failed that block, as the whole network was looking for the longest chain. It was not realistic that you could success to against all of the whole Blockchain network. Because everybody accepted the longest chain principle in common sense, if a perpetrator wanted to change any record, he needed to create another longer chain than the existed one that the whole world miner created, which was impossible. (Except the perpetrator's mining was better than everybody in the world. Since if your chain was not longer than other, your alteration chain wont be accepted.) The longer the time passed, the more robust of the chain (Feng et al. 2021). Generally speaking, if after 6 blocks, your transaction record would be anti-alteration.

4 Blockchain Security Problem-Solving

A ten-year NHS Spine contract awarded to BT involved developing systems and software to support more than 899,000 registered users. It has made transformational healthcare applications available to approximately 1.3 million NHS healthcare staff across England, providing care to circa 50 million UK citizens. In managing overall software delivery lifecycle processes, BT used Rational Unified Process/Systems Engineering (RUP/SE) processes. These were modified to create a methodology that attained Capability Maturity Model Integration (CMMI) Level-3 accreditation.

That BT Global Services methodology is now an internationally recognized standard for complex software development programme delivery (nd; NHS 2020; Lewis 2019). Transaction record statement, block verification, anti-alteration, anti-fake, keep secret and the immunity issues were to be solved.; who is the standard of anti-false and anti-modification, also low scalability, privacy and interoperability PBFT algorithm remain outstanding. IBM deployed the 2nd generation Blockchain technology to some of the NHS nodes (nd; NHS 2020). Moreover, the 3rd generation of Blockchain technology that saved many powers consuming issue was still in progress (De Angelis 2018; Laurence 2019).

5 Evaluation

5.1 Discussion

When working with NHS, a security testing is required. Cyber security testing using Blockchain technology was in demand, in order to ensure the NHS health system guarantee patients safe and secure. During the past years Blockchain technology was deployed in the health sectors to improve authenticity and security using the features of decentralized structure and the distributed ledgers and the proof of work feature and the hash function which make it very difficult to amend (Yang et al. 2017). Although there is not many projects of Blockchain in the NHS, but during the COVID-19 pandemic, according to Reuters, two UK Hospital used the technology to trace and transfer COVID-19 vaccine to monitor the supply and the sensitive temperature the drug need to be save in (Meirino et al. 2019). In UK NHS, before accessing a medical record, the user needed to be authenticated, using keyless infrastructure signature, which is a mechanism to take protection of the integrity of the NHS record into account for make sure it was not modified in any sense. The proposed implementation and experimental finding results could be shown next (Hussein 2021).

Blockchain used hashing and publishing mechanism. Blockchain use keyless signature infrastructure which consists of a sets aggregation server which create a global hash tree. The Blockchain System will be deployed and tested by NHS, and then we would run the above attacks and obtaining data and logs to be analyzed (Laurence 2019).

5.2 Discussion of Blockchain Examples

5.2.1 Blockchain at Barclays

Barclays was a second-largest bank in UK, with branches in Europe. Barclays and companies used Blockchain technology. In reality, they were using Blockchain technology for streamlining fund transfers and Know-Your-Customer processes. To an extent, they even filed for patents against these two features. They will shift from their traditional paper-based records to fully digital and decentralized Vault platform. Thus, their investors could now finally track their money in real-time, although with latency.

5.2.2 Blockchain at Pfizer

Pfizer was another one of the large companies using Blockchain technology. Biogen and Pfizer led organization Clinical Supply Blockchain Working Group (CSBWG), just completed proof of concept for tracking records and managing the digital inventory of pharmaceutical products. In reality, the group also comes with other players such as GlaxoSmithKline, Merck, AstraZeneca, and Deloitte.

5.2.3 Overseas Finance Transfer

In Central American, El Salvador congress has voted, to identify bitcoin as another legal tender of the country in addition to the US dollar. That meant Salvador become the first country in the world to legalize bitcoin. This new law means, unless they could not provide the technology that the exchange needs, all businesses must accept bitcoin as a good or service. The President of El Salvador, Nayib Bukele said, the move would make it more convenient for Salvadorans living abroad to send remittances to their country. He also twitted "This will bring financial inclusion to our country, investment, tourism, innovation and economic development." In country like El Salvador, many people's lives were highly dependent on remittances from overseas. The proportion of remittances from overseas in the country's GDP 20%. 70% of Salvadorans had no bank account. The move to use BTC will open financial services to these people, fast and convenience.

5.3 Evaluation

All these applications technology were Blockchain technology, which recorded all the historic transactions and open to network transparently. Blockchain ledgers contained time, transactions, and related activities. Although excessive delay in reaching consensus and limited throughput still outstanding, scientists have been made effort to improve. Blockchain technology still grown in exponential increase globally. To research Blockchain technology and master the trends impact to big data intelligence for smart applications improve security in the future.

Although excessive delay in reaching consensus and limited throughput still outstanding, scientists have been made effort to improve. Recently, Zhao et al. proposed a method had hierarchical processing and logging of potentially large amount of sensing big data with the Blockchain technology, which could drastically address the issue of limited throughput in Blockchain for IoT of the current distributed ledgers (Zhao et al. 2020).

6 Conclusions and Future Development

6.1 Summary

In this chapter, we had explored the main Blockchain application for NHS outstanding security problems. Since this dealt with different operating systems and many different fields, security issue was quite sophisticated in Blockchain technology applications. Blockchain technology possessed a number of advantages, such as, peer-to-peer network; cryptography; chaining-based data structure; immutability and proof of work that made it was well accepted in the society. Nevertheless, the majority security problem in Blockchain application design, testing and management threat appearance features had been discussed. The security potential was analyzed. Remedy and problem-solving had been explored respectively. An investigation Blockchain technology utilization, testing for decentralized systems to date had been searched. The attacking on authentication system also was looked at. Blockchain was a decentralized, distribution multi-recorded system technology. Its application on AI, process data for health system to realizing cyber security would be achievable and worthwhile. Feng et al. (2021) had pointed out, because the advantages of public Blockchain with cryptography, this technology could also serve for cyberstalking and cyberbullying digital forensics evidence data storage security for law enforcement and society. Blockchain technology might change IoT big data intelligence for smart application. Currently IoT device were managed by supply companies. In the future, centralized management could not support all the devices properly. Blockchain technology could help IoT devices direct secure and reliable to communicating on Blockchain network and avoid intermediary central control disadvantages. There was one more point, the main difference between Blockchain and distributed ledger would be the internal structure system. All Blockchain technologies were a form of the distributed ledger, but not all the distributed ledger systems were Blockchain.

6.2 Suggestion for Future Development

The suggested future development is to explore a more appropriate approach to sort out the multiple Blockchain application diversity issues. Blockchain design and development specific expertise to make use of features of Blockchain technology is taken into account. As currently, it is still an early stage of the project, experimentation for the physical Blockchain would be carried out. The proposed NHS model and experimentation results data would be published soon.

Meanwhile, protect patients' privacy kept confidential. The hashing algorithm could apply more robust algorithms in the society. The deployed system would be tested in non-live environment, by performing several attacks. Alexakos et al. suggested we could use forensics to solve collected sensing Big Data security in smart applications. That was worth to be considered. As these Big Data could be collected from NHS IoT sensors and devices and perform some of the following attacks: adversarial training using Brute Force; adversarial attack for network model; poisoning attack; 51% attack; typo squatting attack and so on (Alexakos et al. 2020; Eze et al. 2020). The newer generalized Blockchain planned to have memory security, thread security and process security, output and input data security, AI security and big data management functions to sort these out. These increased capability of Blockchain decentralized systems, which would have demand for facility, throughput, latency and so on (Laurence 2019; Meirino et al. 2019; Dhillon 2019). Furthermore, although PoW algorithm offered a fair proof for working load to authenticate a qualified block, but it generated months latency for consensus and the impact to eco-environmental protection was not positive. Therefore, further development of alternative PoW would be necessary.

Acknowledgements We sincerely thank the anonymous reviewers and editors for their helpful comments and suggestions on earlier draft of the manuscript.

References

Cambridge, Cambridge Dictionary (2021). https://dictionary.cambridge.org/dictionary/english/blockchain. Accessed 15 June 2021

Builtin, What is Blockchain technology? How does it work? BuiltIn (2018). https://builtin.com. Accessed 15 May 2021

X. Feng, E. Short, A. Asante, Cyberstalking issues. in *International Workshop ACE-2017 IEEE Xplore FL USA* (2017)

W. Zhao, et al., Blockchain-enabled cyber-physical systems: a review. IEEE Xplore (2021). https://doi.org/10.1109/JIOT.2020.30148. Accessed 2 Mar 2021

M. D'Aliessi, How does the blockchain work? One zone (2016). https://onezero.medium.com/how-does-the-blockchain-work-98c8cd01d2ae

W. Zhao, et al., Secure hierarchical processing and logging of sensing data and IoT events with Blockchain, in *ICBCT'2020: Proceedings of the 2020, The 2nd International Conference on*

Blockchain Technology, ACM Digital Library, March 2020 (2020), pp. 52–56. https://doi.org/10. 1145/3390566.3391672. Accessed 4 Mar 2021

IBM, IBM Blockchain. www.ibm.com/blockchain: www.ibm.com/blockchain. (2020). Accessed 16 Feb 2021

X. Feng, et al., A systematic approach of impact of GDPR in PII and privacy. J. IJESI 5 (2021). ISSN 2319-6726

H. Yang, et al., A Blockchain-based approach to the secure sharing of healthcare data. NISK 2017 (2017). ISSN 1894-7735. http://www.nik.no/. Accessed 16 Feb 2021

IBM, IBM Blockchain (2020). www.ibm.com/blockchain: www.ibm.com/blockchain. Accessed 16 Feb 2021

Steemit, How smart contract works. Steem, White paper (2017). https://steem.com

X. Feng, et al., Artificial intelligence and Blockchain for future cyber security application, in *Accepted IEEE 5th International Workshop ACE-2021* (2021).

C. Alexakos, et al., Enabling digital forensics readiness for internet of vehicles, in *23rd Euro WP on Transportation Meeting, EWGT 2020, Sep 2020, Paphos, Cyprus, Transportation Research Procedia*, vol. 52 (2021) pp. 339–346

G. Iredale, Blockchain for healthcare: use cases and applications (2019). https://101blockchains. com/blockchain-for-healthcare. Accessed 25 June 2021, 4 Mar 2021

A. Takyar, "Use Cases of Blockchain in Healthcare" Leeway Hertz (2016). https://www.leeway hertz.com/blockchain-in-healthcare. 15 June 2021

S. De Angelis, Assessing security and performances of consensus algorithms for permissioned Blockchains. Southampton university (2018).

E. Ali et al., A semi-oriented radial measure for measuring the efficiency of decision making units with negative data, using DEA. Eur. J. Oper. Res. **200**(1), 297 (2010)

X. Feng, et al., Artificial intelligence and cyber security strategy, in *IEEE International CyberSciTech Conference, Athabasca University, Canada* (2020)

J. Weisenthal, What'd you miss? Bloomberg Bundesbank together with Deutsche Boerse ledger technology potential cheaper and faster than current settlement mechanisms. Bloomberg (2019). https://www.bloomberg.com/podcasts/material_world1. Accessed 4 June 2021

IBM, Blockchain and the healthcare industry: how to get started. IBM (nd). https://www.ibm.com/ blogs/blockchain/category/blockchain-education/blockchain-explained/. Accessed 25 Feb 2021

BT, Smart city applications on the Blockchain. British Telecom. Springer-professional (nd). Accessed 25 Feb 2021

Udemy, Blockchain Training (2021). http://www.udemy.com. Accessed 25 Feb 2021

NHS, Spine case study. NHS UK (2020). https://business.bt.com/solutions/resources/nhs-spine/. Accessed 12 Feb 2021

C. Lewis, Can Blockchain help to deliver the NHS long-term plan? retrieved from the journal of mhealth (2019). https://thejournalofmhealth.com/can-blockchain-help-to-deliver-the-nhs-long-term-plan/. Accessed 10 Feb 2021

E. Liu, X. Feng, Trustworthiness in the patient centered health care system. Communications in Computer and Information Science, vol. 426 (Springer, Berlin, 2014)

T. Laurence, Introduction to Blockchain technology the many faces of Blockchain technology in the 21st century. Van Hearen Publishing (2019). ISBN: 978-9401804998

M.J. Meirino et al., Blockchain technology applications: a literature review. Braz. J. Oper. Prod. Manag. **16**(4), 672–684 (2019)

K. Hussein, Physical Blockchain design considerations, in *Prepared for the 6th ACE WS* (2021).

E. Eze, S. Zhang, X. Feng X et al., Mobile computing and IoT: radio spectrum requirement for timely and reliable message delivery over internet of vehicles (Springer, Berlin, 2020)

V.J.B. Dhillon, Blockchain in healthcare: innovations that empower patients. Connect professionals and improve care. Pristine Publishing (2019).

Depression Detection from Social Media Using Twitter's Tweet

**Rifat Jahan Lia, Abu Bakkar Siddikk, Fahim Muntasir,
Sheikh Shah Mohammad Motiur Rahman, and Nusrat Jahan**

Abstract Social media is using rapidly for expressing user's opinions and feelings. At present, 'Depression' is propagating in a cursory way which is the reason for the increasing rate of suicides even more. Recently, the status on social media can give the hints of user's mental state along with the situation and activities that are happening with them. Different research works had done for perceiving the mental health of any user through social media which has impressive impact. That is done by analysing expressed opinions, images, sentiments, linguistic style and other activities. Variant studies had proposed variant methods or intelligent systems for detecting depression or state of mental health through social media posts. In this chapter, the posted tweets from users on Twitter will be considered in the sake of detecting depressions. Six machine learning approaches named Multinomial Naive Bayes, Support Vector Classifier, Decision Tree, Random Forest, K-Nearest Neighbor, Logistic Regression had used to differentiate the depressed and non-depressed users. Finally, Support Vector Classifier outperforms among all of the investigated and evaluated techniques with 79.90% accuracy, 75.73% precision, 77.53% recall and 76.61% f1-factor. This study can be a basement for the developers of intelligent systems in the area of users mental conditions detection.

R. J. Lia · A. B. Siddikk · F. Muntasir · S. S. M. M. Rahman (✉) · N. Jahan
Department of Software Engineering, Daffodil International University, Dhaka, Bangladesh
e-mail: motiur.swe@diu.edu.bd

R. J. Lia
e-mail: rifat35-1845@diu.edu.bd

A. B. Siddikk
e-mail: abu35-1994@diu.edu.bd

F. Muntasir
e-mail: fahim35-1900@diu.edu.bd

N. Jahan
e-mail: nusrat.swe@diu.edu.bd

R. J. Lia · A. B. Siddikk · F. Muntasir · S. S. M. M. Rahman
Future Research Lab, Dhaka, Bangladesh

© The Author(s), under exclusive license to Springer Nature Switzerland AG 2022   209
Y. Baddi et al. (eds.), *Big Data Intelligence for Smart Applications*,
Studies in Computational Intelligence 994,
https://doi.org/10.1007/978-3-030-87954-9_9

Keywords Twitter · Depression · Mental illness · Social media · Suicide · Machine learning

1 Introduction

Depression is a mental disorder which can impair many facets of human life. Depression has profound and varied impacts on both physical and mental health (Deshpande and Rao 2017; Coello-Guilarte et al. 2019) and sometimes occurs anxiety or other psychological disorders (Tadesse et al. 2019) such as mood disruption, uncertainty, loss of interest, tiredness, and physical issues (Mustafa et al. 2020). It is also known as a leading cause of disability and is a major contributor to the overall global burden of disease (Burdisso et al. 2019; Cacheda et al. 2019). According to the World Health Organization(WHO), more than 264 million people of all ages suffer from depression globally and 70% of the patients would not even consult doctors, which may sustain their condition to next echelons (AlSagri and Ykhlef 2020; Tong et al. 2019) and in many countries, even fewer than 10% depressed people receive treatment (Gui et al. 2019). It is also entitled the second leading cause of death among 15–29-year-old's people (Tao et al. 2016; Singh et al. 2019; Yang et al. 2020; Wu et al. 2020). Depression is a conventional cause of committing suicide which is increased day by day very rapidly. According to the recent report of World Health Organization(WHO), the number of died people due to suicide is near to 800000 in every year, which means one person in every 40 s. People who are facing less interaction with friends and family use social media as a rule to share their feelings and opinions (Lin et al. 2020). Often, the users express the unuttered feelings for feeling good which drive them on the way of suicide. Some specific words are usually used by the depressed people for expressing their mental stress such as pain, hurt or any other negative words. Through this, the mental conditions of many depressed people had been revealed on different social media (Leiva and Freire 2017). A humongous quantity of data is generated in every second through various social media platforms like Facebook, Twitter, etc. These all data from different platforms have a lot of relevant information for doing behaviour analysis. Twitter's tweet will be used in this research as Twitter is one of the most visited social networking sites where an average of 58 million tweets are generated per day (Lin et al. 2020) and there are over 271 million monthly active users in Twitter (Nambisan et al. 2015). So, it becomes easy to get data for analysing depression with tweets. There are eight basic emotions as features from Twitter posts which helps to identify the type of any tweets even easily (Chen et al. 2018).

In this study, the main goal is to find out which users are depressed. Through detecting those depressed users, there is a chance of decreasing the risk of suicide and save them.

The rest of the chapter is organized as follows: This study has tried to find out depressed twitter users through their posted tweets using six machine learning techniques. In Sect. 2, have reviewed some related works of depression detection with a table and description. In Sect. 3, the methodology of this chapter has shown in figure and described elaborately. The experimental results have shown in Sect. 4 and comparing the results the best algorithm for this topic has also declared. Section 5 mentioned about the limitations as well as talked about future research on it. Finally, this paper ends with Sect. 6 with the conclusions.

2 Literature Review

Several research works had been performed till now for detecting depression through Twitter's tweets as depression is increasing in short order. Several research works have been completed using deep learning algorithms (Guntuku et al. 2017) and several worked with machine learning algorithms for getting the best outcome. Large-scale passive monitoring of social media used for identifying depressed users or any at-risk individuals and Linear Regression and Support Vector Machine(SVM) algorithms were used for this research work (Almeida et al. 2017). A martingale framework which is widely used for detection of changes in data stream settings was implemented in the research work of Vioules et al. (2018). Several algorithms such as Natural Language Processing (NLP), Naive Bayes (NB), Sequential Minimal Optimization (SMO) with a poly kernel, C4.5 decision tree (J48), nearest neighbor classifier (IB1), multinomial logistic regression, rule induction (Jrip), Random Forest, and SMO with a Pearson VII universal kernel function (PUK) were used in the study (Vioules et al. 2018).

Logistic Model Tree (LMT), Sequential Minimal Optimization (SMO) and Random Forests were used in O'dea et al. (2015) and a multipronged approach was designed in this paper that combines results obtained from both Information Retrieval (IR) and Supervised Learning (SL) based systems for detecting depression. Text processing had been used and explained elaborately in several work (Sohan et al. 2018; Rahman et al. 2020, 2018), [?] and this study had gained knowledge from those. Methodology used in Gaikar et al. (2019) this study was an automated computer classifier that could replicate the accuracy of the human coders and this methodology was implemented in this paper. Logistic Regression (LR) and Support Vector Machine (SVM) were the techniques used in Gaikar et al. (2019) for the mentioned methodology.

SVM with NB features (NBSVM) was applied for detecting any depression related words or phrases from Tweets and also classified the type of depression (Lin et al. 2020). A system dubbed SenseMood was used in the paper of Morales et al. (2017). SenseMood demonstrated that the users with depression can be efficiently detected and analyzed by using the proposed system. A novel approach was proposed for identifying users with or at risk of depression by incorporating measures of eight basic emotions features from Twitter posts over time, including a temporal analysis of these features and this approach provided the psychological state successfully (Chen et al. 2018).

The proposed system of Lin et al. (2020) was a hybrid approach of Support vector machine and Naïve Bayes classifier which works well not only with shorter snippets but also with longer snippets. The author of Deshpande and Rao (2017) had also used the Support vector machine and Naive-Bayes classifier for depression detection and got the result by presenting the primary classification metrics including F1-score, accuracy and confusion matrix.

Information about some of the related literature are explained in Table 1. From Table 1. We can acknowledge that most of the literature have worked with very less number of algorithms in order to detect depression from social media.

In this study, we will concentrate on the used words in any tweet by any user and six machine learning algorithms will be applied on the collected data for depression detection in order to get the best result.

Table 1 Summary of some recent literature works on depression detection

Author	Title	Year	Methodology	Algorithm
Sharath Chandra Guntuku, David B Yaden, Margaret L Kern, Lyle H Ungar and Johannes C Eichstaedt	Detecting depression and mental illness on social media: an integrative review	2017	Identify depressed or otherwise at-risk individuals through the large-scale passive monitoring of social media	Linear Regression, Support Vector Machines (SVM)
M. Johnson Vioules B. Moulahi J. Az e S. Bringa	Detection of suicide-related posts in Twitter data streams	2018	Implement a martingale framework, which is widely used for the detection of changes in data stream settings	NLP, NB, Sequential Minimal Optimization, C4.5 J48, Nearest neighbor classifier, Multinomial logistic regression, Jrip, Random Forest
Hayda Almeida, Antoine Briand, and Marie-Jean Meurs	Detecting Early Risk of Depression from Social Media User-generated Content	2017	To detect users in risk of developing depression, we have designed a multipronged approach that combines results obtained from both Information Retrieval(IR) and Supervised Learning (SL) based systems	Logistic Model Tree (LMT), Sequential Minimal Optimization, Random Forests

(continued)

Table 1 (continued)

Author	Title	Year	Methodology	Algorithm
Stephen Wan b, Philip J. Batterham c, Alison L. Calearc, Cecile Paris b, Helen Christensen	Detecting suicidality on Twitter	2015	Design and implement an automated computer classifier that could replicate the accuracy of the human coders	Support Vector Machine(SVM), Logistic Regression(LR)
Jayesh Chavan, Kunal Indore, Mrunal Gaikar, Rajashree Shedge	Depression Detection and Prevention System by Analysing Tweets	2019	Detect any depression related words or phrases from Tweets and also classify the type of depression, if detected.	SVM with NB features (NBSVM)
Chenhao Lin, Pengwei Hu, Hui Su, Shaochun Li Jing Mei,Jie Zhou, Henry Leung	SenseMood: Depression Detection on Social Media	2020	Design a system dubbed SenseMood to demonstrate that the users with depression can be efficiently detected and analyzed by using proposed system	CNN, Bert
Xuetong Chen, Martin D. Sykora, Thomas W. Jackson, Suzanne Elayan	What about Mood Swings? Identifying Depression on Twitter with Temporal Measures of Emotions	2018	Proposed a novel approach for identifying users with or at risk of depression by incorporating measures of eight basic emotions as features from Twitter posts over time, including a temporal analysis of these features	Naıve Bayes, Sequential Minimal Optimization (SMO)

3   Research Methodology

The proposed research methodology of this study provides the complete knowledge about this research work. Through this framework, all detailed works of this study had been shown clearly step by step which ensures that future works based on this topic can take proper ideas from this work. From data collection to prediction's result all are there in this framework. There are several steps in the research methodology of this study. They are:

1. Data Collection and Preparation
2. Data Preprocessing
3. Word Analysis
4. Tokenization
5. Extract Features
6. Depression Detection
7. Output

The purpose of this paper is to detect if any Twitter user is in depression or not just like (Kumar et al. 2019) through following the proposed methodology in Fig. 1. In the first component of this methodology, this study mainly focused on collecting data and preparing the collected data which are the most important for starting the work of this research. After preparing the required data, this study next did data processing for making the data all ready for the consisted dataset. However, there are several steps and processes to follow for preprocessing the data which will be explained later. Then came word analysis which gave an idea of most used words in the dataset. Next component named tokenization breaks the raw data into small tokens consisting of words and sentences. For feature extraction of this study, TF-IDF vectorization was used. Then splitted the data into testing data and training data for further experiments. This study's next component name was depression detection in which several machine learning algorithms had been used for getting better result of detecting depressed users of Twitter which was the main goal of this work alike (Islam et al. 2018; Ziwei and Chua 2019). Then the last component completed with giving the results of different algorithms and gave a clear conclusion of which algorithm is the best for detecting depression.

3.1   Data Collection and Preparation

Dataset collection and preparation is very essential for further work progression of any research. In the research of Gaikar et al. (2019), all the tweets from 18th February,2014 to 23rd April, 2014 which used English words or phrases that are consistent with the vernacular of suicidal ideation were collected. In the training phase of Lin et al. (2020), the labeled Dataset from kaggle were collected and used. A total of 10,000 Tweets were collected from Twitter API for generating the training

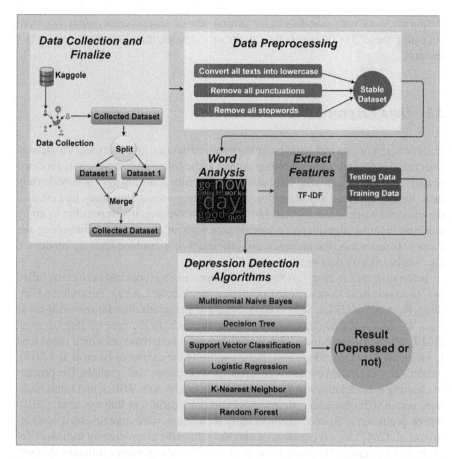

Fig. 1 Research methodology for depression detection in Twitter's tweets

and test dataset by the author of Deshpande and Rao (2017). The dataset used in Tadesse et al. (2019) contained depression-indicative posts of 1293 and standard posts of 548. The data used in Tao et al. (2016); Kumar et al. (2019) were also collected from Twitter. 55,810 tweets from Twitter between Oct 2018 and Feb 2019 were collected for the research work of Singh et al. (2019). Following their shadow, we also collected data from Twitter and will prepare the collected data for future uses. The dataset used in this study was collected from kaggle.com with the title of training.1600000.processed.noemoticon consisting 1600000 million tweets from 6th April,2020 to 16th June,2020 and all of these datas were in English. Two different datasets were splitted with 100000 datas in each dataset with two different names from the collected dataset. The first splitted dataset's data limitations were from index 0 to index 100000 and datas from index 800000 to 900000 million were kept in the second dataset. A new dataset had created concating those two splitted dataset named as PrepareData which had 200000 datas. The new dataset was prepared for testing the work with new dataset for getting new or experimental output. So, from collecting these 200000 tweets though the new dataset, this study had done further research.

3.2 Data Preprocessing

Preprocessing of data just started with the finalized dataset. Preprocessing was needed and crucial before taking any next component as the collected data were not processed at all. Lots of noises, missing values and unstable format are always contained in any data. Before applying any algorithms or next components, this study had cleaned the data and made it suitable for further work. Moreover, it's impossible to apply machine learning algorithms on any dataset without cleaning and formatting the dataset. Besides this, the accuracy and efficiency of a machine learning model are also increased with data preprocessing.

In the study of Lin et al. (2020), several preprocessing types had performen called as Conversion from Uppercase to Lowercase, Apostrophe Lookup, Punctuation Handling, Removal of URLs, Stop word removal and Lemmatization for removing noise from the data and cleaning and simplifying the data. In the case of Tadesse et al. (2019), all the URLs, punctuations and stop words were removed which could lead to erratic results if stayed ignored. Before splitting the dataset of Islam et al. (2018), the dataset was cleaned by removing quotes, extra spacing and symbols. The process of cleaning data includes removing numeric and empty texts, URLs, mentions, hashtags, non-ASCII characters, stop-words† and punctuations in Rahman et al. (2020) before predictions. Sixteen preprocessing techniques were experimented in Guntuku et al. (2017) for giving ideas of choosing the right preprocessing technique for cleaning the dataset. The study (Kumar et al. 2019) also removed different symbols like exclamation marks, punctuations, digits, special characters etc. for cleaning the dataset. However, three types of preprocessing had performed in this study's dataset named as:

1. Convert all texts into lowercase
2. Remove all punctuations
3. Remove all stopwords.

In the finalized dataset, some texts were in lowercase and at the same time some were in uppercase. So, it creates confusion at the time of applying any algorithm for any prediction. All the texts of the finalized dataset were converted into lowercase in order to solve this problem like (Tao et al. 2016; Kumar et al. 2019) also did. The study (Stephen and Prabu 2019) showed an example of the full process of converting texts into lowercase for explaining.

There were lots of punctuations like comma, semicolon, colon, dash, hyphen in the dataset which were not only unwanted but also problematic for this research as punctuations look like just another character to Python. For removing punctuation, the used programming language Python was shown what the punctuations look like by accomplishing using the String package in Python (Singh et al. 2019). So, all these unwanted punctuations were removed from the dataset.

Lastly, this dataset contained lots of stopwords like is, at, which, me, you etc. which are not meaningful and all these stopwords should ignore as these words cas create problems at the time of any word which is meaningful and expressed feeling. So, all these stop words were removed from the dataset. This is done to reduce the work for our algorithms (Singh et al. 2019). In order to remove stop-words from the tweet, Deshpande and Rao (2017) used Nltk library has a set of stop-words which can be used as a reference. The clean and formatted dataset had been obtained by performing all these three types of preprocessing and had applied the clean dataset further.

3.3 Word Analysis

As the dataset of this study had been prepared and cleaned, so now it's possible to do word analysis of this dataset without any problem. However, if word analysis had been performed before data preprocessing, it might give different output from the output of word analysis of cleaned and formatted dataset as any stop word might take place on the position of any important word. Word analysis is called the process of learning more about any word in a meaningful way. Within word analysis, the most used words were shown in large words and less used words were shown in small words which can be witnessed in Fig. 1. So, though word analysis, the idea of the most and less used words were cleared in this research work.

3.4 Tokenization

Tokenization breaks the raw data into small tokens consisting of words and sentences. The tokens help easily in the sake of understanding the context or model developing. There are two types of tokenization. One is called word tokenization and the other is sentence tokenization. If text is splitted into words with any tokenization technique, then that is called word tokenization and if text is splitted into sentences with any tokenization technique, then that is called sentence tokenization. In the case of Deshpande and Rao (2017), the first column of the csv file containing the tweet was extracted and later converted into individual tokens. In the research of Singh et al. (2019), an example was shown for clearing the tokenization's concept. In order to filter the words, symbols and other elements called tokens, Tao et al. (2016) used the TreebankWordTokenizer of Python Natural Language Toolkit (NLTK). However, word tokenization was used in this study. There were 593588 numbers of tokens after tokenization performed on the dataset of this research.

3.5 Extract Features

Feature extraction is a general term for a method for developing blends of the factors to get around these issues while depicting the information with adequate exactness.

The feature extraction method extracts the aspect (adjective) from the improved dataset and the adjective is used to determine the polarity of a sentence. Unigram model extracts the adjective and segregates it Deshpande and Rao (2017).

3.5.1 Term Frequency-Inverse Document Frequency (TF-IDF)

TF-IDF is a statistical measure for evaluating how relevant a word is to a document in a collection of documents. It consolidates two measurements, Term Frequency and Inverse Document Frequency. The initial measurement considers the occasions that a term shows up in a record and gives higher position. The term frequency measures the number of words in any dataset and document frequency measures the provided information of those words. According to Singh et al. (2019), Term Frequency-Inverse Document Frequency, oftenly referred as TF-IDF, created a document-term matrix, where there was still one row per tweet and the column still represented a single unique term. In this research, the vectorization was performed utilizing the Tf-Idf Vectorizer like (Singh et al. 2019). The Vectorizer is also answerable for sifting through pointless words or stop words, for example, "a", "an", "the", "in", "i", "you", "them", "has", "been" which have uninformed worth.

Here,

$$TF(t) = (Number\ of\ times\ term\ t\ appears\ in\ a\ document)/(Total\ number\ of\ terms\ in\ the\ document).$$

$$IDF(t) = log_e(Total\ number\ of\ documents/Number\ of\ documents\ with\ term\ t\ in\ it).$$

TF-IDF for a word in a document is calculated by multiplying two different metrics named term frequency(tf) and inverse document frequency(idf). If the word is very common and appears in many documents, this number will approach 0. Otherwise, it will approach 1.

$$tfidf(t, d, D) = tf(t, d).idf(t, D)$$

After feature extraction, the full dataset was splitted into Testing Data and Training Data. 30% data were kept in Testing Data and rest 70% data were in Training Data.

3.6 Depression Detection

After splitting the data into testing and training, now it's time to work on those data. The 30% testing data had been used in order to get accuracy, precision, recall, f1-

score from several machine learning algorithms in favor of depression detection. On the base of test data, this study got different outputs for different machine learning algorithms. The applied machine learning algorithms for detecting if any user is on depression or not and how many percentage of users are in depression or not are:

1. Multinomial Naive Bayes (MNB)
2. Decision Tree (DT)
3. Support Vector Classification (SVC)
4. Logistic Regression (LR)
5. K-Nearest Neighbor (KNN)
6. Random Forest (RF).

3.7 Output

Several machine learning algorithms were applied for depression detection from the tweets in order to find out how many users are in depression and how many users are not. After applying all those machine learning algorithms, this study saw that different algorithms gave different accuracy. However, in spite of some poor output of some algorithms, the bright side was that though that, this research work came to a conclusion of which algorithm is the best for depression detection from tweets to find out if any user is in depression or not in AlSagri and Ykhlef (2020), the study got 70% accuracy from Support Vector Machine and 71% accuracy from Naive Bayes. 79% accuracy was gained in Deshpande and Rao (2017) for depression detection.

However, in this study, it will be shown which algorithm is best in terms of providing accuracy, precision, recall and f1-score for getting the best machine learning algorithm for depression detection.

4 Experimental Results and Discussions

The effectiveness of proposed research methodology had been evaluated in this section of this study. Several machine learning algorithms had been applied in this research work for getting a better result of which algorithm gave better results for detecting depression from Twitter's tweets. In this section, all the performance of different applied algorithms will be discussed and find out the best one. Performance metrics used in this study will also be discussed in this section.

4.1   Performance Metrics

There are several metrics which can be used in order to evaluate the machine learning algorithms. However, for choosing any performance metric, the basic things need to be remembered are the performance of the machine learning algorithms, dependency, and importance of various characteristics which will influence the result of the algorithms. Performance metrics used for this study were Confusion matrix, Accuracy, Precision, Recall and F1-score.

Confusion Matrix
Confusion matrix is the most used and easiest way for measuring the performance with two or more types of classes in any classification problem. Basically the confusion matrix is a table which contains two dimensions called actual and predicted which helps to get any prediction. There are four terms in the confusion matrix named True Positives(TP), True Negatives (TN), False Positives (FP), False Negatives (FN).

$$\text{Accuracy Formula} = (TP+TN)/ (TP+TN+FP+FN)$$

$$\text{Precision Formula} = TP/ (TP+FP)$$

$$\text{Recall Formula} = TP/ (TP+FN)$$

$$\text{F1-Score Formula} = 2*(Precision*Recall)/(Precision+Recall)$$

So, this study used all these performance metrics on all the machine learning algorithms which were used in this research.

4.2   Depression Prediction Performance

Six types of machine learning algorithms were used in this research work for finding out which algorithm provides the best performance for detecting depression with tweets. The finalized and preprocessed dataset was used in those algorithms after splitting that dataset into 30% in testing and 70% in training data. On the basis of test data, the predictions were performed and found out the accuracy, precision, recall and f1-score for all of the machine learning algorithms. All the results of accuracy, precision, recall and f1-score for all applied machine learning algorithms were judged and compared among each other on account of declaring the best machine learning algorithm.

Here goes the accuracy, precision, recall and f1-score of all applied machine learning techniques-

In Fig. 2., there is a comparison graph for accuracy among all the applied algorithms. From the graph, it's clear that Support Vector Classifier had the best accuracy among all the used machine learning algorithms which was 76.90% . At the same

Fig. 2 Accuracy of all used algorithms

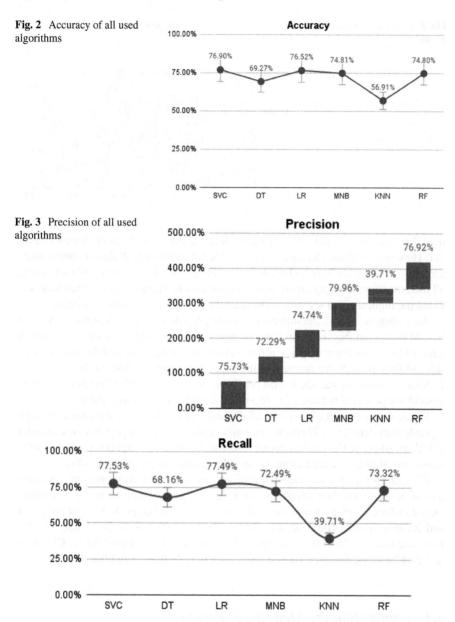

Fig. 3 Precision of all used algorithms

Fig. 4 Recall of all used algorithms

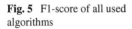

Fig. 5 F1-score of all used algorithms

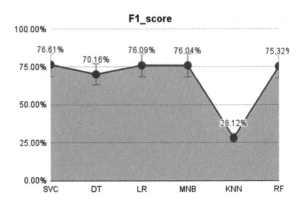

time, Logistic Regression had only 0.38% less accuracy than Support Vector Classifier. However, without Decision Tree, all other algorithms had almost approximate accuracy as Decision Tree had the lowest accuracy. So, Fig. 2. indicated that among all used machine learning algorithms in this research, Support Vector Classifier provided the highest accuracy where Decision Tree provided the lowest accuracy.

According to Fig. 3., the highest precision provided machine learning algorithm was Multinomial Naive Bayes for this research work. Multinomial Naive Bayes gave 79.96% precision which was quite effective for depression detection. On one side, Multinomial Naive Bayes had the highest precision, while on the other side, K-Nearest Neighbor had the lowest precision. Rest of the applied machine learning algorithms provided between 72–76%precision which was also standard.

Figure 4. indicates the recall rate of all used machine learning algorithms through a graph with data level. Through the graph, it shows that Support Vector Classifier (SVC) resulted in the highest recall rate and K-Nearest Neighbor resulted in the lowest recall rate. The recall rates of rest algorithms are also shown in Fig. 4.

F1-score for all the machine learning algorithms which were used in this study had been picturized beautifully in Fig. 5. In the Fig. 5., Support Vector Classifier, Decision Tree, Logistic Regression, Multinomial Naive Bayes, K-Nearest Neighbor and Random Forest had gradually 76.61, 70.16, 76.09, 76.04, 28.12 and 75.32% F1-Score rate. So, 76.61% is the highest F1-score which is Support Vector Classifier and 28.12% is the lowest F1-score rate which is K-Nearest Neighbor.

4.3 Comparison the Algorithms' Results

In the previous component, accuracy, precision, recall and f1-factor of all applied algorithms had been described elaborately. Though that description, it had been stated that Support Vector Classifier (SVC) gave the highest accuracy, Multinomial Naive Bayes(MNB) provided the best precision and Support Vector Classifier (SVC) provided the highest recall and f1-factor both.

Table 2 Comparison of all applied algorithms' results

Algorithms' name	Accuracy (%)	Precision (%)	Recall (%)	F1-factor (%)
Support Vector Classifier (SVC)	76.90	75.73	77.53	76.61
Decision Tree (DT)	69.27	72.29	68.16	70.16
Logistic Regression (LR)	76.52	74.74	77.49	76.09
Multinomial Naive Bayes (MNB)	74.81	79.96	72.49	76.04
K-Nearest Neighbor (KNN)	75.40	69.39	78.86	73.82
Random Forest (RF)	74.80	76.92	73.32	75.32

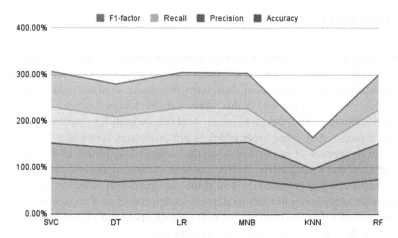

Fig. 6 Comparison of all applied algorithms' results

All applied algorithms provided accuracy, precision, recall and f1-factor had shown in Table 2. Though Table 2., it's clear that Support Vector Classifier (SVC) had the highest accuracy, recall and f1-factor. At the same way, Multinomial Naive Bayes(MNB) had the highest precision. However the precision rate of Support Vect(SVC) is little less than Multinomial Naive Bayes(MNB) and Random Forest(RF). So, Support Vector Classifier (SVC) is the best algorithm for depression detection according to this thesis work (Fig. 6).

5 Limitations and Future Research Direction

This research had only worked with the data collected from Twitter. Some studies had worked on anxious depression prediction (Rahman et al. 2020), depression level detection. However, this study worked on depression detection based on Twitter's tweet. So, by this it's clear that this study only worked for detecting if any Twitter user is depressed or not but not for all social media users. This research also mentioned that this work was only for English language. Moreover, this thesis only detected depression of any user thought the tweet but couldn't provide the status or level of that depressed user. These are some limitations for this study.

In future, this study will be worked for identifying the types of depression in depth so that it will be able to inform about the depressed person's depression state. Moreover, it will also work with the data from all other social media and in many more languages.

6 Conclusions

This study presented a generic depression detection model with the use of six machine learning algorithms. The objective of this study is depression detection from any tweets which represent depression though any depressive word and reduce depression & suicides out of depression. In this work, a methodology has been followed and through following this methodology, any tool can be built for detecting depression easily in future. Any other tweet which is pre-processed, it can be told if the data indicates depression or not by following the methodology of this study. This study also resulted that among all the applied machine learning algorithms, Support Vector Classifier(SVC) had provided the best result for depression detection which means if any other study wants to work on depression detection based on machine learning algorithms, then that study can use Support Vector Classifier(SVC) without any doubt for getting the best result for depression detection.

References

M. Deshpande, V. Rao, Depression detection using emotion artificial intelligence, in *2017 International Conference on Intelligent Sustainable Systems (ICISS)* (IEEE, 2017), pp. 858–862

L. Coello-Guilarte, R.M. Ortega-Mendoza, L. Villaseñor-Pineda, M. Montes-y-Gómez, Crosslingual depression detection in Twitter using Bilingual word alignments, in *International Conference of the Cross-Language Evaluation Forum for European Languages* (Springer, Cham, 2019), pp. 49–61

M.M. Tadesse, H. Lin, B. Xu, L. Yang, Detection of depression-related posts in reddit social media forum. IEEE Access **7**, 44883–44893 (2019)

R.U. Mustafa, N. Ashraf, F.S. Ahmed, J. Ferzund, B. Shahzad, A. Gelbukh, A multiclass depression detection in social media based on sentiment analysis, in *17th International Conference on Information Technology–New Generations (ITNG 2020)* (Springer, Cham, 2020), pp. 659–662

S.G. Burdisso, M. Errecalde, M. Montes-y-Gómez, A text classification framework for simple and effective early depression detection over social media streams. Expert. Syst. Appl. **133**, 182–197 (2019)

F. Cacheda, D. Fernandez, F.J. Novoa, V. Carneiro, Early detection of depression: social network analysis and random forest techniques. J. Med. Internet Res. **21**(6), e12554 (2019)

H.S. AlSagri, M. Ykhlef, Machine learning-based approach for depression detection in twitter using content and activity features. IEICE Trans. Inf. Syst. **103**(8), 1825–1832 (2020)

L. Tong, Q. Zhang, A. Sadka, L. Li, H. Zhou, Inverse boosting pruning trees for depression detection on Twitter (2019). arXiv:1906.00398

T. Gui, L. Zhu, Q. Zhang, M. Peng, X. Zhou, K. Ding, Z. Chen, Cooperative multimodal approach to depression detection in Twitter, in *Proceedings of the AAAI Conference on Artificial Intelligence*, vol. 33, no. 01 (2019), pp. 110–117

X. Tao, X. Zhou, J. Zhang, J. Yong, Sentiment analysis for depression detection on social networks, in *International Conference on Advanced Data Mining and Applications* (Springer, Cham, 2016), pp. 807–810

R. Singh, J. Du, Y. Zhang, H. Wang, Y. Miao, O.A. Sianaki, A. Ulhaq, A framework for early detection of antisocial behavior on Twitter using natural language processing, in *Conference on Complex, Intelligent, and Software Intensive Systems* (Springer, Cham, 2019), pp. 484–495

X. Yang, R. McEwen, L.R. Ong, M. Zihayat, A big data analytics framework for detecting user-level depression from social networks. Int. J. Inf. Manag. **54**, 102141 (2020)

M.Y. Wu, C.Y. Shen, E.T. Wang, A.L. Chen, A deep architecture for depression detection using posting, behavior, and living environment data. J. Intell. Inf. Syst. **54**(2), 225–244 (2020)

M.J. Vioules, B. Moulahi, J. Aé, S. Bringay, Detection of suicide-related posts in Twitter data streams. IBM J. Res. Dev. **62**(1), 1–7 (2018)

C. Lin, P. Hu, H. Su, S. Li, J. Mei, J. Zhou, H. Leung, Sensemood: depression detection on social media, in *Proceedings of the 2020 International Conference on Multimedia Retrieval* (2020), pp. 407–411

V. Leiva, A. Freire, Towards suicide prevention: early detection of depression on social media, in *International Conference on Internet Science* (Springer, Cham, 2017), pp. 428–436

P. Nambisan, Z. Luo, A. Kapoor, T.B. Patrick, R.A. Cisler, Social media, big data, and public health informatics: ruminating behavior of depression revealed through twitter, in *2015 48th Hawaii International Conference on System Sciences* (IEEE, 2015), pp. 2906–2913

X. Chen, M.D. Sykora, T.W. Jackson, S. Elayan, What about mood swings: identifying depression on twitter with temporal measures of emotions, in *Companion Proceedings of the The Web Conference 2018* (2018), pp. 1653–1660

S.C. Guntuku, D.B. Yaden, M.L. Kern, L.H. Ungar, J.C. Eichstaedt, Detecting depression and mental illness on social media: an integrative review. Curr. Opin. Behav. Sci. **18**, 43–49 (2017)

H. Almeida, A. Briand, M.J. Meurs, Detecting early risk of depression from social media user-generated content, in *CLEF (Working Notes)* (2017)

B. O'dea, S. Wan, P.J. Batterham, A.L. Calear, C. Paris, H. Christensen, Detecting suicidality on Twitter. Internet Interv. **2**(2), 183–188 (2015)

M. Gaikar, J. Chavan, K. Indore, R. Shedge, Depression detection and prevention system by analysing tweets, in *Proceedings 2019: Conference on Technologies for Future Cities (CTFC)* (2019)

M. Morales, S. Scherer, R. Levitan, A cross-modal review of indicators for depression detection systems, in *Proceedings of the Fourth Workshop on Computational Linguistics and Clinical Psychology–From Linguistic Signal to Clinical Reality* (2017), pp. 1–12

C.S.A. Razak, M.A. Zulkarnain, S.H. Ab Hamid, N.B. Anuar, M.Z. Jali, H. Meon, Tweep: a system development to detect depression in twitter posts, in *Computational Science and Technology* (Springer, Singapore, 2020), pp. 543–552

A. Kumar, A. Sharma, A. Arora, Anxious depression prediction in real-time social data, in *International Conference on Advances in Engineering Science Management & Technology (ICAESMT)-2019* (Uttaranchal University, Dehradun, India, 2019)

M.R. Islam, M.A. Kabir, A. Ahmed, A.R.M. Kamal, H. Wang, A. Ulhaq, Depression detection from social network data using machine learning techniques. Health Inf. Sci. Syst. **6**(1), 1–12 (2018)

B.Y. Ziwei, H.N. Chua, An application for classifying depression in tweets, in *Proceedings of the 2nd International Conference on Computing and Big Data* (2019), pp. 37–41

S. Symeonidis, D. Effrosynidis, A. Arampatzis, A comparative evaluation of pre-processing techniques and their interactions for twitter sentiment analysis. Expert. Syst. Appl. **110**, 298–310 (2018)

J.J. Stephen, P. Prabu, Detecting the magnitude of depression in twitter users using sentiment analysis. Int. J. Electr. Comput. Eng. **9**(4), 3247 (2019)

S.S.M.M. Rahman, K.B.M.B. Biplob, M.H. Rahman, K. Sarker, T. Islam, An investigation and evaluation of N-Gram, TF-IDF and ensemble methods in sentiment classification, in *International Conference on Cyber Security and Computer Science* (Springer, Cham, 2020), pp. 391–402

M.F. Sohan, S.S.M.M. Rahman, M.T.A. Munna, S.M. Allayear, M.H. Rahman, M.M. Rahman, NStackSenti: evaluation of a multi-level approach for detecting the sentiment of users, in *International Conference on Next Generation Computing Technologies* (Springer, Singapore, 2018), pp. 38–48

M.M. Rahman, S.S.M.M. Rahman, S.M. Allayear, M.F.K. Patwary, M.T.A.A. Munna, Sentiment analysis based approach for understanding the user satisfaction on android application, in *Data Engineering and Communication Technology* (Springer, Singapore, 2020), pp. 397–407

S.S.M.M. Rahman, M.H. Rahman, K. Sarker, M.S. Rahman, N. Ahsan, M.M. Sarker, Supervised ensemble machine learning aided performance evaluation of sentiment classification, in *Journal of Physics: Conference Series*, vol. 1060, no. 1 (IOP Publishing, 2018), p. 012036

A Conceptual Analysis of IoT in Healthcare

Muhammad Azmi Umer, Muhammad Taha Jilani, Asif Rafiq, Sulaman Ahmad Naz, and Khurum Nazir Junejo

Abstract There are numerous health problems associated with human life. Usually, they are not instantaneous. There are significant parameters based on which any disease or health problem arises. Nowadays hospitals are using state of the art infrastructure to keep track of patients' health conditions. It is possible using Wireless Body Area Network (WBAN). This chapter proposes a machine learning model to discover hidden facts and to predict the future behavior of patients based on current health conditions. Further, a conceptual analysis is presented using a heart disease dataset.

1 Introduction

Recent advancements in information technology have enabled significant improvements in healthcare. The Internet of things (IoT) is playing a vital role in today's modern society. IoT is a system based on machine to machine (M2M) communication. Nowadays hospitals are using the latest hardware gadgets to monitor the health condition of patients. Among other technologies, the Wireless Body Area Network (WBAN) is playing an important role in hospitals. Hardware gadgets used in WBAN

M. A. Umer (✉)
DHA Suffa University, and KIET Karachi, Karachi, Pakistan
e-mail: azmi.umer@dsu.edu.pk

M. T. Jilani
Karachi Institute of Economics and Technology, Karachi, Pakistan
e-mail: m.taha@pafkiet.edu.pk

A. Rafiq · S. A. Naz
DHA Suffa University, Karachi, Pakistan
e-mail: asif.rafiq@dsu.edu.pk

S. A. Naz
e-mail: sulaman.ahmed@dsu.edu.pk

K. N. Junejo
DNNae Inc., Karachi, Pakistan

© The Author(s), under exclusive license to Springer Nature Switzerland AG 2022
Y. Baddi et al. (eds.), *Big Data Intelligence for Smart Applications*,
Studies in Computational Intelligence 994,
https://doi.org/10.1007/978-3-030-87954-9_10

Fig. 1 Connectivity of
Wireless Body Area
Network over the Internet

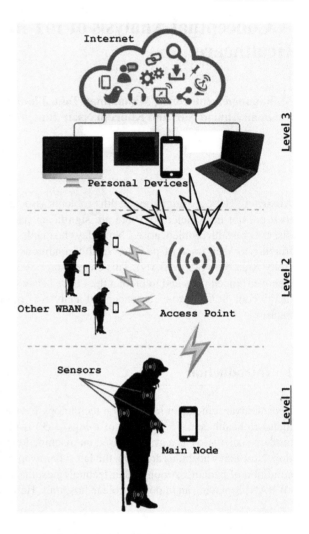

continuously update the patients' health parameters in the centralized system. The patient does not need to be physically under supervision in a ward or hospital. The mode of supervision is remote while visits are conducted virtually. The interface or gadget at the patient's end collects the data with the help of sensors like pulse rate, heartbeat, blood pressure, oxygen level, temperature, sleep, etc. Thus patient's care workflow becomes automated and the data collected is viewable at the doctor's end. Mere collection from patient's and viewing of data at doctor's portal helps in analyzing and drawing predictions about the health of patient and hence many calamities can be avoided. But this process can become even more productive if analysis is also automated and doctor receives the trend summary reports along with the actual historical data. This historical data can become very bulky over the time and manual analysis becomes impossible. However, by using data mining and machine learning techniques, regression functions can be constructed and hence trends can be found predicting the possible future behavior of patients including heart attacks, stroke, etc. The proposed model uses machine learning algorithms for this purpose.

WBAN is an active research area with great potential for step up in the field of personal health care (Negra et al. (2016)). It consists of some strong sensing devices which can sense vital signs as well as emotions like fear, happiness, anxiety level, etc. These sensors can be rooted under patient's skin or as detachable. Some of the examples include Wearable devices like bands, gloves, etc. Basic working model of communication of some devices is explained in Fig. 1. It shows a three step process. Level 1 refers to the communication within WBAN which consists of some wireless sensing devices and the main node of WBAN which may be patient's smart phone. In level 2, the main node of WBAN is connected with any personal device like laptop etc. Last level is concerned with the connection of laptop with the internet and the beyond. This chapter presents a model to use the information provided by WBAN to predict the future behavior of patients. The following are some main research questions that motivated the current study.

RQ1: Does the patient suffering from specific disease have some specific medical traits?

RQ2: Can machine learning be used to predict the future behavior of patients?

Contributions: We answer the above research questions through following contributions (a) A model to predict the future behavior of patient (b) A conceptual analysis of the proposed model.

Organization: Sect. 2 presents the related work in this domain. Section 3 describes preliminaries related to WBAN. The proposed model is described in Sect. 4. A conceptual analysis of proposed model is presented in Sect. 5. The discussion on research questions stated earlier is provided in Sect. 6. Conclusion and possible future work is discussed in Sect. 7.

2 Related Work

Machine Learning algorithms are being applied in health-care Ghassemi et al. (2018); Qayyum et al. (2020), Cyber Security Ahmed et al. (2021); Umer et al. (2020), Surveillance Hunain et al. (2021), Weather Forecasting Umer et al. (2021), and several other domains. It has robust performance in predictive analytics. It has been applied in healthcare analytics like in Chen et al. (2017) Convolutional Neural Network was used for disease risk prediction. The latent factor model was used to overcome the problem of missing data. Similarly, a study reported in Abdelaziz et al. (2018) used the machine learning algorithms (Linear Regression and Neural Networks) to predict chronic kidney disease. Linear Regression was used to study the factors influencing chronic kidney disease while Neural Network was used to predict the disease. Zhang et al. (2014), discusses a healthcare system in which Wireless Sensor Network (WSN) was used to monitor the health condition of the patient. It has divided the architecture into three layers where layer 1 includes sensor node devices, layer 2 includes the access network and the Internet while layer 3 includes the medical server and the telemedicine. A study reported in Tennina et al. (2014) has proposed an e-health system through which a patient at home could be tracked by the doctor using WSN. In Yang and Logan (2006), the 1999 National Inpatient sample data sets (which was used for Healthcare cost and utilization project) was analyzed using data mining techniques. The results obtained by data mining techniques were compared with the expert knowledge survey of 97 gastrointestinal tract surgeons. This analysis showed that the comparison of these two results was consistent in some factors which are highly associated with paraesophageal hernia. Moreover, the data mining approach also revealed some other related disorders which were not known by the expert at that time and neither reported in the literature. In Ferreira et al. (2012), data mining techniques were used to improve the diagnosis of neonatal jaundice. To perform experiments, 227 healthy newborn infants with 35 or more weeks of gestation were studied. For this purpose, more than 70 variables were collected and analyzed. Machine learning algorithms like Decision tree (J48) and neural networks (multi-layer perceptron) were applied in this study. The accuracy of prediction was 89%. Naive Bayes (NB), Multi Layer Perceptron (MLP) and simple logistics algorithms provided the best results.

Nexleaf Thomas (2020) offers wireless remote temperature monitoring system helps to regulate the temperature inside the refrigerators to avoid vaccine spoiling. Similarly ample work has also been done to ensure environment safety by emitting less carbon dioxide. This process is controlled by a cloud-based remote monitoring system. SystemOne Briodagh (2017) uses IoT services of Vodafone to endure real-time diagnostic information to the concerned health professionals. In Pfizer taps ibm for research collaboration to transform parkinson's disease care (2016), the consultant-patient communication-gap is bridged my making real-time modifications in the prescribed amount of certain medicines on the basis of data collected by sensors that collect the changes in patient's behavior at a very minute and macro level. This was implemented in patients having a brain disorder that leads to shaking, stiffness, and difficulty with walking, balance, and coordination. The loop wearable device in (2021) can be helpful in monitoring vital signs of COVID patients as it was designed to remotely measure the heart rate, measuring oxygen saturation in blood of patients with any chronic obstructive pulmonary disease. More applications of IoT in healthcare can be found in Thomas (2020).

New experimental techniques were introduced in Moser et al. (1999) for Data Mining Surveillance System (DMSS) using an electronic healthcare database to monitor the emerging infections and antimicrobial resistance. Information provided by DMSS can be a good indicator of an important shift in infection and antimicrobial resistance patterns in ICU. In a study reported in Kaur et al. (2012), an initiative was taken to increase the availability of abusive drug usage data to provide better drug recovery services. Observational study of cases registered in the California Department of Alcohol and Drug Program was utilized for this purpose. Initially, the Minitab diagnostic tool was used to diagnose the above-mentioned cases. Further it provided access to the Casemix databases to retrieve the hidden information using data mining tools. K-means clustering was used to determine the existence of patients admitted and discharged on the usage of abusive substances between 2 and 3 consecutive years. Data were classified into educated and non-educated classes for a categorized race along with correlation age at the time of admission in the hospital. During the analysis phase, it was made sure to include the treatments provided to patients during their stay at the hospital and also the discharge status of the final medical diagnosis. Results showed that the incidence rate of admission cases in the 45–49 age group was on a rising edge. With more than 40% probability, patients who were acquiring the maximum number of abusive substance had obtained post-graduate education Kaur et al. (2012).

Data mining was used in Cao et al. (2008) to extract complex data from the major clinical trial repository. This data was summarized and visualized to make informed decisions about future cancer vaccine clinical trials. Data mining techniques were used in Setoguchi et al. (2008) to determine in which situations analyses based on various types of exposure propensity score (EPS) models produce unbiased and/or efficient results. Recursive partitioning, logistic regression, and neural networks were used in this study. Logistic regression provided the robust results while the neural network provided the least numerically biased estimates.

3 Wireless Body Area Network

WBAN consists of small wireless sensor nodes that are interconnected into/on/off a human body (up to some cm) via Radio Frequency (RF) as shown in Fig. 2. It is used for collecting both the medical and general purpose data from the body. Nowadays, WBAN is playing a vital role in medical sciences, specifically for human healthcare systems. It is a state of the art technology that has reshaped the medical practices including diagnostics, investigations, prescriptions, etc. It provides a new mechanism for medical practitioners and doctors to diagnose disease in an efficient manner by providing quick and effective solution. Apart from medical, it has a significant contribution in the field of entertainment and consumer electronics. Despite this, it is an active research area for researchers in clinical advancements Rathee et al. (2014).

With the recent advancements in IoT, WBAN has acquired a crucial status in the field of computation, diagnosis, and implementation. The diagnostic of health parameters requires power and energy sustainability, therefore some recent techniques like fog computing are provisioned to overcome such issues in IoT based WBAN. We use to cater millions of terabytes of data during the human health condition diagnostics, therefore specific machine learning (ML) algorithms are required for the next level optimization in fog computing techniques Amudha and Murali (2020).

Another important issue related to WBAN is "Coexistence". In this problem, WBAN has to deal with interference due to the same channel simultaneous occupancy of multiple WBANs. A study reported in Sun et al. (2018) used the ML algorithms namely decision tree (DT) and naïve Bayes classifier (NBC) to tackle the above

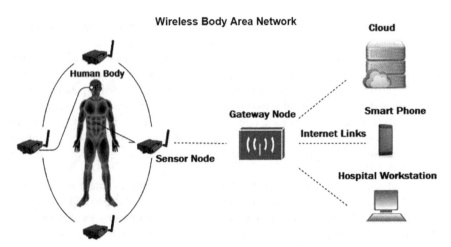

Fig. 2 Wireless Body Area Network

mentioned problem. It describes how the coexistence state in multiple WBANs be predicted in efficient and timely manner.

In contrast to WSN, a typical framework is required to be standardized for the robust protocol structuring of WBAN. For this purpose, a sub task group under IEEE 802 named 802.15.6 has already been established and managing the same. Both Physical and MAC layers' specifications have been taken into account for efficient use of Bandwidth Kwak et al. (2010).

4 Proposed Model

WBAN can be used to monitor the patients' health condition. All this data gets recorded in the system, therefore this data could be used to extract meaningful results and to predict the behavior of patients in the future. To achieve this goal, this chapter has proposed a model using machine learning techniques as described in Fig. 3. The proposed model is further described in the following sub sections.

4.1 Data Sets and Feature Selection

Data available in the system would be used to form data sets. Relevant attributes/ features would be selected from these data sets. This requires an understanding of the dataset at the initial step. It helps in selecting the important attributes for training the model. For this purpose, several statistical techniques can be applied to deeply study the insights of data. It includes but not limited to finding the correlation among

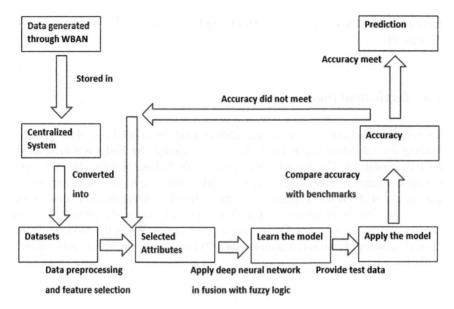

Fig. 3 Work Flow

the attributes, finding the noise in the dataset using box plot, covariance, deviations, central tendency, etc.

4.2 Selection of Algorithm

The selection of an algorithm does not depend only on the type of problem but also on the type of dataset which is being used to train the model. There are several machine learning algorithms which could be used for this purpose. For example, Neural networks can be used for this purpose. They have the capability of approximating a function that depends on a large number of inputs. This particular system may have to deal with a large number of inputs to cover every possible aspect of patients' health condition. Futhermore, Neural networks learn from the experience which could be helpful in the prediction of a patients' medical behavior.

4.3 Fusion of Deep Neural Networks and Fuzzy Logic

Fuzzy logic has been used in conjunction with deep neural networks (DNN) in Leke et al. (2015) to perform the missing data imputation tasks. The proposed model is focused on critical issues related to patient's health condition, therefore it is recommended to use DNN in conjuction with fuzzy logic to handle the missing values in the dataset.

4.4 Implementation

To train the machine learning model, dataset must be divided into three sets i.e. training set, validation set, and test set. At the begining, the model is trained using the training dataset. The trained model is then evaluated on validation set to observe the performance. This validation step is used to tune the model based on certain parameters of algorithm. Futhermore, it also helps in avoiding the overfitting of the model. This step is an iterative step as it is required to repeat the feature selection process and data processing for making a good model. When the model gets the desired accuracy, then it is tested on test data. This step gives the overall performance of the model.

4.5 Parallelization of Machine Learning Algorithms

It is important to parallelize the above-mentioned techniques. Real-time data has high velocity, therefore the system should be capable of dealing with big data. A study reported in Liu et al. (2007) has discussed several issues of parallelization of data mining algorithms. It includes workload balance, inter-process communication, and other factors.

5 Conceptual Analysis of Proposed Model

The study reported here has performed a conceptual analysis of the proposed model using the heart disease dataset available at UCI repository Uci (2019). This dataset consists of the following 14 attributes:

- Age
- Sex
- Chest Pain Type
- Resting Blood Pressure
- Serum Cholestrol
- Fasting Blood Sugar
- Resting Electrocardiographic Results
- Maximum Heart Rate Achieved
- Exercise Induced Angina
- ST Depression Induced By Exercise Relative to Rest

Fig. 4 Box Plots (1 of 3)

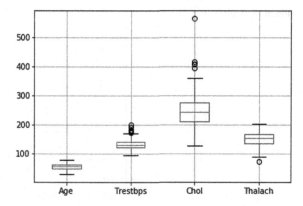

Fig. 5 Box Plots (2 of 3)

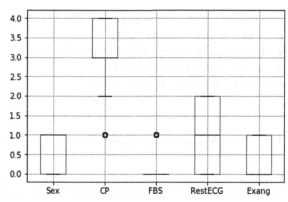

- The Slope Of The Peak Exercise ST Segment
- Number Of Major Vessels (0-3) Colored By Flourosopy
- Thal
- Angiographic Disease Status (Class Variable).

Boxplots of different features are shown in Figs. 4, 5, and 6 to identify the features containing outliers.

5.1 Algorithms

Following machine learning algorithms were used to train the model on the heart disease dataset:

- Logistic Regression (LR)
- Linear Discriminant Analysis (LDA)
- KNeighbors Classifier (KNN)
- Classification And Regression Tree Classifier (CART)

Fig. 6 Box Plots (3 of 3)

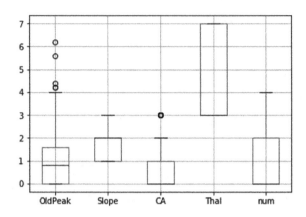

- Support Vector Machine (SVM)
- Multi Layer Perceptron Classifier (MLP)
- Random Forest Classifier (RF)
- Gradient Boosting Classifier (GBC)

5.2 Results and Discussion

Four iterations were run using the above-mentioned algorithms. The actual score (accuracy) of these four iterations are shown in Fig. 7. In order to determine the algorithms deviating on each iteration, mean and the standard deviation of each algorithm was calculated, as shown in Figs. 8 and 9. It is evident from Fig. 9 that there are three algorithms that are deviating from their scores in each iteration and these are CART, MLP, and RF. Among which MLP was found to be a highly deviating algorithm on each iteration with the same set of parameters configuration.

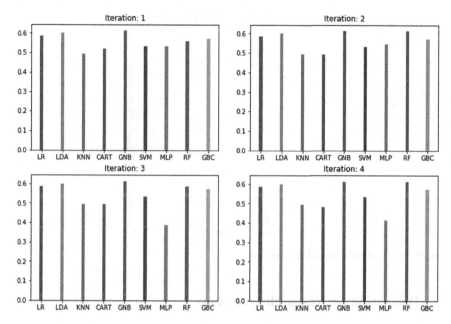

Fig. 7 Results of four iterations

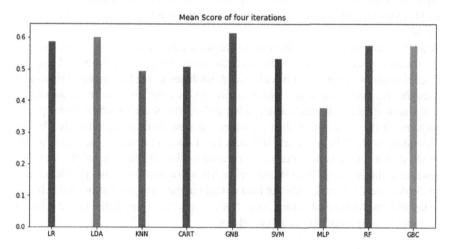

Fig. 8 Mean score of four iterations

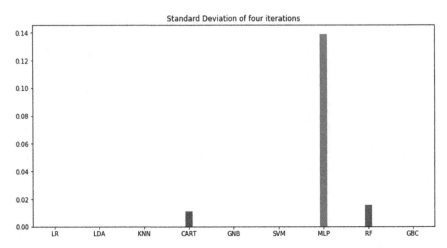

Fig. 9 Standard deviation of four iterations

6 Discussion on Research Questions

Now we would discuss the research questions stated in the introductory section.
RQ1: Does the patient suffering from specific disease have some specific medical traits?
RQ2: Can machine learning be used to predict the future behavior of patients?

The answer to both questions is yes, because the study reported in Sect. 5 used the dataset consists of various medical related attributes to predict the heart disease. It used the machine learning techniques to predict the heart disease. There are various machine learning techniques which work differently. Some of which finds specific patterns in the dataset and then classifies the new instance accordingly. As ML algorithms were able to classify different heart disease, it means there are specific medical traits in the patient that could be used to classify the patients' disease. This leads us to the answer of RQ1 stated earlier. ML algorithms finds the specific patterns in the dataset of patients. This can lead to find out the changes in structural behavior using certain attributes of patient, thus they are capable to predict the future behavior of patient based on current health conditions.

7 Conclusion and Future Work

This chapter has proposed a model to predict the patients' medical behavior in future. Success in the proposed model could be extremely beneficial for the human life and opens a research direction for the scientist in the related fields. This chapter has presented a conceptual analysis using a heart disease dataset. Complete Practical implementation of the proposed model using proper clinical data is still needed to be done. Hopefully, novel aspects of the proposed model could lead to some significant improvements in healthcare.

References

R. Negra, I. Jemili, A. Belghith, Wireless body area networks: applications and technologies. Procedia Comput. Sci. **83**, 1274–1281 (2016)

M. Ghassemi, T. Naumann, P. Schulam, A.L. Beam, R. Ranganath, Opportunities in machine learning for healthcare (2018). arXiv:1806.00388

A. Qayyum, J. Qadir, M. Bilal, A. Al-Fuqaha, Secure and robust machine learning for healthcare: a survey (2020). arXiv:2001.08103

C.M. Ahmed, M.A. Umer, B.S., S.B. Liyakkathali, M.T. Jilani, J. Zhou, Machine learning for cps security: applications, challenges and recommendations, in *Machine Intelligence and Big Data Analytics for Cybersecurity Applications* (Springer, 2021), pp. 397–421

M.A. Umer, A. Mathur, K.N. Junejo, S. Adepu, Generating invariants using design and data-centric approaches for distributed attack detection. Int. J. Crit. Infrastruct. Prot. **28**, 100341 (2020)

M. Hunain, T. Iqbal, M.A. Siyal, M.A. Umer, M.T. Jilani, A framework using artificial intelligence for vision-based automated firearm detection and reporting in smart cities, in *Artificial Intelligence and Blockchain for Future Cybersecurity Applications* (Springer, 2021), pp. 237–255

M.A. Umer, M.T. Jilani, K.N. Junejo, S.A. Naz, C.W. D'Silva, Role of machine learning in weather related event predictions for a smart city, in *Machine Intelligence and Data Analytics for Sustainable Future Smart Cities* (Springer, 2021), pp. 49–63

M. Chen, Y. Hao, K. Hwang, L. Wang, L. Wang, Disease prediction by machine learning over big data from healthcare communities. IEEE Access **5**, 8869–8879 (2017)

A. Abdelaziz, M. Elhoseny, A.S. Salama, A. Riad, A machine learning model for improving healthcare services on cloud computing environment. Measurement **119**, 117–128 (2018)

Y. Zhang, L. Sun, H. Song, X. Cao, Ubiquitous WSN for healthcare: recent advances and future prospects. IEEE Internet Things J. **1**(4), 311–318 (2014)

S. Tennina, M. Di Renzo, E. Kartsakli, F. Graziosi, A.S. Lalos, A. Antonopoulos, P.V. Mekikis, L. Alonso, Wsn4qol: a WSN-oriented healthcare system architecture. Int. J. Distrib. Sens. Netw. **10**(5), 503417 (2014)

J. Yang, J. Logan, A data mining and survey study on diseases associated with paraesophageal hernia, in *AMIA Annual Symposium Proceedings*, vol. 2006 (American Medical Informatics Association, 2006), p. 829

D. Ferreira, A. Oliveira, A. Freitas, Applying data mining techniques to improve diagnosis in neonatal jaundice. BMC Med. Inf. Decis. Mak. **12**(1), 143 (2012)

M. Thomas, 6 iot in healthcare applications leading to an improved industry (2020). https://builtin.com/internet-things/iot-in-healthcare

K. Briodagh, Systemone and vodafone tackle medical challenge with iot (2017). https://www.iotevolutionworld.com/iot/articles/432161-systemone-vodafone-tackle-medical-challenge-with-iot.htm

Pfizer taps ibm for research collaboration to transform parkinson's disease care (2016). https://www.pfizer.com/news/press-release/press-release-detail/pfizer_taps_ibm_for_research_collaboration_to_transform_parkinson_s_disease_care

The promise of clinical wearables for people with chronic conditions. https://spryhealth.com/the-loop-monitoring-solution/. Accessed 30 June 2021

S.A. Moser, W.T. Jones, S.E. Brossette, Application of data mining to intensive care unit microbiologic data. Emerg. Infect. Dis. **5**(3), 454 (1999)

H. Kaur, R. Chauhan, S.M. Aljunid, Data mining cluster analysis on the influence of health factors in casemix data. BMC Health Serv. Res. **12**(1), O3 (2012)

X. Cao, K.B. Maloney, V. Brusic, Data mining of cancer vaccine trials: a bird's-eye view. Immunome Res. **4**(1), 7 (2008)

S. Setoguchi, S. Schneeweiss, M.A. Brookhart, R.J. Glynn, E.F. Cook, Evaluating uses of data mining techniques in propensity score estimation: a simulation study. Pharmacoepidemiol. Drug Saf. **17**(6), 546–555 (2008)

D. Rathee, S. Rangi, S. Chakarvarti, V. Singh, Recent trends in wireless body area network (wban) research and cognition based adaptive wban architecture for healthcare. Health Technol. **4**(3), 239–244 (2014)

S. Amudha, M. Murali, Deep learning based energy efficient novel scheduling algorithms for body-fog-cloud in smart hospital. J. Ambient Intell. Humaniz. Comput. 1–20 (2020)

Y. Sun, T. Chen, J. Wang, Y. Ji, A machine learning based method for coexistence state prediction in multiple wireless body area networks, in *EAI International Conference on Body Area Networks* (Springer, 2018), pp. 203–217

K.S. Kwak, S. Ullah, N. Ullah, An overview of ieee 802.15. 6 standard, in *3rd International Symposium on Applied Sciences in Biomedical and Communication Technologies (ISABEL 2010)* (IEEE, 2010), pp. 1–6

C. Leke, T. Marwala, S. Paul, Proposition of a theoretical model for missing data imputation using deep learning and evolutionary algorithms (2015). arXiv:1512.01362

Y. Liu, W.-k. Liao, A.N. Choudhary, J. Li, Parallel data mining algorithms for association rules and clustering (2007)

Uci (machine learning repository). heart disease dataset (2019) https://archive.ics.uci.edu/ml/datasets/Heart+Disease

Securing Big Data-Based Smart Applications Using Blockchain Technology

Rihab Benaich, Imane El Alaoui, and Youssef Gahi

Abstract Nowadays, Big Data is the most salient paradigm. It has become a game-changer in the current technology and has aroused various industries and research communities worldwide. Big Data-based applications have shown much potential in data-driven decisions that have changed all business operations, including cost optimization, online reputation, and customer loyalty. Nevertheless, Big Data also brings many security and privacy issues. For this reason, Big Data applications require accurate methods to ensure reliable data storage, sharing, and decision-making. Among these methods, the Blockchain is known as the clever combination of distributed exchange, consensus, and cryptography mechanisms. It has been brought to the forefront to resolve security challenges in various fields such as Healthcare, banking, smart cities, etc. This paper first presents the considerable difficulties faced when adopting Big Data that can be resolved using Blockchain. Afterward, we overview the Blockchain technology by projecting its components, workflows, classification, and related characteristics. Finally, we present the importance of combining Big Data and Blockchain through reviewing the novel implementations proposed by researchers in great domains.

Keywords Blockchain · Big data · Security · Healthcare · Banking · Game theory · IoT · Smart cities · VANETS

R. Benaich (✉) · I. E. Alaoui · Y. Gahi
Laboratoire Des Sciences de L'Ingénieur, Ecole Nationale Des Sciences Appliquées, Ibn Tofail University, Kenitra, Morocco
e-mail: rihab.benaich@uit.ac.ma

I. E. Alaoui
e-mail: Imane.el.alaoui@uit.ac.ma

Y. Gahi
e-mail: gahi.youssef@uit.ac.ma

241

Y. Baddi et al. (eds.), *Big Data Intelligence for Smart Applications*,
Studies in Computational Intelligence 994,
https://doi.org/10.1007/978-3-030-87954-9_11

1   Introduction

With the increase of people using the Internet, the amount of produced data is practically exploding by the day. This explosion opens up incredible processing opportunities for organizations, such as data sharing and data-driven decisions (Faroukhi et al. 2020) and proposing customizable services. However, this noteworthy and extensive usage also comes with the necessity of ensuring the security and protection of data storage and transmissions (Gahi et al. 2016).

It is worth noting that this vast data, also known as Big data, is, most of the time, very sensitive (Gahi et al. 2016) as it represents a direct projection of our day-to-day activities. Big data is closely related to our personal life as it mainly relies on our transactions, interactions, and pattern mining. As Big data becomes a grand reality, it has to be clearly defined. Since the 2000s, many characteristics have been associated with Big data (El Alaoui et al. 2019), such as Volume, Variety, Velocity, Value, Veracity, Variability, and Visualization. Also known as 7Vs (El Alaoui and Gahi 2019), see Fig. 1.

Because of these varied characteristics, ensuring security using traditional methods becomes a complex task (Hu et al. 2014). For example, many conventional security techniques cannot handle the scale and velocity required by Big Data-based applications.

To overcome the scalability and velocity aspects, distributed approaches should be used. These approaches are designed to run over large clusters, which are a set of interconnected machines. Big Data systems' common distribution aspect makes Blockchain technology an ultimate solution for Big Data security challenges. It allows coping with security issues related to sharing and storing Big Data by applying cryptography, decentralization, consensus principles, and ensuring transaction trust.

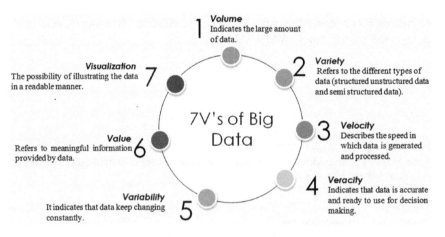

Fig. 1 The seven V's of big data

Moreover, today, Blockchain technology has become an ideal fit for Big Data problems through its high capacity to prevent the significant malicious activities caused by cybercriminals. The Blockchain not only restricts the intruders but also ensures access to multiple nodes in the network. This safeguarding feature of Blockchain strengthens and boosts the security level compared to many unique technologies used in various enterprises such as the cloud or virtual private network, or any other technology.

For this reason, many researchers have taken a keen interest in proposing attractive and diverse Blockchain models and frameworks to share and securely store Big Data in various sectors, namely Healthcare, Banking, IoT, gaming, and smart applications.

In this contribution, we focus on reviewing the recent studies that have tackled Big Data and Blockchain and their alliance. Furthermore, we discuss the advantage and the limits of the proposed models.

The following research questions will guide this contribution:

- What are the application domains of blockchain technology in the context of Big data?
- How does Blockchain technology address the challenge of Big data security?
- What are the significant challenges remaining in the usage of Blockchain as a secure solution for Big data intelligent applications?

This paper is organized as follows: in the second snd third sections, we present the research methodology, and we overview the concept of Blockchain technology by showing its components, workflows, and related characteristics. Afterward, in the fourth section, we highlight exciting studies based on Blockchain brought to various domains such as the healthcare field, smart cities, IoT, Vanets, 5G networks, Banking, and Game theory. Then in Sect. 5, we discuss the importance of using Blockchain in the context of Big Data. Finally, we conclude the paper and show some future works.

2 Research Methodology

To carry out this paper, we have used a systematic review methodology to identify and summarize the relevant research contributions that answer the above questions. According to (Tranfield et al. 2003) standards, we have conducted our review to develop a review plan regarding the methods to be used and the main research questions to be outlined and reports to be extracted.

This literature review highlighted the critical role of the outstanding Blockchain technology in coping with Big data challenges in numerous domains.

We took three stages to perform this study:

1. First of all, we researched and identified multiple primary papers related to our topic.
2. Secondly, we filtered the primary chosen papers.
3. Finally, we validated the filtered articles for data synthesis and analysis.

The primary studies were chosen based on the inclusion and exclusion criteria applied to the reviewed publications, the study's scope, and verifications done by subject and title and reading the abstract and findings.

This research method enables the investigation to be focused on papers published in renowned academic journals such as IEEE Xplore, Springer, Google Scholar, Wiley online library, ACM digital library, web of science, and so forth.

In our research approach, we covered various papers and magazines that were published in recent years. This allowed us to illustrate the progression of the thoughts. Therefore, we selected particular keywords such as "Big Data," "Blockchain," "Security," and "Blockchain applications" to assure a superior approach in literature research.

A considerable number of articles on the order of 300 were found through the selection of source research. A relevant and rigorous study of cross-referenced and redundant keyword combinations was used to refine this first bundle. A total of 180 articles were chosen, which were then subjected to a more in-depth study.

The filtered papers were carefully studied and assessed to ensure that they were relevant to the subject of our research. The last group consisted of 100 articles.

3   An Overview of Blockchain Technology

Blockchain is undoubtedly a brilliant invention, which has been introduced in 2008 by the pseudonym Satoshi Nakamoto (Nakamoto). It is a register where data is stored and kept by a decentralized system of computers. The Oxford dictionary is described as "A system in which a record of transactions made in Bitcoin or another cryptocurrency is maintained across several computers that are linked in a peer-to-peer network." This technology, based on trust, redefines our relationship to information and value transfers.

Blockchain architecture is composed of a set of functional components. The major ones are:

- **Nodes**: nodes are an essential component of Blockchain that allows access to data. They act as a communication point and perform multiple tasks such as ensuring the reliability of stored data (2021) in the Blockchain network.
- **Block**: it contains the block header and many transactions. The block header is hashed repeatedly to generate proof of work for mining rewards. Also, the block header serves to locate a specific block on an entire Blockchain.
- **Transactions**: defined as an exchange of assets, processed and managed under some specific rules of the entity service.
- **Smart contract**: is an autonomous agent unfalsifiable and permanent. Smart contracts are logical rules in coded text embedded in a file Blockchain to manage transactions.

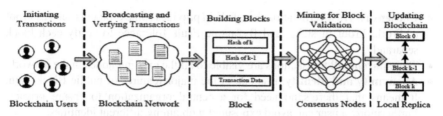

Fig. 2 Blockchain workflow

- **Merkle tree**: also known as a binary hash tree, it is a tree in which each leaf node is marked by the hash value of the block transaction data. Merkle trees are used to effectively and securely encode Blockchain data.

  An overview of the Blockchain workflow is shown in Fig. 2.
  The procedure shown in Fig. 2 works as follows:

- The first step consists of initiating and broadcasting a transaction to the distributed network via a node.
- After that, all Blockchain network nodes verify the transaction (the node that performs the broadcast).
- More than one node can bundle different subgroups of the newly verified transactions in their candidate constituencies and broadcast them over the entire network in the third step.
- In the fourth step, some nodes or the entire network nodes validate blocks by executing functions defined by consensus protocol.
- The last step consists of the attachment of the verified block to Blockchain, and finally, all nodes update their local replica.

The Blockchain can be classified into the following categories depending on its usage (Viriyasitavat and Hoonsopon 2019):

- **Public**: public Blockchains are decentralized and visible by anyone, and they don't have a specific owner, for example (Bitcoin, Ethereum).
- **Private or Permissioned**: they require some rules and privileges to manage and control who can read and write to the Blockchain.
- **Hybrid or Consortium**: these Blockchains are public only to a specific group. The consensus process is controlled by known, privileged servers using all parties' rules.

It is also important to mention that Blockchain technology is acknowledged with these main characteristics (Yaga et al. 2018):

- **Decentralized**: a key Blockchain feature. It means that Blockchain does not have to depend on location nodes anymore. Data can be recorded, stored, and updated for distribution.
- **Immutable**: every time a record of transactions is added, it's permanent and cannot be changed.

- **Consensus**: Trust verification, Blockchain provides some protocols such as Proof of Work, Proof of stake, and Byzantine Fault Tolerance to verify each block separately.
- **Transparent**: all transactions are visible to all the existing nodes in the network.
- **Anonymity and identity**: it is one of the major features of public Blockchain. It means that there's no need for a central organization to ensure privacy. Furthermore, a user can avoid exposure by obtaining different identities.
- **Tamper-proof ledger**: the usage of cryptography ensures the security of all transactions done in the network. They cannot be modified or altered until all nodes are compromised.

Combining these inherent attributes has made Blockchain a viable solution for Big data issues such as security. As a result, researchers have proposed attractive models based on big data and Blockchain to ensure the security of Big Data-based applications.

The following section presents this noteworthy healthcare, banking, smart cities, IoT, VANETs, and 5G.

4 Blockchain in the Service of Big Data

Researchers use Blockchain for different security aims in the context of Big Data. In this section, we first emphasize the security aims by reviewing recent researches in various domains. Then, we compare these studies by showing their benefits and limits.

4.1 Healthcare Field

Healthcare is a critical and essential domain that has attracted massive attention from researchers. As a result, several studies have proposed and enhanced Blockchain models to cope with several healthcare challenges in the context of Big Data.

Blockchain technology redefines the data modeling and management embedded in many health care systems (Liang et al. 2017b). This is mainly due to its adaptability and differentiation capabilities, security, and sharing of information and medical services unprecedentedly. One of the most crucial usages of Big Data and Blockchain is securing and sharing personal health data obtained from wearable devices such as smartwatches and activity trackers via the creation of a user-centric health data sharing based on decentralized and permissioned Blockchain. This solution offers data security via the deployment of a channel formation scheme. Furthermore, it ensures the enhancement of identity management via the membership service provided by Blockchain (Liang et al. 2017b). It also allows the prevention of any alteration of the patients' health data through applying a proof of integrity and validation that is continuously reachable from a cloud database then linked to the Blockchain network. Another potential of Blockchain that should not be overlooked is its capability to surmount vulnerabilities in the old sharing methods, such as the inappropriate access of data and the integration of malicious programs.

Also, Blockchain is used in pharmacy (Clauson et al. 2018) and helps to control and secure the medication return process. In this regard, the Blockchain has shown its potential toward the product (medicines) tracing by allowing manufacturers, distributors, and pharmacists to submit tracing information in a shared ledger, with automated verification of critical data. Besides, Blockchain has demonstrated a potential toward detection and response and enables public and private actors to report and identify medicines suspected of imitation, illegal, or dangerous. Therefore, the safety of medications increases without disclosing sensitive information.

It is also important to mention that the interest in the Blockchain has drastically increased during the pandemic of COVID-19 (Abd-alrazaq et al. 2021). According to the European Parliamentary Research Service, the Blockchain was considered among the ten powerful technologies which have reduced COVID-19 challenges (Mihalis Kritikos 2020). It is used to track public health information, particularly outbreaks of infectious diseases. Combining both Blockchain and Big Data enables real-time data processing, further preventing the spread of the disease to epidemic levels. It also allows obtaining more accurate reporting and effective responses. However, these effective responses are insufficient without safeguarding the patient's information, which requires a high level of security. Hence, by health information exchange based on the Blockchain, the privacy of the patient is ensured. The aspect of securing each patient's medical data is obtained by deploying an electronic medical record Blockchain that uses on and off-chain storage verification records (Ahir et al. 2020).

4.2 Banking Field

Blockchain is widely used in the banking sector and has many advantages in the Big Data context. This innovation reduces the risk of fraud and prevents some potential scams in all banking environments, such as financial and online markets (Hassani et al. 2018) via monitoring and recording every change of the data within the blocks for every transaction in real-time. Furthermore, blockchain technology is considered a trusted network for the digital banking sector due to its encryption capabilities and public key infrastructure. Additionally, the distributed ledger "Blockchain" offers another aspect that should not be neglected, relying on the difficulty of operating the newly added data by various units in real-time. This difficulty results in enhanced privacy of the massive data transiting.

Blockchain also addresses various issues such as operational risk and administrative costs. Furthermore, integrating Blockchain and Big Data into the financial sector optimizes multiple processes, such as creating new financial services and ensuring the integration of the performance and profitability of actors (microcredit, micro-payment transactions almost free of charge, and so forth. These benefits could be obtained through using a cost-cutting technique, "transaction commitment," that drastically decreases transaction time and storage for small amounts of money (Rezaeibagha and Mu 2019).

Furthermore, Blockchain and Big Data allow enhancing the security of banking transactions highly. They would generate new revenue models, capacity gains, cost reduction of millions, and significant losses across the industry. It is also worth noting that many analysts predict positive growth of the level of trust of customers due to the combination of Blockchain and Big Data. This combination will change customers and banks, insurers, and other financial institutions by identifying suspicious transactions by tracking customer transactions and activities in real-time.

4.3   Smart Applications

In recent years, the notion of "smart city" has emerged as a new paradigm to enhance citizen's daily life, by providing personalized and adapted services. This paradigm is based on personal data that are closely related to our daily basis. However, they are several challenges related to security and storage in smart city systems. For this reason, the idea of securing the data aroused the interest of many researchers.

Many studies have tackled the smart city field by proposing new methods and practices. The major one is the combination of Blockchain and Big Data. This combination provides secure communication between physical devices in a heterogeneous environment through real-time data monitoring. It also provides the quality of service (QoS) by minimizing traffic rate fluctuations and the diversity of new devices (Alam). Furthermore, Blockchains can be used to create a smart city system that enables devices to securely and reliably transfer currency and data between all smart city devices.

Also, the combination of Big Data and Blockchain is widely used in smart transportation, especially in the automotive industry. This combination has revolutionized intelligent transportation systems (ITS) by building safe, dependant and self-sufficient ITS ecosystems to provide services such as remote software-based vehicle operation (Deepa et al. 2021).In the same context of smart transportation security, it is noteworthy that (Wang and Qu 2019) have created a secure critical management architecture based on Blockchain technology to provide network security. This solution is based on using security managers to record vehicle departure data, encapsulate the blocks to transfer keys, and then deploy rekeying to the cars within the secured domain. This framework provides a key management system for critical transfers among security administrators in a heterogeneous vehicle network. Other researchers have proposed an interesting system that ensure the security of vehicles using Blockchain (Hîrțan et al. 2020). This system is based on an offline Blockchain storage system. All confidential data gathered from users is kept and then shared with the help of unique encryption keys applicable to a specific vehicle cluster. The solution comprises two applications, the first one is installed on the client's smartphone, and the second is installed on the server. It has shown the capacity to ensure privacy policies sent to the client's application to offer the precise transit routes.

Furthermore, the combination of Big Data and Blockchain allows managing the security issues such as protocol vulnerabilities, privacy, eavesdropping, and attacks on internet-connected devices and vehicle communication networks (Ismagilova et al. 2020).

4.4 Game Theory

Today, most gaming platforms are hosted on centralized servers, and transactions are frequently conducted on mobile phones or desktop computers that lack adequate security (Jacob 2020). In fact, because of this centralization aspect, servers could be easily hacked. Furthermore, this lack of server security could lead to enormous damages such as data loss or interruption of services. Therefore, many studies have tackled the combination of Blockchain and Big Data in gaming to cover these risks. This combination has brought the potential to change the vulnerability of centralized servers. Hackers will be unable to damage a decentralized Blockchain network since (i) there is no server to destroy, (ii) nodes share the maintenance of distributed databases, (iii) each node has complete information in the database. As a result, players can safely store digital collectibles purchased in Blockchain-based games in their crypto wallets (Aran Davies 2020).

Furthermore, due to the decentralization aspect of Blockchain, trust is established between all nodes; players are unable to modify the data. This means that trust is built between all players in the industry, from developers to actors (Gainsbury and Blaszczynski 2017). Also, the Blockchain permit gamers to make their payments in a secure way using digital crypto coins.

Besides the security aspect, the Blockchain also offers new features to the gaming world, such as digital trade assets between players (Casino et al. 2019). The use of intermediary sites obtains this ability of digital trading. In addition, to provide user-friendly functions and interfaces for running shared data (Robin8 2019).

Another aspect of Blockchain technology resides in solving data instability in peer-to-peer games by providing data authentication and permanent storage via a new proof-of-play consensus methodology underpins this serverless turn-based strategy game (Wu et al. 2020). This solution is based on three key steps: matchmaking, gaming session, and gameplay.

4.5 Internet of Things

The Internet of things (IoT) refers to the billions of physical devices deployed worldwide connected to the Internet (Atzori et al. 2010). These devices collect and share massive data. Thus, they allow conducting data analysis for better and faster decision-making. However, there are many security issues, such as data leaks and vulnerabilities concerning physical equipment; hackers can easily alter that.

For this reason, Blockchain was introduced to ensure privacy and to face challenges related to IoT security. Using Blockchain as the basis of devices reduces the possibility of hacking by decreasing malicious programs and spyware (Banerjee et al. 2018). In this context, the Blockchain was introduced to trace the history of the firmware. If a corrupted firmware is identified, it will be driven to revert to its prior version. Also, the Blockchain was brought to the front line to maintain the Refence Integrity Metric (RIM) of the datasets by storing membership informations such as the address, owner, and sharing policies.

Furthermore, Blockchain technology allows protecting IoT networks against potential entry points of intruders without the need of a central authority. Therefore, Blockchain is considered a leading solution for organizations involved in auditing, tracing a supply chain, or securing connected streetlights in intelligent cities due to the encryption capability of Blockchain, its dispersed storage, and its irrefutable, tamper-proof record (Plummer and Writer 2018).

Besides the security aspect, Blockchain creates a marketplace that allows data providers (IoT sensors) and data consumers (IoT application owners) to trade in real-time via smart contracts and makes value from collected data (Jain 2021).

Another breakthrough of blockchain use in IoT is embodied in a reliable and flexible solution designed for IoT services, specifically the adoption of drones in various IoT scenarios, namely agriculture, delivery, and so forth. In fact, (Liang et al. 2017b) have proposed a solution that consists of using a public Blockchain and the standard cloud server in which the hashed data records collected from drones are anchored to the Blockchain network. After that, a Blockchain receipt for each data record stored in the cloud is generated to reduce the burden of moving drones with limited battery and processing capability while gaining improved data security assurance. In addition, this usage provides trust, data capability integrity audit, data convenience, and scalability.

4.6   Big Data and Blockchain in the Service of VANETs

VANETs are one of the most brilliant inventions of the century. However, they still include many challenges that could be resolved with Blockchain. Such as data management and security in a vehicular environment. For example, the usage of Blockchain inVANETs enables essential functionalities such as ensuring traffic safety and parking space management through promoting decentralized system management (Peng et al. 2020).

Furthermore, Blockchain played a pivotal role in resolving critical message dissemination issues in VANETs via storing the history of vehicle trust level in Blockchain and event messages (Shrestha et al. 2020). Thus, when a car experiences an event, such as an accident, the event message with various parameters is broadcasted to nearby vehicles in the Blockchain network. Also, Blockchain allows the preservation of the circulation of untrustworthy information sent by malicious vehicles. In addition to these advantages, using Blockchain and Big Data's combination as a security feature of the VANET network allows strengthening network security and facilitating data transmission. Lu et al. (2018) have implemented a Blockchain-based anonymous reputation system that builds a trust paradigm for VANETs. This paradigm enables the transparency of certificates and revocations to be efficiently accomplished through proofs of presence and absence based on enhanced Blockchain technology and conditional anonymity via public keys. This level of protection is guaranteed via a trust-based Blockchain design that effectively mitigates multiple network attacks such as (DoS attack and Sybil attack) (Khan et al. 2019). These attacks become almost non-existent since the potential of altering centralized data storage has become remarkably decreased through the application of consortium Blockchain-based data sharing for VANETS (Zhang and Chen 2019).In other words, given the tamper resistance property of Blockchain, the operation of data sharing become more controlled. The deployment of consortium Blockchain resides in sending each vehicle data to the nearest roadside units (RSU). Then, the RSUs function as pre-selected nodes that construct blocks based on data from the connected vehicles. At this stage, the Blockchain that contains the vehicle data is created by obtaining consensus among RSUs. Consequently, the RSUs control the data sharing process via intelligent contract technology provided by the distributed ledger technology.

However, despite the critical role of Blockchain in VANETs, research is still lacking, mainly concerning real-time and scalability. Moreover, the mobility of VANETs creates difficulty in proof of work in the Blockchain because of the continuous dynamicity of Blockchain nodes (Kim 2019).

4.7 The Fifth Generation Based Applications (5G)

The novel features of 5G allow supporting new business models and services, including mobile operators, businesses, call providers, government executives, and infrastructure providers. However, they also face many challenges, such as data mobility and network privacy (Nguyen et al. 2019).

For this reason, several studies have proposed various Blockchain models to address 5G challenges in the context of Big Data. Moreover, blockchain characteristics offer promising solutions to 5G's challenges. One of these essential Blockchain features is decentralization, which allows for banishing external authorities' requirements in the 5G ecosystem.

In other words, decentralization of 5G networks enables avoiding single-point failures and ensuring data availability (Nguyen et al. 2019). It is also important to note that Blockchain offers high security for 5G networks due to smart contracts that support the services of 5G, such as preservation of 5G resources against alteration and data authentification (Nguyen et al. 2019).

Furthermore, Blockchain provides immutability to the 5G network performs multiple tasks such as (usage information for billing, resource utilization, and trend analysis).

Besides the security and immutability aspects, the Blockchain allows to ensures transparency of the 5G services, enabling service providers and users to fully access, verify, and monitor transaction activities over the network with equal rights (Nguyen et al. 2019). Furthermore, as mentioned above, Blockchain reveals a sequence of security features that remove the need for centralized network infrastructure or third-party authorities and decrease the single point failure. Furthermore, the security of device-to-device (D2D) communication may be achieved by constructing a peer-to-peer network using Blockchain. This converts each D2D device as a Blockchain node to maintain a ledge replica of validating and monitoring transactions to improve the system's transparency and dependability (Nguyen et al. 2019).

All the above studies have designed exciting solutions based on Blockchain and Big data to face many fields. The table below summarizes these solutions by presenting the significant covered challenges and showing their limits and advantages (Table 1).

5   Discussion

Through this paper, we highlighted how security challenges related to Big Data could be resolved with Blockchain. Researchers have proposed Blockchain models and frameworks to face many security challenges in diverse Healthcare, banking, IoT, and so forth.

By reviewing these studies, we have noticed that Blockchain played a pivotal role in securing and preserving data privacy through its decentralization, tamper-proof, and immutability attributes. Moreover, far beyond, the combination of Big Data and Blockchain has assured the security of the data by creating partitioning models based on smart contracts. Accordingly, the data sharing models are generated without the need for a reliable third party.

Besides the security aspect provided by Blockchain, many outstanding advantages are emerging in the implementation of Blockchain claimed by several studies. For instance, transparency and a remarkable reduction of the costs of data storage. This significant minimization of the cost is achieved via the Blockchain storage capacity, which allows storing a large amount of data for long periods instead of using shared data holding platforms, which require many resources. In this process, Blockchain relies on the automatic performance of transactions that are supervised by smart contracts.

Moreover, due to the transparency and decentralization aspects of Blockchain technology, users can look through the record of all transactions.

Despite the advances brought by the usage of both Blockchain and Big Data. A lot of issues still unreliable and require more research. Scalability is an example of these problems; it is known as a critical requirement of Big Data. Companies and businesses

Table 1 Summary of blockchain and big data applications

Field	Reference	The faced challenges	Used methods	Advantages	Limits
Healthcare	Liang et al. (2017b)	The privacy issues and vulnerabilities existing in personal health data and storing system	Implementing an access control scheme by utilizing the Hyperledger Fabric membership service component, and a tree-based data processing	– Preservation of the integrity of health data within each record – Protection and Validation of personal health data	Difficulty in combining both personal health data and medical data
	Wang et al. (2018)	The challenge of sharing medical data	A medical data sharing platform based on permissioned Blockchains Smart contracts - Concurrent Byzantine Fault Tolerance-Consensus mechanism	– The use of encryption technology ensures that users have complete autonomy in their medical data – Smart contract technology is used to enable users to set up different access permissions of medical data	– High cost of centralized data storage – Lack of security and protection of personal privacy

(continued)

Table 1 (continued)

Field	Reference	The faced challenges	Used methods	Advantages	Limits
	Ahir et al. (2020)	Virus detection during a pandemic	– Mathematical modeling – The QR Code system – Artificial intelligence	– Deep learning to detect the symptoms of the virus – AI signals robots to keep social distancing – Blockchain technology to maintain patient records – Big Data to detect the spread of the virus	The high cost of the material
	Clauson et al. (2018)	The challenge of managing supply chain complexity in the healthcare field	Recognizing stakeholders engaged in finding solutions using Blockchain for the health supply chain	– Preserving integrity of the health supply chain – Medical devices become secured – Increasing the function of the Internet of health things(IoHT)	– Proof of concept – Requirement of a pilot phase

(continued)

Table 1 (continued)

Field	Reference	The faced challenges	Used methods	Advantages	Limits
Banking	Hassani et al. (2018)	Expanding the research and development of Blockchain in the banking field	– Blockchain-based KYC (know your customer) process – Smart contracts – Filtering technique and signal extraction	– The use of Blockchain prevents the piracy of historical information of banking data	– The high cost of developing a Blockchain-enabled system – Problem of currency stability
	Bandara et al. (2018)	The challenge of managing high transactions in the current public Blockchains	– Each node in the Blockchain runs its own Cassandra node – All services are built as docker containers and deployed via Kubernetes	– High scalability, high availability, and full-text search features It makes Big Data more secure, structured and allows data analytics to be more easily performed on big data – Support to store large data payloads with Cassandra	The challenge to introduce the designed application for an online 3D printing marketplace
Smart field	Biswas and Muthukkumarasamy (2016)	Overcoming security challenges in smart cities	They are integrating blockchain technology with smart devices	– Improved reliability – Better fault tolerance capability – Faster and efficient operation – Scalability	Interoperability issue

(continued)

Table 1 (continued)

Field	Reference	The faced challenges	Used methods	Advantages	Limits
	Liang et al. (2017a)	Securing drone communications during data collection and transmission	Compiling hash data records collected from drones to Blockchain network then creating a Blockchain receipt Containing each data recorded is stored in the cloud	– Enhanced security of the data – Provisioning data integrity and cloud auditing	Nodes don't require any permission to participate
	Xu et al. (2018)	The lack of cooperation between edge devices to share voluminous data	– Consensus mechanism none as Proof-of-Collaboration (PoC) – Filter algorithm for transaction offloading (FTF)	– Reducing storage resources occupied by the Blockchain	Framework in a green and efficient manner is still an open issue
	Ismagilova et al. (2020)	Security in smart cities	– Fog computing characteristics – Protocol of privacy-preserving authentication (PPA) – Fully Privacy-Preserving and Revocable Identity-BasedBroadcastEncryption (FPPRIB) – Piracy Zones	– Assuring data privacy – The content of the data is not revealed – The increased privacy will encourage the adoption of smart cities	Smart cities require the employment of many emergent technologies (IoT, sensors, GPS) remain notable threats related to security
Game theory	School of Computer Science and Electronic Engineering, University of Essex, U.K et Dey (2018)	Securing attacks in Blockchain using game theory	Using intelligent software agents to monitor the activity of stakeholders in the Blockchain networks to detect any sort of attack	Prevention of attacks and risk of alteration	The need for the application layer of the network to make a decision

(continued)

Table 1 (continued)

Field	Reference	The faced challenges	Used methods	Advantages	Limits
	Gainsbury and Blaszczynski (2017)	The necessity of a third party to prevent frauds to ensure the security of the game	– Using cryptocurrencies – Smart contracts – Blockchain-based domain name system	– Players can operate outside of regulatory jurisdictions – Transactions are verified	The issue of classification of Bitcoin as currency, money, or an item
	Wu et al. (2020)	The vulnerability remaining in a single cluster	– Proof-of-Play consensus model – Proof-of-work – Proof of stake – P2P turn-based strategy game	– The joining players have equal roles – Global synchronization – Offers gaming sessions	The difficulty to manage a large number of players
Internet of things	(Alam)	Introducing a new Blockchain architecture with Big data analytics to increase connectivity performance throughout the smart cities	The Integration system for smart objects using cloud and Blockchains on the IoT (IoT nodes, P2P network)	– The physical devices are allowed to communicate securely with other physical devices in heterogeneous environments – Big Data analytics perform an increasingly significant role in strategic planning	– Discovering the IoT nodes is a challenge across all smart devices – The privacy issue – Scalability issues

(continued)

Table 1 (continued)

Field	Reference	The faced challenges	Used methods	Advantages	Limits
	Dorri et al. (2017)	Providing security for IoT devices in a smart Home	BC-based smart home framework	Protection against DDOS and linking attacks	– Low scalability – Low latency and high requirement of resources
	Liang et al. (2017a)	The challenge of ensuring data resilience in cloud-based IoT applications	The Blockchain collects data from drones and commands from control systems – Cloud server – Cloud database	– Ensuring the security of drones – the solution offers the chance to store a large amount of data	– Scalability issue – The cloud operating systems are vulnerable
Vehicular ad hoc networks (Vanets)	Shrestha et al. (2020)	The problem of security in traditional VANETs	Building a local Blockchain for exchanging real-world event messages across vehicles within a country's borders	– The messages of vanets become secured – Vehicles ensure a secured and distributed database	– Scalability issues – Storage and message overhead
	Lu et al. (2018)	Preserving vehicles from attacks	– Using a privacy-preserving trust model forVANETs known as blockchain-based anonymous reputation system (BARS) – Public keys	– Prevention of vehicles from fraudulent messages – The solution provides an efficient and robust trust model for Vanets	Scalability (the solution can't support a large number of vehicles)

(continued)

Table 1 (continued)

Field	Reference	The faced challenges	Used methods	Advantages	Limits
	Tan and Chung (2020)	The problem of data interference and the lack of security caused by the large group of vehicles in heterogeneous VANETS	– Securing authentification – Key management – Edge computing infrastructure – Using Consortium Blockchain	– Real-time arrangement – Security and resistance against attacks	Cost of computation and communication
	Iqbal et al. (2020)	The traditional cloud data centers are inadequate for vehicles. A tremendous amount of bandwidth is consumed, which negatively influences delay-sensitive applications	– Edge computing – Fog computing – Using distributed ledger-based decentralized trust management scheme to solve the offloading issue in VANETs – Consensus-based on techniques such as Byzantine fault tolerance and Raft	– Blockchain-based social reputation framework at roadside unites permits the decision model to choose from a pool of trusted vehicles for any arriving assignments. As a result, the level of privacy expands	Overloaded fog resources
The fifth generation (5G)	Chaer et al. (2019)	How can Blockchain offers opportunities for the fifth generation(5G)	– Smart contract with the Service Level Agreements (SLA) – Home server subscriber (HSS) – Dynamic Spectrum Sharing Decentralized application (DApp)-DLT	– Network slicing – 5G infrastructure sharing – International roaming – 5G infrastructure crowdsourcing	– Scalability issues – Interoperability issues – Smart contracts – Data privacy complexity – Standardization and Regulations – Transaction and Cloud Infrastructure Costs

(continued)

Table 1 (continued)

Field	Reference	The faced challenges	Used methods	Advantages	Limits
	Nguyen et al. (2019)	Challenges in 5G network (risk of data interoperability)	Using the Blockchain Framework to store immutable ledgers	– Customized and advanced user-centric value – D2D communication – Software-defined networking(SDN)	– Scalability – latency problem

cannot efficiently handle and process the exploding volume of data without using data scaling techniques. Moreover, the lack of standardization is another drawback of combining Blockchain and Big Data. In other words, standardization denotes a minimal degree of interoperability, such as (different Blockchain networks and different consensus models).

The interoperability problem leads to increased resources used in transactions generated by users in various blockchain networks. Also, without standardization, sharing data and value between participants in divergent Blockchain networks becomes complex or almost impossible. Furthermore, another limit of the adoption of Blockchain in the service of Big Data is redundancy. This limit highlights the fundamental issue of decentralization attributes brought by Blockchain. In other words, redundancy means that every node in the Blockchain network must traverse and process each intermediate node independently (Mudrakola 2018) to target the objective node in the network. As a result, the redundancy inherent in Blockchain technology has an impact on its efficiency.

Additionally, Blockchain faces another limit related to signature verification. The employment of this signature verification in Blockchain technology means that each transaction made in the Blockchain network should be verified and signed digitally using private or public keys. Consequently, the computation of this signature verification process becomes highly complex and takes a long time. This limit hampers the alliance of Blockchain with Big Data.

6 Conclusion

The technology of Blockchain is considered a hot topic given its robust features and good advantages. Today, Blockchain's importance has extended far beyond the financial sector and has shown notable improvements, especially toward security concerns. On the other side, Big Data must be handled securely and accurately to avoid any interruption or loss of the data. Therefore, Blockchain has been brought to the front line in Big Data contexts. In this paper, we have presented the alliance of Blockchain with Big Data to cope with Big Data storing and sharing security issues. For this, we have overviewed the multiple advantages and limits of different Blockchain researches that have tackled Big data issues in many sectors ranging from health and technology (IoT, 5G, VANETs, Smart field) to entertainment concerns.

Despite the advancement brought by Blockchain technology, there are still plenty of challenging issues concerning the use of Blockchain in the context of big data, such as interoperability and scalability. However, distributed ledger technology is still maturing. Moreover, it can be linked to new-age technologies, such as artificial intelligence and IoT, to build platforms and infrastructures that ensure advanced data privacy and security.

As future work, we propose a Blockchain model that copes with unsolved Big data security issues such as scalability and standardization.

References

A.A. Abd-alrazaq, M. Alajlani, D. Alhuwail, A. Erbad, A. Giannicchi, Z. Shah, M. Hamdi, M. Househ, Blockchain technologies to mitigate COVID-19 challenges: a scoping review. Comput. Methods Programs Biomed. Updat. **1** (2021). https://doi.org/10.1016/j.cmpbup.2020.100001

S. Ahir, D. Telavane, R. Thomas, The impact of artificial intelligence, blockchain, big data and evolving technologies in coronavirus disease—2019 (COVID-19) curtailment, in *2020 International Conference on Smart Electronics and Communication (ICOSEC)*. (IEEE, Trichy, India, 2020), pp. 113–120

T. Alam, Blockchain-based big data analytics approach for smart cities. Kansai University **62**(9), 17

A. Davies, How Blockchain Could Redefine the Gaming Industry, in DevTeam. Space (2020). https://www.devteam.space/blog/how-blockchain-could-redefine-the-gaming-industry/. Accessed 18 Mar 2021

L. Atzori, A. Iera, G. Morabito, The internet of things: a survey. Comput. Netw. **54**(15), 2787–2805 (2010). https://doi.org/10.1016/j.comnet.2010.05.010

E. Bandara, W.K. Ng, K. De Zoysa, N. Fernando, S. Tharaka, P. Maurakirinathan, N. Jayasuriya, Mystiko—blockchain meets big data, in *2018 IEEE International Conference on Big Data (Big Data)*. (IEEE, Seattle, WA, USA, 2018), pp. 3024–3032

M. Banerjee, J. Lee, K.-K.R. Choo, A blockchain future for Internet of things security: a position paper. Digit. Commun. Netw. **4**(3), 149–160 (2018). https://doi.org/10.1016/j.dcan.2017.10.006

K. Biswas, V. Muthukkumarasamy, Securing smart cities using blockchain technology, in *2016 IEEE 18th International Conference on High Performance Computing and Communications; IEEE 14th International Conference on Smart City; IEEE 2nd International Conference on Data Science and Systems (HPCC/SmartCity/DSS)*. (IEEE, Sydney, Australia, 2016), pp. 1392–1393

F. Casino, T.K. Dasaklis, C. Patsakis, A systematic literature review of blockchain-based applications: current status, classification and open issues. Telemat. Inform. **36**, 55–81 (2019). https://doi.org/10.1016/j.tele.2018.11.006

A. Chaer, K. Salah, C. Lima, P.P. Ray, T. Sheltami, Blockchain for 5G: Opportunities and Challenges, in *2019 IEEE Globecom Workshops (GC Wkshps)*. (IEEE, Waikoloa, HI, USA, 2019), pp. 1–6

K.A. Clauson, E.A. Breeden, C. Davidson, T.K. Mackey, Leveraging blockchain technology to enhance supply chain management in healthcare: an exploration of challenges and opportunities in the health supply chain. BHTY (2018). https://doi.org/10.30953/bhty.v1.20

N. Deepa, Q.-V. Pham, D.C. Nguyen, S. Bhattacharya, B. Prabadevi, T.R. Gadekallu, P.K.R. Maddikunta, F. Fang, P.N. Pathirana, A survey on blockchain for big data: approaches, opportunities, and future directions (2021). arXiv:200900858 [cs]

A. Dorri, S.S. Kanhere, R. Jurdak, P. Gauravaram, Blockchain for IoT security and privacy: the case study of a smart home, in *2017 IEEE International Conference on Pervasive Computing and Communications Workshops (PerCom Workshops)*. (IEEE, Kona, HI, 2017), pp. 618–623

I. El Alaoui, Y. Gahi, The impact of big data quality on sentiment analysis approaches. Procedia Comput. Sci. **160**, 803–810 (2019). https://doi.org/10.1016/j.procs.2019.11.007

I. El Alaoui, Y. Gahi, R. Messoussi, Big data quality metrics for sentiment analysis approaches, in *Proceedings of the 2019 International Conference on Big Data Engineering* (Association for Computing Machinery, New York, NY, USA, 2019), pp. 36–43

A.Z. Faroukhi, I. El Alaoui, Y. Gahi, A. Amine, Big data monetization throughout big data value chain: a comprehensive review. J. Big Data **7**(1), 3 (2020). https://doi.org/10.1186/s40537-019-0281-5

Y. Gahi, M. Guennoun, H.T. Mouftah, Big data analytics: security and privacy challenges, in *2016 IEEE Symposium on Computers and Communication (ISCC)*. (IEEE, Messina, Italy, 2016), pp. 952–957

S.M. Gainsbury, A. Blaszczynski, How blockchain and cryptocurrency technology could revolutionize online gambling. Gaming Law Rev. **21**(7), 482–492 (2017). https://doi.org/10.1089/glr2.2017.2174

H. Hassani, X. Huang, E. Silva, Banking with blockchain-ed big data. J. Manag. Anal. **5**(4), 256–275 (2018). https://doi.org/10.1080/23270012.2018.1528900

L.-A. Hîrțan, C. Dobre, H. González-Vélez, Blockchain-based reputation for intelligent transportation systems. Sensors **20**(3), 791 (2020). https://doi.org/10.3390/s20030791

H. Hu, Y. Wen, T.-S. Chua, X. Li, Toward scalable systems for big data analytics: a technology tutorial. IEEE Access **2**, 652–687 (2014). https://doi.org/10.1109/ACCESS.2014.2332453

S. Iqbal, A. Malik, A.U. Rahman, A. Waqar, Blockchain-based reputation management for task offloading in micro-level vehicular fog network. IEEE Access **8**, 52968–52980 (2020). https://doi.org/10.1109/ACCESS.2020.2979248

E. Ismagilova, L. Hughes, N.P. Rana, Y.K. Dwivedi, Security, privacy and risks within smart cities: literature review and development of a smart city interaction framework. Inf. Syst. Front. (2020). https://doi.org/10.1007/s10796-020-10044-1

N. Jacob, How blockchain is making digital gaming better, in *Blockchain Pulse: IBM Blockchain Blog* (2020). https://www.ibm.com/blogs/blockchain/2020/02/how-blockchain-is-making-digital-gaming-better/. Accessed 22 Mar 2021

S. Jain, Can blockchain accelerate internet of things (IoT) adoption, in *Deloitte Switzerland* (2021). https://www2.deloitte.com/ch/en/pages/innovation/articles/blockchain-accelerate-iot-adoption.html. Accessed 25 Mar 2021

A.S. Khan, K. Balan, Y. Javed, S. Tarmizi, J. Abdullah, Secure trust-based blockchain architecture to prevent attacks in VANET. Sensors **19**(22), 4954 (2019). https://doi.org/10.3390/s19224954

S. Kim, Impacts of mobility on performance of blockchain in VANET. IEEE Access **7**, 68646–68655 (2019). https://doi.org/10.1109/ACCESS.2019.2918411

X. Liang, J. Zhao, S. Shetty, D. Li, Towards data assurance and resilience in IoT using Blockchain, in *MILCOM 2017—2017 IEEE Military Communications Conference (MILCOM)*. (IEEE, Baltimore, MD, 2017a) pp 261–266

X. Liang, J. Zhao, S. Shetty, J. Liu, D. Li, Integrating Blockchain for data sharing and collaboration in mobile healthcare applications, in *2017 IEEE 28th Annual International Symposium on Personal, Indoor, and Mobile Radio Communications (PIMRC)*. (IEEE, Montreal, QC, 2017b), pp. 1–5

Z. Lu, W. Liu, Q. Wang, G. Qu, Z. Liu, A privacy-preserving trust model based on blockchain for VANETs. IEEE Access **6**, 45655–45664 (2018). https://doi.org/10.1109/ACCESS.2018.2864189

M. Kritikos, Ten technologies to fight coronavirus **28** (2020)

S. Mudrakola, Blockchain limitations: this revolutionary technology isn't perfect—and here's why, in *TechGenix* (2018). https://techgenix.com/blockchain-limitations/. Accessed 28 Mar 2021

S. Nakamoto, Bitcoin: a peer-to-peer electronic cash system **9**

D.C. Nguyen, P.N. Pathirana, M. Ding, A. Seneviratne, Blockchain for 5G and beyond networks: a state of the art survey (2019). arXiv:191205062 [cs, eess, math]

C. Peng, C. Wu, L. Gao, J. Zhang, K.-L. Alvin Yau, Y. Ji, Blockchain for vehicular internet of things: recent advances and open issues. Sensors (Basel) **20**(18) (2020). https://doi.org/10.3390/s20185079

L. Plummer, T. Writer, Blockchain, IoT & security, in *Intel* (2018). Accessed 21 Mar 2021

F. Rezaeibagha, Y. Mu, Efficient micropayment of cryptocurrency from blockchains. Comput. J. **62**(4), 507–517 (2019). https://doi.org/10.1093/comjnl/bxy105

Robin8, Why is trading digital assets via Blockchain the best solution? in *Medium* (2019). https://medium.com/@robin8/why-is-trading-digital-assets-via-blockchain-the-best-solution-6897db75faff. Accessed 22 Mar 2021

School of Computer Science and Electronic Engineering, University of Essex, U.K., Dey S, A proof of work: securing majority-attack in blockchain using machine learning and algorithmic game theory. IJWMT **8**(5), 1–9 (2018). https://doi.org/10.5815/ijwmt.2018.05.01

R. Shrestha, R. Bajracharya, A.P. Shrestha, S.Y. Nam, A new type of blockchain for secure message exchange in VANET. Digit. Commun. Netw. **6**(2), 177–186 (2020). https://doi.org/10.1016/j.dcan.2019.04.003

H. Tan, I. Chung, Secure authentication and key management with blockchain in VANETs. IEEE Access **8**, 2482–2498 (2020). https://doi.org/10.1109/ACCESS.2019.2962387

D. Tranfield, D. Denyer, P. Smart, Towards a methodology for developing evidence-informed management knowledge by means of systematic review. Br. J. Manag. **14**(3), 207–222 (2003). https://doi.org/10.1111/1467-8551.00375

W. Viriyasitavat, D. Hoonsopon, Blockchain characteristics and consensus in modern business processes. J. Ind. Inf. Integr. **13**, 32–39 (2019). https://doi.org/10.1016/j.jii.2018.07.004

R. Wang, W.-T. Tsai, J. He, C. Liu, Q. Li, E. Deng, A medical data sharing platform based on permissioned blockchains, in *Proceedings of the 2018 International Conference on Blockchain Technology and Application—ICBTA 2018*. (ACM Press, Xi'an, China, 2018), pp. 12–16

S. Wang, X. Qu, Blockchain applications in shipping, transportation, logistics, and supply chain (2019), pp. 225–231

F. Wu, H.Y. Yuen, H.C.B. Chan, V.C.M. Leung, W. Cai, Infinity battle: a glance at how blockchain techniques serve in a serverless gaming system, in *Proceedings of the 28th ACM International Conference on Multimedia*. (ACM, Seattle WA USA, 2020), pp. 4559–4561

C. Xu et al. Making big data open in edges: a resource-efficient blockchain-based approach. IEEE Trans. Parallel Distrib. Syst. **30**(4), 870–882 (2018)

D. Yaga, P. Mell, N. Roby, K. Scarfone, Blockchain technology overview. (National Institute of Standards and Technology, Gaithersburg, MD, 2018)

X. Zhang, X. Chen, Data security sharing and storage based on a consortium blockchain in a vehicular ad-hoc network. IEEE Access **7**, 58241–58254 (2019). https://doi.org/10.1109/ACCESS.2018.2890736

Overview of Blockchain-Based Privacy Preserving Machine Learning for IoMT

Rakib Ul Haque and **A. S. M. Touhidul Hasan**

Abstract The rapid growth of the internet of medical things (IoMT) in practical life causes privacy concerns for data owners at the time of data analysis. There are many methods available to examine the IoMT data in a privacy-preserving manner but a study is needed to understand the best practical method for real life. So, this study will not propose a novel method but will explore the state-of-the-art method to address the most realistic method for privacy-preserving data analysis. This considered two widely used technologies are Cryptography and Differential privacy. Performances are considered in terms of accuracy and time consumptions for the latest published research. Based on the analysis it is clear that both methods have flaws and can be used depending on the constraints of domains. Moreover, Differential privacy is more practical as it has almost similar time consumptions just like standard unsecured methods.

1 Introduction

IoMT Joyia et al. (2017) is rapidly expanding in the healthcare sector, and enormous volumes of data continuously gather from these devices. Diversified Machine Learning (ML) Goodfellow et al. (2016) algorithms are used to train these vast amounts of data to build the prediction model. The data analyst conducts ML training on the data collected from various data owners. However, data owners often disagree to share

R. U. Haque
School of Computer Science & Technology, University of Chinese Academy of Sciences, Shijingshan District 100049, Beijing, China
e-mail: rakibulhaqueraj@mails.ucas.ac.cn

Institute of Automation Research and Engineering, Dhaka 1205, Bangladesh

A. S. M. T. Hasan (✉)
Department of Computer Science and Engineering, University of Asia Pacific, Dhaka 1205, Bangladesh
e-mail: touhid@uap-bd.edu

Institute of Automation Research and Engineering (IARE), Dhaka 1205, Bangladesh

© The Author(s), under exclusive license to Springer Nature Switzerland AG 2022
Y. Baddi et al. (eds.), *Big Data Intelligence for Smart Applications*,
Studies in Computational Intelligence 994,
https://doi.org/10.1007/978-3-030-87954-9_12

their sensitive IoMTs data for training, as there are privacy issues. This study focuses on the scenario where data analysts and data owners share their data in a Blockchain platform.

ML method, where participants are unaware of each other sensitive information in a Blockchain platform is known as Blockchain-based Privacy-Preserving Machine Learning (BPPML) Al-Rubaie and Chang (2019). BPPML makes sure that sensitive data and ML models are private to the owners and results are disclosed to the designated participants. Most of the studies focused on cryptography Haque et al. (2020); Haque and Hasan (2021) and differential privacy Abadi et al. (2016). Blockchain is consolidated in order to record the transaction's information at the time of data sharing and develop a private network. Homomorphic cryptography is often used in BPPML in terms of the cryptographic method, where arithmetic operations are done on encrypted data rather than the original text. On the other hand, in Differential privacy-based BPPML methods, original data are added with noise before the ML training. These are the two most popular methods often used for privacy-preserving ML training. Each of these methods has advantages and limitations depending on the use case.

The main focus of this article is to describe the best feasible method in terms of accuracy and time complexity for BPPML in between cryptography and differential privacy methods. Again, both BPPML techniques will be described to illustrate the process of protecting data privacy. This study intends to help professionals in data privacy, Blockchain, and machine learning fields by showing a comparative study between Blockchain-based data privacy-preserving methods (i.e., cryptography and differential privacy) on ML for IoMT data.

The rest of the paper is articulated as follows. Section 2 and 3 represent Related works and Preliminaries. Section 4 represents System overview and Model construction and Sect. 5 represents Experimental setup and analysis of results. Finally, in Sect. 6 we conclude our papers.

2 Related Works

Numerous research has focused on privacy issues. i.e., cryptographic Bost et al. (2015), differential privacy Abadi et al. (2016), and privacy-preserving data publishing Hasan et al. (2018), Hasan et al. (2018, 2017, 2016). Whereas cryptographic and differential privacy is time-consuming and provides less data utility, respectively. Again, ML training is not a concern of privacy-preserving data publishing. In addition, these methods cannot keep data owner and data analyst information at the time of data sharing. Some recent solutions for protecting the data privacy of data owners at the time of training a ML methods are secure k-nearest neighbor (k-nn) Haque et al. (2020), secure support vector machine (SVM) Shen et al. (2019), and secure linear regression (LR) Haque and Hasan (2021). All these methods consolidate Blockchain to keep the transaction information of data owner and data analyst at the time of ML training with encrypted IoMT data and achieve the best potential accuracy like regular SVM, k-nn, and LR. However, secure SVM, secure k-nn, and secure LR need several comparisons and calculations that result in higher space and time complexity for analyzing the health data.

Research on cryptographic ML methods mostly focused on any specific domain Sakuma and Kobayashi (2010), some do not consider Blockchain Saranya and Manikandan (2013) and only a few sets of research utilizing Blockchain Zhu and Li (2020). None of them are versatile as they are based on a specific setup and address all the privacy concerns related to data integrity, authenticity, and privacy. Some of the previous research discuss the flaws of PPML Al-Rubaie and Chang (2019), which do not cover BPPML and also lack experimental analysis. This study examines both cryptography and Differential privacy-based BPPML methods with an experimental analysis and also suggest the best possible use for them. In this study, Blockchain and a Cryptosystem (partially Homomorphic: Paillier) are applied to address the above concerns when employing ML classifiers with IoMT data of the owners. A distributed ledger is used to record the encrypted (Paillier) IoMT data of each data owner. Encrypted data can be obtained from data owners by Data analysts for employing the ML classifier. In the Privacy-Preserving DP-based ML method, gaussian and laplacian noises are employed in this study. This study utilized two security definitions to achieve security goals: modular sequential composition and secure two-party computation Goldreich (2021); Canetti (2000).

3   Preliminaries

All background technologies are discussed in this section.

3.1   Machine Learning

Machine Learning is a computer science domain where the system can learn without any isolated programming Al-Rubaie and Chang (2019). ML methods' prime goal is to learn the process of achieving particular responsibilities by stereotyping from data. Some of the tasks are prediction analysis and pattern recognition in data. ML algorithms can be categorized depending on the learning style, such as supervised, semisupervised, unsupervised learning. ML applications can be achieved in any of the three mentioned above and do not rigidly belong to any methods, such as recommender systems and dimensionality reduction. Figure 1 illustrates the three types of ML methods.

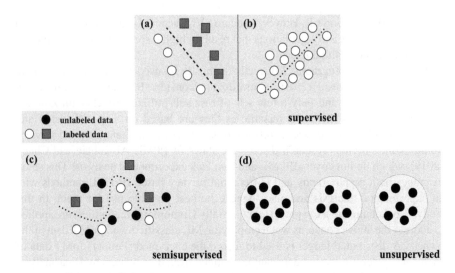

Fig. 1 Machine Learning Tasks Overview. **a** Classification: determining a dividing line within the two classes, "rectangle" and "circles", **b** Regression: implementing a predictive design, **c** Classification: finding a separating dashed line among the three classes, "circles (black and white)", "rectangle" classes, where black circles are unsupervised data and white circles represent supervised data, **d** Clustering: classifying sample sets into several clusters

3.1.1 Supervised Learning

It uses labeled datasets. An output value (i.e., continuous value for regression and class label for classification) is related to the individual feature vector. In the training phase, that identified dataset is utilized in order to develop a model. That model can forecast the label of the brand-new feature vectors at the testing phase. Classification categorizes feature vectors into more than one class, and identifying the class of the new sample is the main aim of the ML classification algorithm. For example, Fig. 1a shows that a hyperplane is separating two classes. Various classification methods can be utilized for categorization applications such as Logistic Regression, Neural Networks, and Support Vector Machines, etc. If the label is dependent or a continuous value in discrete value, then this situation is known as regression. The feature vectors are known as independent variables. The main goal of the regression model is to fit a forecasting model (i.e., line) with a labeled dataset in a way such that the gaps within the line and the observed data points are lessened, which is shown in Fig. 1b.

3.1.2 Semisupervised Learning

Most real-world data are unlabeled as data labeling needs special devices and human experts, making it costly. Scientific observation says that a tiny volume of labeled data can significantly increase the training mechanism's performance. Figure 1c shows the process of training a model with the label and unlabeled data set.

3.1.3 Unsupervised Learning

It does not require any labeled dataset and each feature vector has no response variable associated with it. This type of learning aims to find an internal structure from the dataset. Figure 1d shows the working process of unsupervised learning. Clustering (K-means) is one of the most well-known unsupervised methods. It categorizes feature vectors into various clusters. All similar samples should belong to a cluster, and any distance measuring method (i.e., euclidian, manhattan, etc.) can be utilized for the similarity test.

3.2 Homomorphic Cryptosystem

Three methods combinedly develop Cryptosystems: key generation ($KeyGen$), encryption of data (Enc), decryption of data (Dec). ($SK; PK$) are representing (private key; public key) used in public-key cryptosystems. These key pairs are used for encryption and decryption. A cryptosystem can be Homomorphic if and only if features of it map with ciphertext calculation to the respective plaintext being unaware about the decryption key. It is defined in Definition 1.

Definition 1 (HomomorphicAl-Rubaie and Chang (2019)) Homomorphic property can be achieved by public-key encryption (Gen, Enc, Dec) only if for all n and all ($PK; SK$) output by Gen (1^n), it is possible to define groups M, C (depending on PK only) such that:

- M is the message space and C's elements are all ciphertexts results, which are encrypted by Enc_{pk}.
- Dec_{sk} ($\sigma(c_1, c_2)$) = $\sigma(m_1, m_2)$ is held for any (m_1, m_2) $\in$ M, any c_1 output by $Enc_{pk}(m_1)$, and any c_2 output by Enc_{pk} (m_2).

3.3 Differential Privacy

Differential Privacy based methods can preserve privacy by combining arbitrary noise to the input datasets or redundancies in a specific algorithm or the algorithm result. DP approaches can be divided into 2 categories (i.e., Global DP and Local DP). Global DP requires a trusted aggregator in order to add noise, and local DP does not require any trusted server because it enables individual input parties to add the noise locally. Differential privacy concept is defined in Dwork (2006). $\epsilon-$

differential privacy can be achieved by a randomized function K if all datasets D_1 and D_2 varying on at most unit component, and all $S \subseteq Range(K)$,

$$\frac{Pr[K(D_1) \in S]}{Pr[K(D_2) \in S]} \leq e^\epsilon$$

DP makes sure any order of reply to inquiries is approximately fairly acceptable to occur, whether a dataset has a particular record involved or not. One of the important properties of DP is composition. It facilitates the scheme and interprets the complicated DP method from the more simplistic building blocks of DP. The composition of K mechanisms sequence in which the i-th mechanism affords ϵ_i-DP is ($\sum_{i=1}^{k} \epsilon_i$) Dwork (2006). It is a common phenomenon that the privacy budget's effectiveness may lessen due to the repeated utilization of the method. However, this privacy loss can be quantified with DP. In order to yield (ϵ, δ)-DP, δ is added to the right-hand side of the previous equation, which will generate a relaxation with a more delicate privacy budget than the pure ϵ-DP. DP can be used in a distributed learning method, where multiparty input is necessary to protect data privacy. Besides, it is protected from postprocessing, Which means that an adversary will never enhance the privacy loss even in the appearance of supplemental knowledge. De-anonymization, which is used for linkage attack, can be neutralized by DP. It is already clear from the previous definition of DP that it is a randomized algorithm. DP can be categorized based on the application of randomness.

- **Input perturbation**: In the input perturbation approach, noise is combined with the raw data. Computation is done on the perturbed data, and the outcomes will be differentially private.
- **Algorithm perturbation**: It is a method wherein an iterative method noise is added in intermediate data.
- **Output perturbation**: Here noise is added to the generated model after running a public training model. The exponential mechanism is quite useful, where added perturbation may lead to misclassification.
- **Objective perturbation**: In order to minimize empirical risk, objective perturbation is used, where noise is added to the objective function at the of training the algorithms.

ϵ- global DP is mostly relaid on a centralized authority, known as database maintainer, as noise is added in the output. This article mainly focused on the area where there is no trusted third party. A robust and privacy-preserving private system can be achieved with Local DP. Generally, data analysts want to train their ML model from different data owners, then one of the best possible privacy-preserving ways is Local DP. In Local DP, individual data owners will add noise with their raw data and send the data analyst's perturb data. Blockchain with local DP is used in different manners. Only the transaction can be recorded at the time of data sharing from data owners to data analysts. In another application, individual participants can have their smart contracts. Smart contracts of data owners will apply local DP, and data analysts' smart contracts will hold the ML model. A data owner can upload their raw data to its own smart contract, where the smart contract will apply Local DP on the data set and send them to the smart contract of the data analyst. For better privacy, participants can use private networks like Hyperladger.

3.4 Blockchain

Blockchain is a ledger of Indelible transactions maintained in a distributed network of peer nodes Aleksieva et al. (2020). These nodes maintain a copy of the ledger by applying transactions in each block validated by a consensus protocol, grouped into blocks where a hash is included that binds each block to the previous block. Basically, the blocks are cryptographically linked with their antecedent block and form a chain of irreversible records that are distributed and shared across the P2P consensus network. If any member creates a new block in the chain, the new block is sent to everyone in the network. Each node then verifies the block that the block is not tampered with. If there is no exertion of illegal influences and everything is ok the new block is added to the chain. A Blockchain-based system can assist as a confirmation clearinghouse for these documents by providing official and specific evidence regarding the authenticity of the documents.

Hyperledger Fabric is a permissioned distributed ledger technology (DLT) platform. This framework is used for implementing a personal Blockchain where only authenticated participants can access the network. It follows some governance rules that ensure the validation of the information. It has high modularity, which enables versatility. Instead of an open unauthorized system that allows anonymous identities to participate in the network, protocols such as "proof of work" are needed to legitimize the transaction and "secure the network" Aleksieva et al. (2020). Members of the HyperLedger Fabric Network are listed through a Trusted Membership Service Provider (MSP). The structure of Hyperledger Fabric allows only authorized users to participate in the network by membership service. The membership service verifies a member by performing some task that includes enrollment, issuance, and authentication of the access token between a user, a client, and an endorser.

4 System Overview & Model Construction

System model, the threat model, and model construction are discussed in this section.

4.1 System Model

The employed systems aim to make sure privacy and protected sharing of data between data analysts and data owners. Data analyst gets encrypted or DP-based

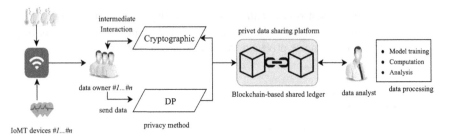

Fig. 2 System model: In this model data analysts want to train its ML model (with Cryptographic and DP based privacy method) over the IoMT dataset of the data owners

IoMT data from each data owner. All the shared data are recorded in a distributed ledger of Blockchain by forming transactions in order to keep authenticity. Data analysts assemble its Ml model by acquiring recorded data from the public ledger of Blockchain. Data analysts erect a protected method utilizing protected building blocks (Secure Comparison and Secure Polynomial operation) for the Cryptographic method and use the noise data for the DP-based method. At the time of employing the secure ML model, it is important to have interactions among data owners and data analysts in order to share results for the Cryptographic method. Figure 2 shows the system model.

- **IoMT Devices**: are accountable for collecting and transferring important IoMT data by wired and wireless networks.
- **Data Owners**: gather every data from the IoMT devices.
- **Data Analyst**: want to train a ML model upon the data-set gathered from multiple Data Owners:.

Security Goals: π, which is a training privacy-preserving protocol for ML training satisfies the following requirements.

- Data analyst cannot learn any sensitive data.
- Each data owner should be unaware of the ML model parameters of data analyst.
- The Sensitive information of each data owner should be private.

4.2   Threat Model

In the employed model, participants do not esteem anyone, but themselves, this type of participants can be formally termed as legitimate but inquisitive adversaries. To be more specific, all participants in the system are legitimate in following the protocol of the model but also strive to know the private data of other participants. The thread model is discussed in this section. Again, the ML training protocol π, which is pre-designed and data analyst is fair in obeying it. He has an interest in the private data of other participants and also tries to gain further information from their data by investigating. On the other hand, there might be an attempt from the data owner side to recognize the model parameters of data analysts from the intermediate results.

- **Ciphertext Pattern Recognition**. data analysts only receive the information (encrypted) registered in Blockchain. Intermediate outputs for example iteration steps can be recorded by the data analyst at the time of training the ML method.
- **Backend Method Recognition**. data analyst is assumed to be well aware of the type of the dataset than the knowledge, which can be distinguished from ciphertexts. Various Data owners and Data analysts can jointly scheme against other participants to stive their private data.

4.3 Data Sharing via Blockchain

This research assigned each data instance to the corresponding feature vectors, which are preprocessed locally. To record encrypted or noisy datasets a transaction is maintained in the Blockchain. The employed transaction has two domains: inputs and outputs. Input field has:

- Sender address
- Encrypted or DP-based data space
- Identity of source device (IoMT)

  Output terminal (corresponding) has:

- Receiver address
- Encrypted or DP-based data space
- Identity of source device (IoMT)

This study employs the employed model in the Hyperledger Fabric platform. It is known as a permissioned Blockchain platform. Sender and receiver's addresses are hash values. Paillier determines the encrypted data. The length of each entry is 128 bytes. The segment length is 4 bytes for the type of IoMT device. The sender node assembling the transaction and broadcast it in the Peer to Peer (P2P) Angrish et al. (2018) system of the Blockchain network. Operation's correctness is validated by the Miner nodes and transactions are packaged in a new block by a particular miner node. Each block may record various transactions. Classical consensus protocols are used to add blocks to the chain. i.e. the Proof of Work (PoW) mechanism Vukolić (2015) or Byzantine Fault Tolerance (BFT) Stifter et al. (2019).

4.4 Model Construction

This section presents the construction details of the employed k-nn model. Note that we are just analysing proposing any novel method but analyzing the existing Cryptographic and Differential Privacy based methods in order to identify the best suitable algorithm for real-life. The goal these methods are to protect the privacy of distinct data owner and data analyst while training an ML model over multiple private datasets from various data owner.

Protected polynomial addition and subtraction are employed in order to train the ML method in a cryptographic manner. Homomorphic property of Pailler is derived as: $[[m_1 + m_2]] = [[m_1]] \times [[m_2]] \bmod N^2$ and subtraction is derived as: $[[m_1 - m_2]] = [[m_1]] \times [[m_2]]^{-1} \bmod N^2$.[1] Secure polynomial multiplication of Paillier is represented in Eq. 1.

$$[[am_1 + bm_2]] = [[m_1^a]] \times [[m_2^b]] \bmod N^2 \tag{1}$$

Statistical indistinguishability of Paillier makes polynomial addition, subtraction, and multiplication secure. Again, for comparing among encrypted numbers secure comparison is employed. Consider, A and B want to compare their values $[[m_1]]$ and $[[m_2]]$ privately by maintaining following protocol π, where their original values are private m_1 and m_2.

The most commonly used ϵ-LDP formula is explained below:

$$x' = x + Lap(\frac{\delta}{\epsilon})$$

where, x' noise data, x plain text, δ predefined sensitivity, ϵ privacy budget, and function Laplace $Lap(\ )$.

5 Experimental Setup & Result Analysis

Tools used for this study are: MacBook Pro (DDR3, 1600 MHz, 4 GB), 2.5 GHz, processor: Intel Core i5, which acting as P and A concurrently. Secure Polynomeal Operation and Comparison are developed in Google's Collaboratory platform in python 3 language. Breast Cancer Wisconsin data set (BCWD), Diabetes data set (DD), and Heart disease data set (HDD) are the three health datasets are utilized in this study Haque et al. (2020). BCWD represents breast mass and cell nuclei from the images. Malignant and benign are used to label the dataset. On the other hand, HDD and DD datasets have 9 numeric and 13 discrete attributes, respectively. For

---

[1] $[[m]]^{-1}$ represents the modular multiplicative inverse. It can calculate $[[m]] \times [[m]]^{-1} \bmod N^2 = 1$ in Paillier. $[[m]]^{-1}$ can be calculated by $\phi(N)$ function, $[[m]]^{-1} = [[m]]^{\phi(N)-1}$.

Table 1 Statistics of datasets

Datasets	Instances	Attribute	Discrete	Numerical
	Number	Number	Attributes	Attributes
BCWD	699	9	0	9
HDD	303	13	13	0
DD	768	9	0	9

the employed system training and test, sets are divided into 80 and 20% respectively. Table 1 illustrates the statistics of datasets.

Cryptosystems do not allow any floating-point operation so all the operations must be done in whole numbers as a result format conversion from float to integer become compulsory.[2] One of the important aspects of asymmetric cryptosystem's security is the key length.

Three most popular metric, i.e., accuracy (2), precision (3) , recall (4).

$$accuracy = \frac{t_p + t_n}{t_p + t_n + f_p + f_n} \tag{2}$$

$$precision = \frac{t_p}{t_p + f_p} \tag{3}$$

$$recall = \frac{t_p}{t_p + f_n} \tag{4}$$

Here, the positive or relevant classes has represented as t_p. These classes are precisely labeled. The negative or irrelevant class that are labeled correctly are represented as f_p. f_n and t_n represents the number of relevant but mislabeled and the number mislabeled but irrelevant, respectively in the test result. The outcomes are shown in Table 2.

Cryptographic methods provide robust privacy and also maintain the accuracy of the dataset than any other algorithm. However, cryptographic methods' main limitations are that they are costly in terms of space and time complexity. These methods work on integer formate, and in order to maintain that, lots of preprocessing are required for formate conversion. Rakib Haque et al. (2020) applied BPPML (k-nn) based on cryptographic method on three data sets containing less than 500 records and found that it took nearly one hour for each of the datasets. As a result, in the case of a larger dataset, these methods become impractical. Table 2 shows the performance comparison based on accuracy and time consumption between standard k-nn and secure k-nn (i.e., Cryptographic and LDP). On the other hand, LDP-based methods for BPPML can provide robust privacy based on privacy budget. The main

---

[2] According to the global standard IEEE 754 representation of a floating-point binary number is D is $D = (-1)^s \times M \times 2^E$, where the sign bit is s, a significant number is M and exponential bit E.

Table 2 Summary of performance for k-nn ($t = 8$). t stands for the threshold value of k-nn

				Privacy budget of LDP		
Measures	Dataset	Standard	Cryptographic	$\epsilon = 0$	$\epsilon = 1$	$\epsilon = 3$
Accuracy	BCWD	96.96%	97.80%	93.55%	94.84%	96.54%
	HDD	83.50%	82.33%	77.90%	79.50%	82.95%
	DD	79.00%	78.00%	72.34%	76.12%	78.34%
Precision	BCWD	96.54%	96.26%	87.84%	90.64%	95.94%
	HDD	83.85%	82.30%	76.35%	79.15%	82.55%
	DD	77.00%	76.00%	68.21%	72.54%	76.56%
Recall	BCWD	96.85%	96.67%	89.33%	92.64%	95.85%
	HDD	83.85%	82.66%	76.33%	79.55%	82.32%
	DD	75.90%	75.10%	67.49%	71.75%	74.59%
Time	BCWD	0.7 s	3357.2 s	2.1 s		
	HDD	0.4 s	2534 s	1.2 s		
	DD	0.2 s	3709 s	0.6 s		

advantage of these kinds of methods is that they are not as expensive as cryptographic methods in space and time complexity. A small amount of time is required to apply LDP in the dataset, and the time consumption for model training is the same as the original dataset. The main limitation is accuracy. LDP-based dataset never provides the same accuracy as the original dataset because of the privacy budget. Less accuracy confirms higher privacy, and higher privacy provides lesser accuracy. Moreover, DP is more practical than cryptographic methods. Professional institutions like Apple, Google, etc., already employed their own LDP methods.

6 Conclusion

Though the above-mentioned methods are available to train an ML model privately, still non-private Ml algorithms are widely utilized. Privacy laws are not utilized properly in the real world. Privacy laws should force companies to announce their activities related to user data and also provide users the right to keep their sensitive data private. Existing BPPML methods employed by researchers are not utilized in the real-world for some limitations. This study mainly focused on two important BPPML methods used by the researchers and some other methods like zero-knowledge proof, etc. The uncovered methods are up growing but not as versatile as the covered method. The main concerns are lack of public awareness about their privacy and also stick policies from governments.

Acknowledgements Authors thanks the school of computer science and technology of the University of Chinese Academy of Science, Beijing, China, and the Department of Computer Science and Engineering of University of Asia Pacific, Dhaka, Bangladesh for their support towards this study.

References

G.J. Joyia, R.M. Liaqat, A. Farooq, S. Rehman, Internet of Medical Things (IOMT): applications, benefits and future challenges in healthcare domain. J. Commun. **12**(4), 240–247 (2017)

I. Goodfellow, Y. Bengio, A. Courville, Machine learning basics. Deep Learn. **1**, 98–164 (2016)

M. Al-Rubaie, J.M. Chang, Privacy-preserving machine learning: threats and solutions. IEEE Secur. Priv. **17**(2), 49–58 (2019)

R.U. Haque, A.S.M.T. Hasan, Q. Jiang, Q. Qu, Privacy-Preserving K-nearest neighbors training over blockchain-based encrypted health data. Electronics **9**, 2096 (2020). https://doi.org/10.3390/electronics9122096

R.U. Haque, A.T. Hasan, Privacy-Preserving multivariant regression analysis over blockchain-based encrypted IoMT data. Artif. Intell. Blockchain Future Cybersecur. Appl. **45**

M. Abadi, A. Chu, I. Goodfellow, H.B. McMahan, I. Mironov, K. Talwar, L. Zhang, Deep learning with differential privacy, in *Proceedings of the 2016 ACM SIGSAC Conference on Computer and Communications Security* (ACM, New York, NY, USA, 2016), pp. 308–318

R. Bost, R.A. Popa, S. Tu, S. Goldwasser, Machine learning classification over encrypted data, in NDSS, vol. 4324, p. 4325 (2015)

A.S.M. Hasan, Q. Qu, C. Li, L. Chen, Q. Jiang, An effective privacy architecture to preserve user trajectories in reward-based LBS applications. ISPRS Int. J. Geo-Inf. **7**(2), 53 (2018)

A.S.M. Hasan, Q. Jiang, H. Chen, S. Wang, A new approach to privacy-preserving multiple independent data publishing. Appl. Sci. **8**(5), 783 (2018)

A.S.M. Hasan, Q. Jiang, C. Li, An effective grouping method for privacy-preserving bike sharing data publishing. Future Int. **9**(4), 65 (2017)

A.T. Hasan, Q. Jiang, J. Luo, C. Li, L. Chen, An effective value swapping method for privacy preserving data publishing. Secur. Commun. Netw. **9**(16), 3219–3228 (2016)

M. Shen, X. Tang, L. Zhu, X. Du, M. Guizani, Privacy-preserving support vector machine training over Blockchain-based encrypted IoT data in smart cities. IEEE Int. Things J. **6**(5), 7702–7712 (2019)

J. Sakuma, S. Kobayashi, Large-scale $k-$means clustering with user-centric privacy-preservation. Knowl. Inf. Syst. **25**(2), 253–279 (2010)

C. Saranya, G. Manikandan, A study on normalization techniques for privacy preserving data mining. Int. J. Eng. Technol. (IJET) **5**(3), 2701–2704 (2013)

Y. Zhu, X. Li, Privacy-preserving $k-$means clustering with local synchronization in peer-to-peer networks. Peer-to-Peer Netw. Appl. **13**(6), 2272–2284 (2020)

O. Goldreich, *Foundations of Cryptography: Volume 2, Basic Applications* (Cambridge University Press, 2009)

R. Canetti, Security and composition of multiparty cryptographic protocols. J. Cryptol. **13**(1), 143–202 (2000)

V. Aleksieva, H. Valchanov, A. Huliyan, Implementation of smart-contract, based on hyperledger fabric Blockchain, in *2020 21st International Symposium on Electrical Apparatus and Technologies (SIELA)*, Bourgas, Bulgaria, pp. 1–4, (2020). https://doi.org/10.1109/SIELA49118.2020.9167043

C. Dwork, Differential privacy, in *Automata, Languages and Programming*. ed. by M. Bugliesi, B. Preneel, V. Sassone, I. Wegener (Springer, Berlin, Heidelberg, 2006), pp. 1–12

A. Angrish, B. Craver, M. Hasan, B. Starly, A case study for Blockchain in manufacturing:"FabRec": a prototype for peer-to-peer network of manufacturing nodes. Procedia Manuf. **26**, 1180–1192 (2018)

M. Vukolić, The quest for scalable Blockchain fabric: Proof-of-work vs. BFT replication, in *International Workshop on Open Problems in Network Security* (pp. 112–125) (Springer, Cham, Oct. 2015)

N. Stifter, A. Judmayer, E. Weippl, Revisiting practical byzantine fault tolerance through Blockchain technologies, in *Security and Quality in Cyber-Physical Systems Engineering* (Springer, Cham, 2019), pp. 471–495

Big Data Based Smart Blockchain
for Information Retrieval
in Privacy-Preserving Healthcare System

Aitizaz Ali, Muhammad Fermi Pasha, Ong Huey Fang, Rahim Khan, Mohammed Amin Almaiah, and Ahmad K. Al Hwaitat

Abstract In digital healthcare systems, the patients face significant problems identifying an optimal vacant and available slot for their appointment as the number of patient requests directly correlates with the slot availability. Existing smart healthcare systems facilitate the patient to reserve a particular time slot and attain real-time healthcare information. However, most of these systems need sensitive information about patients, i.e., desired destination of the healthcare providers. Moreover, existing systems utilize a centralized system which makes these systems vulnerable to numerous intruders' attacks and security breaches (particularly related to service providers) and results in single-point failure of the entire system. In this paper, Ring Signature for permissioned Block-chain-based Private Information Retrieval scheme is proposed to improve privacy-preserving in the smart healthcare system. Our proposed Scheme initially utilizes an improved multi-transaction mode consortium block-chain constructed by different numbers of requests to the healthcare providers to achieve maximized appointment offers based on availability, transparency, and security. Our proposed Scheme is useful for information retrieval from multiple domains. The simulation results verify that the proposed Scheme provides a predominant performance in ensuring maximized patient privacy with reduced computation and communication overheads.

Keywords Digital healthcare · Consortium blockchain · Ring signature · Authentication · Privacy

A. Ali · M. F. Pasha · O. H. Fang
School of Information Technology, Monash University, Subang Jaya, Malaysia

R. Khan
Department of Computer Science, Abdul Wali Khan University, Mardan, Pakistan
e-mail: rahimkhan@awkum.edu.pk

M. A. Almaiah (✉)
Department of Computer Networks and Communications, College of Computer Sciences and Information Technology, King Faisal University, Al-Ahsa 31982, Al Mubarraz, Saudi Arabia
e-mail: malmaiah@kfu.edu.sa

A. K. Al Hwaitat
Department of Computer Science, University of Jordan, Amman, Jordan

279
Y. Baddi et al. (eds.), *Big Data Intelligence for Smart Applications*,
Studies in Computational Intelligence 994,
https://doi.org/10.1007/978-3-030-87954-9_13

1   Introduction

In the recent past, the rapid growth in digital healthcare in the big cities has introduced the central issue for the patients to determine the vacant appointment space for patient examination (Garson and Adams 2008). This process of identifying an optimal space of booking in smart healthcare systems is considered to contribute to about 30% of the traffic congestion (Griffiths 2005). Moreover, one report presented that nearly millions of security breaches happened to patient health records each day. This security breach is considered to be increased due to an unsecured access control system (Gronwald and Zhao 2018). Hence, the network transaction time, jitter, and traffic congestion need to be prevented by integrating technology that could aid them in determining the suitable vacant slots in the smart healthcare system (Almaiah and Al-Khasawneh 2020). In this context, the smart healthcare system is emerging as a predominant solution for suitably identifying the appointment slot with the new advancements happening in the Internet of Things (IoT) and wireless communications. IoT devices are deployed in every individual hospital in the smart healthcare system, such that ultrasonic sensors are used to identify the availability of the appointment slot (Adil et al. 2020). Thus, it facilitates the healthcare service providers with the status of occupancy associated with appointment spaces (Adil et al. 2020). The healthcare service provider enables the option of verifying whether the appointment space is available for online reservations, such that the patient can determine the availability of the vacant appointment slots. Despite the benefits mentioned above, implementing a smart healthcare system imposes a diversified number of challenges that need to be handled before its wide implementation (Adil et al. 2020).

One of the major issues in the smart healthcare system is the privacy of information associated with the patients (Khan et al. 2020). Most current approaches require the patients to disclose sensitive information to the service providers about their reservation times, destinations, and real identities (Al Hwaitat et al. 2020). This disclosure of information by the patients aids the healthcare service provider to perceive and interpret life patterns and daily activities such as income level, health condition, home address, and work address depending on the data analysis attained with the background information (Adil et al. 2020). The majority of the current smart appointment systems are generally centralized and are identified to suffer from many limitations (Almaiah et al. 2020). First, the current smart appointment systems are vulnerable to data loss and security breaches when sensitive information such as name, phone number, email address, and daily appointment information associated with the patients are stored in the database (Qasem et al. 2021). Second, they are highly and inherently prone to the problem of single-point failure. In addition, they also have the risk of remote hijacking attacks and distributed denial of service (DDoS) attacks that can convert the appointment services unavailable (nd).

In this paper, Ring Signature and Improved Multi-Transaction Mode Consortium Blockchain-based Private Information Retrieval (RS-IMTMCB-PIR) Scheme is proposed for Privacy-Preserving in Smart Appointment System. This RS-IMTMCB-PIR Scheme is proposed for preventing the attacks that threaten appointment service availability. It prevents a single point of failure by incorporating an improved multi-transaction mode consortium blockchain that helps the appointment lot owners in attaining transparent, secure and available appointment offers with maximized reliability. It also inherited an improved private information retrieval approach for protecting the location privacy of patients and secretly retrieving the offers of appointment from the improved multi-transaction mode consortium blockchain. It also utilized Ring signature for anonymous authentication of patients during the online reservation of available appointment slots. The proposed RS-IMTMCB-PIR scheme's simulation experiments are conducted to determine their predominance with respect to the maximized degree of patients' privacy, computation overhead, and communication overheads on par with the compared benchmarked schemes.

2 Related Work

In this section, the Blockchain-based smart appointment systems explained in the literature in recent years are presented with the merits and limitations.

Due to the emergence of crypto technology such as Ethereum and Bitcoin, such research related to blockchain has gained most of the attraction for the researcher. Blockchain has the capability to stores and share data in a decentralized manner, immutable and trusted. It removes intermediate parties, requiring any central entity to check the transactions (Adil et al. 2020). To achieve trust within a network and among peers, blockchain is considered to be a less complex method for sharing PHR. It combines diverse computing powers from several nodes in the network, making it more applicable for high computational power and speed (Adil et al. 2020). Blockchain platform provides numerous applications and processes, including consensus Protocol, Hashing, P2P topology, Immutable Ledger, and mining. In Fig. 2 is clearly shown the working of a blockchain network which consist of multiple nodes and protocols. These protocols are called smart contracts. Yup et al. (nd) used the blockchain as a tool for healthcare intelligence keeping in view users' privacy. Data access control for privacy was proposed by them and devised the healthcare digital gateway. The main issue with this system is that it doesn't support cross-organization and fine-grained access control. Zhang et al. (Halpin and Piekarska 2017) proposed a healthcare framework based on PSN to secure the digital health system. The author has designed two novel techniques for validation and allocation of clinical data within a distributed network. The overall complexity of the said framework is high, and also it has security vulnerabilities to attacks. Xia et al. (2019) proposed a cloud-based framework using blockchain. The author in Ghani et al. (2019) designed a cross-domain access control system prototype to provide efficient security to clinical records. This system is called coarse-grained access control. Coarse-grained access control is lacking in precision. A coarse-grained access control framework affects the performance of access control. It also affects the security of a system. Mobile-based healthcare was proposed by Liang et al. (2019). This is also called a record-sharing framework using BC through an approach based on user-centric security to limit unprivileged users' access and enhance privacy via a channel formation scheme. The issue in this approach is the computational cost due to the complex cryptographic mechanism. In their research work, Jain and Singh (2020) designed a clinical data exchange system based truly on the blockchain and later developed a series of verification mechanisms for improved security and privacy. Li et al. (Ghanbari 2019) suggested a healthcare framework for data privacy and integrity through novel algorithms for memory management that help data management. Fan et al. (2020) devised blockchain-based health information of patients through efficient consensus mechanisms to provide an efficient security and privacy framework. The problem with this approach was a lack of flexibility and a cross-organizational approach. Pandey et al. (2020) provide better security to record and share patient clinical information using an attribute-based encryption scheme. In this approach, they integrated SC to ensure reliability and to provide a facility to monitor the PHR sensitive data. Guo et al. (2020) designed an attribute-based signature (ABS) method for clinical users to manage medical health records truly based on blockchain. In this technique, the objective was to achieve optimum privacy of the delivered model through a P2P technique for efficient privacy. Uddin et al. (2019) proposed a framework to monitor and trace patient history-related data. This system relies on a node controlled by authorized patients in the main module of the system; through this approach, the author got better security of the system through various experiments and variation in data sets. This approach has computational overhead due to its complex mechanism for monitoring. Ghanbari et al. (2019) introduced a distributed attribute-based signature scheme for medical systems founded on peer-to-peer ideas. They also developed P2P-based records sharing protocol that support algorithms. But the main problem is that this model uses a signature system that can't predict collision and social engineering attacks. Choo et al. (2020) highlighted recent research challenges and developed an AC strategy for electronic health records (EHR) through a fine-grained access control system. Pandey et al. (2020) Proposed an EHR system to provide security to clinical records using distributed ledger technology (DLT). Further, the author has justified an improved cross-organization sharing of health records using access control policies. Thakkar et al. (2020) designed a novel approach to measure the framework's efficiency using blockchain. Performance optimization related to caching and configuration authorization policy were achieved through this approach. Sukhwani (2019) evaluated the performance parameters in a novel way for the hyperledger fabric framework. Gorenflo et al. (2020) provided a detailed method for analyzing and evaluating the optimization for the provenance and performance of the blockchain framework and developed a framework by configuring it to reduce input/output. But this approach was considered complex computationally. They achieved enhanced performance by reducing the computation time. Last but not least, Chen et al. (2019) proved a novelty in designing the secure, searchable encryption (SSE) model for electronic healthcare records using blockchain. The algorithm takes the index of healthcare records as an input and then provides a related search using SSE. Comparative analysis of the state-of-the-art blockchain-based approaches for PHR systems is carried out in Table 1 points using blockchain for the clinical environment. The approach support is detecting security issues without affecting scalability. The problem with

Table 1 Response time and the transaction time of the proposed approach

	Number of users	Proposed framework		Medrec		SHealth		ECC-SS	
		Time	PR$_{(i)}$	Time	PR$_{(i)}$	Time	PR$_{(i)}$	Time	PR$_{(i)}$
Size of the Block-10000	2	102	1	154	1	279	1	712	1
	4	104	1.019	156	1.012	281	1.007	714	1.002
	6	105	1.028	157	1.019	283	1.014	715	1.005
	8	107	1.049	159	1.032	284	1.017	718	1.008
	10	108	1.077	160	1.038	286	1.025	719	1.009
	12	110	1.098	161	1.045	287	1.028	721	1.012
	14	112	1.006	162	1.051	288	1.032	723	1.015
	16	114	1.117	164	1.064	290	1.039	724	1.016

Irving et al. is that its security and computational cost are high, and its performance is poor in case of security breaches such as collusion and phishing attack. Yazdinejad et al. proposed a blockchain-based method to facilitate verification to 5G networks. This method was based on software-defined networking (SDN). The main advantage of this technique is that they eliminate the cost for Re-authentication when devices usually exchange in between cells in 5G networks.

A Smart Appointment systems-based on Non-fungible token was proposed for construction of appointment pool that introduces transparent platform for the patients and the healthcare service providers in determining the available space with reduced amount of time (Viale Pereira 2017; Manjula and Rajasekaran 2020; Niranjana-murthy et al. 2019; Kumar and Mallick 2018). This appointment system included non-fungible token for introducing maximized transparency into the complete system (Hassanien et al.2018; Pavithran et al. 2020; Sultan et al. 2019; Karthikeyyan and Velliangiri 2019). It was proposed as a method to derive added advantage of revenue generated from the vacant space with the benefit of smart contracts. However, the overheads of communication and computation were considered to be comparatively compared to the conventional appointment approaches. A Consortium Blockchain-based Smart Appointment and booking was proposed for achieving maximized avail-ability, transparency and security by storing appointment slots inside theblockchain nodes (Alotaibi 2019; Khan and Salah 2018; Wattenhofer 2016; Yu et al. 2019; nd). It was proposed with the techniques of cloaking for hiding the locations of the patients in order to protect their location privacy. The cloaking mechanism provided the blockchain with the potential of validating and returning the available appoint-ment offers that possibly exists the cloaked area. It provided the patients with the option of selecting the optimal appointment offers such that direct reservation can be attained. It was developed for preventing the time incurred during the search of available appointment space performed a blockchain-based systematic review of IoT application and its belonging security problems. Depending on a regulated, assessed, and useful content analysis of the academic literature, they evaluated the extended categorization of blockchain-depended IoT applications via diverse parts like smart health care, industrial, smart home, smart city, electronic business model, ad-hoc vehicle networks, and supply chain management. Further, they demonstrated the security difficulties of discussed IoT applications, basically restrictions of the IoT technology. Given that, their study was not done systematically. Therefore, this limitation has provided the context for this research.

Further, Elliptic Curve Cryptography (ECC)-based smart appointment Systems (ECC-SMS) was proposed for protecting patients privacy based on the cumulative prevention of attacks imposed over confidential information exchange (Hur and Noh 2010). This ECC-SMS facilitated users' privacy based on the adoption of zero knowledge proofs integrated with the implementation of ECC algorithm. The performance of this ECC-SMS was determined to reduce the network overhead and execution based on their investigation attained with real world test-bed prepared with the available hardware platform. An appointment space sharing mechanism using blockchain was proposed with the name SHealth for sharing and publishing the availability of appointment slots by the service providers without disclosing the sensitive information of patients (Jalan 2018). This SHealth was developed with incentives that are provided to the patients depending on the rates and the appointment time with anonymity. It provides data provenance using blockchain in order to handle the privacy and security issues that are most possible during smart appointment of patients. The blockchain is used in this SHealth for enabling transaction between the patient and doctor that share sensitive information with proofs of images captured during appointment and leaving in a method of tamper-proof. The access control mechanism provided by this SHealth prevented the inclusion of any third party that has the possibility of data utilization. Bloom filters-based privacy preservation retrieval Scheme (BF-PPRS) was proposed with the merits of specific application navigation result in order to prevent attacks launched over the information shared by the patients during appointment process (Pflanzner and Kertész2018). This BF-PPRS included the cloud resources in identifying theavailable appointment spaces with maximized anonymity. In specific, Bloom filters were incorporated for supporting and retrieving the navigation results of the interaction of a specific application or entity with the service providers. It facilitated the patients to access the cloud in estimating the available appointment slots with anonymity in accessing the generated query. It is considered to retrieve the encrypted navigation results effectively based on the utilization of certain units and metrics. The results of this BF-PPRS were determined to facilitate potential appointment navigation with maximized probability of retrieval with low communication and computation overhead.

Most of the researchers surveyed and reviewed the basic security problems of the IoT. They categorized these problems depending on the low-level, intermediate-level, and high-level layers of the IoT. They briefly discussed the procedures offered in the literature to leverage the security of IoT at diverse stages. Additionally, a parametric analysis of IoT attacks and the belonging feasible remedies have been supplied. They considered the attack indications and mapped them to feasible solutions suggested in the literature. They also discussed the way blockchain is able to be utilized for solving some relevant IoT security issues. They outlined and identified security in IoT, but the role of blockchain and the importance of its security in the IoT is discussed very little.

3   Proposed Framework

In this section, the complete details about the Improved multi-transaction mode consortium blockchain constructed by appointment slot owners, improved private information retrieval approach utilized for secretly retrieving appointment offers of the improved multi-transaction mode consortium blockchain and Ring signature potential anonymous authentication process during the online reservation of available appointment slots reservation is presented.

3.1   Design of Multi-transaction Mode Consortium Blockchain

The Multi-Transaction Mode Consortium Blockchain (MTMCB) is designed based on the different characteristics derived from the current technologies of Blockchain that correspond to secure encryption, point-to-point encryption, distributed storage, and relative decentralization (Choo et al. 2020). These features of MTMCB help in generalizing the 'transaction of smart healthcare systems that could be achieved through blockchain technology into 'processing and results of patient-doctor space allocation', and reconstructs 'the mechanism of verifying the transaction' through 'block distribution and storage'. Figure 1 presents the view of the transaction and

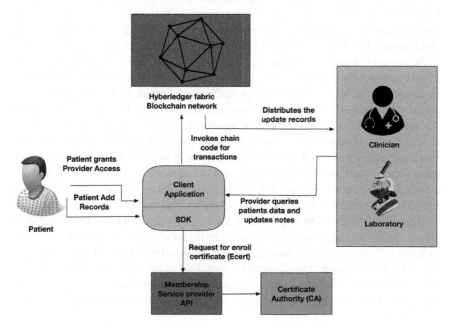

Fig. 1 Proposed framework

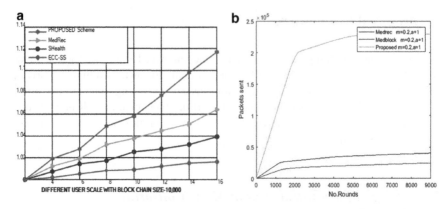

Fig. 2 Part **a** The comparative analysis of our proposed Schemes communication overhead with the benchmark part **b** Comparative analysis of proposed framework with the benchmark models having number of rounds −9000

regulatory node system used in the design of MTMCB used in the proposed scheme. The regulatory node system is included in the server-side of the healthcare service providers of the smart appointment system for achieving the process of initialization, auditing, and transaction processing. However, the operation of initialization is performed only once.

On the other hand, transaction nodes can represent the mobile devices, PCs, and payment machines that could be used to check the availability of appointment space, perform online reservations, and complete payment to the service providers for the appointment space allocated to the Patients (Pandey et al. 2020). In this context, the block name file related to the MTMCB blockchain represents the specific structure of the system file managed by the regulatory node. In the block name file, the Name of the block and the associated addresses of the complete set of patients performing a transaction for online reservation of appointment space with the help of the block are stored simultaneously. Figures 2 and 3 present the user set and block name set logical storage structure. The performance of reading over the disk storage cannot satisfy the requirements of the fast retrieval when the rapid retrieval frequency of the user file and block name files are considered. When the system is initiated, the user and the block name files are read, parsed semantically, and stored in the Redis cache. This Redis is a NoSQL database technology that is developed based on the structure of memory key value. The data can be cached in the memory, and master–slave replication can be significantly realized by constructing the Redis database cluster. This realization allows identifying the frequently accessed data from the memory to minimize the time incurred in the data query. Redis maps the logical association between one or multiple key-value pairs. Rather, the classical relational database maps the logical mode to the table's series. Thus, the selection concerning key structure is vital in the Redis cache. A potential design of Redis cache is determined to enhance the query efficiency

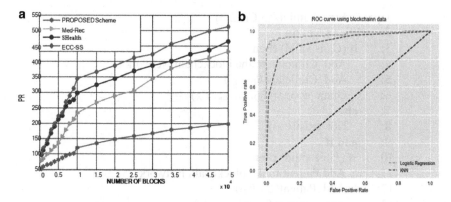

Fig. 3 **a** Proposed RS-IMTMCB-PIR-Response Time and delay with different number of blocks, **b** ROC curve for our proposed blockchain based privacy preserving framework

with reduced memory overhead. However, the computational complexity of the Redis cache-based private information retrieval may increase with a corresponding increase in the number of users and transactions performed by the users. Hence, B+-tree is integrated with Redis cache for handling computational overhead by reconstructing the block name structure, and user set logical structure with the view to enhance the response time and retrieval rate of the Private information in smart appointment systems.

3.2 Proposed Blockchain Based Privacy Algorithm

Admin Node.

Our proposed algorithm for admin node is mentioned below: It consist of three main modules i.e. Input, Output and Initialization.

Input: Enrolment Certificate (E_C) requested from Certification Authority (C_A).

Output: Access to P_{HL}, C_{HL} and L_{HL} transactions for all (P_{HL}, C_{HL}, L_{HL}) B_N.

Initialization: N_{Admin} should   be   valid   node. N_{Admin} can Write/Read/Update/Remove nodes C_{ID}, P_{ID}, L_{ID}.

```
1:procedure Admin(P_ID,C_ID,L_ID)
2:  while (True) do
3:    if (C_IDis valid) then
4:       Add_Clinician to the blockchain Network
5:       Add_Clinician(B_N, C_ID)
6:       Grant_access(C_ID, U_Name, P_K)
7:    else
8:       Not_exist(C_ID)
9:    end if
10:   if (P_IDis valid) then
11:      Add Patient to the blockchain Network
12:      Add_Patient(B_N,P_ID)
13:      grant_access(P_ID, U_Name, P_K)
14:   else
15:      Not_exist(P_ID)
16:   end if
17:   if (L_IDis valid) then
18:      Add Lab to the blockchain Network
19:      Add_Lab(B_N,L_ID)
20:      grant_access(L_ID, U_Name, P_K)
21:   else
22:      Not_exist(L_ID)
23:   end if
24: end while
25:    int N; {0 means bad behavior, 1 means good behavior}
```

3.3 Enhanced Storage Model of Improved Redis Cache

The private information retrieval method is initiated on the MTMCB from the user name (patient's Name) and the block files associated with the user name for the party to perform the appointment transaction. In order to achieve the Enhanced Storage Model of Improved Redis Cache, the Name of the user is replaced with User Transaction Identifier" as the key, and the combination of names corresponding to block files is set to the value of the key. The logical storage structure of this User Block Collection is depicted in Fig. 1.

3.3.1 Redis Cache Initialization-Based Storage

The steps involved in Redis Cache initialization-based storage are presented as follows.

Step 1: When the smart appointment system starts its operation, read and explore function is imposed over "Block Name File" by initializing the "User Block Collection" in the Redis cache based on Name of the Block and User Transaction Identifier existing in the block name file.

Step 2: Generate the B+-Tree index associated with "User Block Collection" to accomplish the relation between the "User Block Collection" and the generated index.

Step 3: Another new B+-Tree index is created to represent the user's files as presented in step 2 with the keyword as "Name of the User", when the user collection increases with low potential sequential retrieval process.

3.3.2 Construction of User Block Collection

In this phase, information related to the Blockname is recorded in the "User Block Collection" based on information derived from the transaction participants. The steps involved in the construction of User Block Collection are presented as follows.

Step 1: Generate a block file for recording the current block contents when it is stored in the block file, and it is named based on the blocks' hash value.

Step 2: Choose a single participant of the transaction and identify its user transaction identifier from the B+-Tree index with the user's name derived from the user collection when one or more users participate in the appointment allocation process.

Step 3: Identify the key from the user block file based on the user transaction identifier of the participating users. If the user transaction identifier exists, then the value is extended. Further, the block's name is written at the end of the value to sort them based on the order of timestamp. On the other hand, if it does not exist, then key-value metadata is appended. Finally, the key is the user transaction identifier, file name, and updated B+-tree index.

Step 4: Until the complete set of users complete their processing and perform the extension operation, repeat steps 2 and 3 for processing the successive transaction user.

3.3.3 Application of Improved B+-Tree and Redis Cache-Based Private Information Retrieval for Private Information Retrieval

Once the user block collection is defined, then the algorithm of block-based private information retrieval in the proposed framework completely prevents Block Name File and tries to accomplish the relation between the "User Block Collection" and "User Set." For instance, let us consider that a smart appointment system requires to query the complete block files when two patients participate in the online reservation of appointment space. In this scenario, the service provider can rapidly access all block files associated with the participating patients (users), then the collection of block file names related to the two patients addresses can be determined from the user block file as long as the associated addresses of associated patients are identified through the user file.

Step 1: The retrieval method identifies the user transaction identifier from the user's name by the user set accessing from the name of one or more users participating in the smart appointment system.

Step 2: Utilize B+-Tree to identify the key location that stores the user address derived from the user block set for determining the complete records of block file name associated with the users.

Step 3: Repeat Step 2 until the complete set of block file names relevant to the transaction's participants existing in the records are identified.

Step 4: A block file associated with the users is obtained as the target file output of this query based on the block name file record set determined in steps 2 and 3, respectively.

4   Experimental Evaluations of the Proposed RS-IMTMCB-PIR Scheme

The performance investigation of the proposed framework and the benchmarked Medrecord blockchain, Smart Health solution, and Elliptic Curve Cryptography (ECC)-based Smart solution approaches are conducted using the necessitated cryptographic operations through Raspberry Pi 3 equipment that runs Python charm cryptographic library in the system configuration of 1 GB RAM and 1.2 GHz Processor. In the first part of the research, our proposed framework and the benchmarked Medrec, Smart health Solution, and ECC-SS approaches are evaluated based on communication overhead in private information retrieval with different appointment allocations available in each cell and number of blockchain nodes. In this experiment, communication overhead refers to the transmitted message size (in bytes) exchange between the patients and the blockchain nodes in the retrieval phase and between the patients and the appointment lot service provider in the reservation phase. Demonstrates the communication overhead of our proposed framework in private information retrieval with a different number of appointment allocation available in each cell. The communication overhead of our proposed framework is considerably minimized even when the number of appointment allocation in unit cell increases since it is capable in handling the required number of retrieval by storing in the B+-Tree indexing data structure that aids better performance. Hence, the communication overhead of the proposed framework in private information retrieval with a different number of appointment allocation available in each is reduced by 5.24%, 6.84%, and 7.54%, excellent to the benchmarked SHealth, MedRec, and ECC-Smart solution approaches. Demonstrates the communication overhead of the proposed framework in private information retrieval with a different number of blockchain nodes. Even when the numbers of blockchain nodes are increased, the communication overhead of the proposed framework scheme is considerably minimized as Redis cache-based indexing played a vital role in eliminating the factors that delay the retrieval of private information users when required by the service providers. The communication overhead of the proposed RS-IMTMCB-PIR in remote information retrieval with a different number of blockchain nodes. In the first part of the investigation, the proposed RS-IMTMCB-PIR and the benchmarked MBO-SMS, CB-SMS, and ECC-SMS approaches are evaluated based on communication overhead in private information retrieval with different parking numbers allocation available in each cell and number of blockchain nodes. In this experiment, communication overhead refers to the transmitted message size (in bytes) exchange between the driver and the blockchain nodes in the retrieval phase and between the driver and the parking lot service provider in the reservation phase. Illustrates the communication overhead of the proposed framework in in private information retrieval with different numbers of appointment allocation available in each cell. The communication overhead of the proposed framework is considerably minimized even when the number of parking allocations in the unit cell increases since it can handle the required number of retrieval by storing in the B+-Tree indexing data structure that aids better performance. Hence, the communication overhead of the proposed scheme in private information retrieval with a different number of appointment allocation available in each is reduced by 5.24, 6.84, and 7.54%, excellent to the benchmarked Medrec, Shealth, and ECC-SS approaches. Demonstrates the communication overhead of the proposed framework in private information retrieval with a different number of blockchain nodes. Even when the numbers of blockchain nodes are increased, the communication overhead of the proposed framework is considerably minimized as Redis cache-based indexing played a vital role in eliminating the factors that delay the retrieval of private information users when required by the service providers.

In Table 1 is shown clearly the response time and the transaction time of the proposed approach and the bench mark framework.

In Fig. 2 is shown the simulation results of our proposed framework. It's very clear that for Blockchain size 10,000 the efficiency of our proposed framework is better than the benchmark models (Table 2).

Further, Figs. 3, 4, 5 present the relative proportion of retrieval of our proposed framework and the benchmarked Medrec, SHealth and ECC-SS approaches with different users and block sizes of 2500, 5000, 7500 and 10,000, respectively. The relative proportion of retrieval achieved by our proposed framework is considered to be highly improved, even when the number of blocks in the blockchain gets increased from 2,500 to 10,000.

In Figs. 2 and 3 is shown the simulation for the attribute of Blockchain having size 7,500.

From Figs. 2 and 3 it's very obvious that the efficiency of the proposed approach is better than the bench mark model. The efficiency is calculated according to the response time (R_T) and delay jitter (D_t). As compared to the bench mark model the efficiency of the proposed framework increases as the number of nodes increases up to 7500. The limit for our proposed framework is up to 7500 nodes (Fig. 6).

In Figs. 4 and 5 are explained the simulations results of the number of blocks and the communication overhead that are carried out during the transaction from

Table 2 Retrieval time of proposed system with different users and block size-7500

	Number of users	Proposed framework		Medrec		SHealth		ECC-SS	
		Time	PR$_{(i)}$	Time	PR$_{(i)}$	Time	PR$_{(i)}$	Time	PR$_{(i)}$
Size of the Block-7500	2	94	1.000	135	1.000	248	1.000	612	1.000
	4	96	1.021	137	1.014	250	1.008	615	1.004
	6	97	1.032	139	1.029	253	1.02	616	1.011
	8	99	1.053	140	1.037	255	1.028	619	1.014
	10	100	1.064	142	1.051	258	1.04	621	1.017
	12	102	1.085	143	1.059	259	1.044	623	1.018
	14	104	1.006	145	1.074	261	1.052	624	1.019
	16	105	1.117	147	1.088	262	1.056	625	1.017

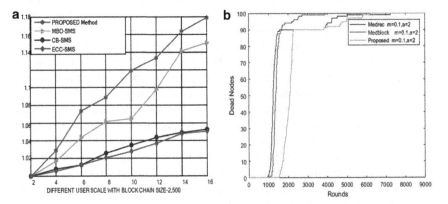

Fig. 4 **a** Efficiency of the proposed system with different users and block size-7500, **b** Comparative analysis of our proposed framework with the benchmark models

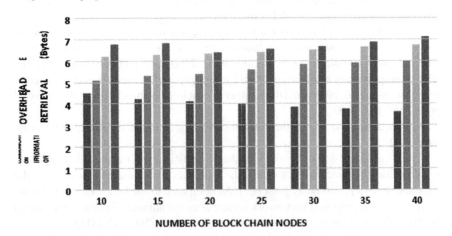

Fig. 5 Communication overhead with number of blockchain nodes

one node and to another nodes respectively. In Figs. 4 and 5 the number of nodes that we have simulated ranges to 5000 respectively. From the representation of the color i.e. Blue bar, green bar, yellow bar and red bar respectively represents different framework and the overhead time. The blue color represents our proposed framework which has least communication overhead as compared to the benchmark models.

Figure 7 represents the simulation of our proposed framework for the number of nodes up to 5000. From this simulation diagram its very clear that the efficiency of our proposed framework is more as compared to the benchmark modes. The number of nodes are 5000 in this case.

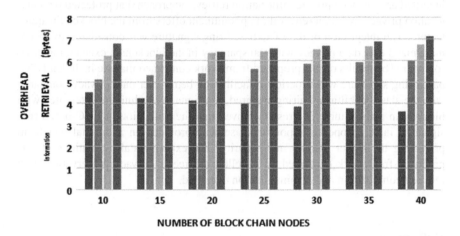

Fig. 6 Proposed framework-communication overhead with number of blockchain nodes

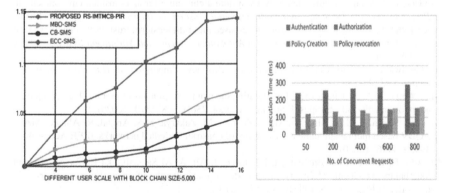

Fig. 7 Efficieny of the proposed framework with the number of blockchain modes (5000)

5 Conclusion

The proposed RS-IMTMCB-PIR Scheme was presented as a privacy-preserving smart appointment system based on private information retrieval and blockchain utilization. Our proposed framework uses a ring signature approach based on ECC to integrate with the constructed Improved Multi-Transaction Mode Consortium Blockchain architecture to achieve maximum transparency. The number of appointment lot services constructed this improved multi-transaction mode consortium blockchain architecture provides for transparency, availability, and security. It also inherited an enhanced private information retrieval approach that protected patients' location privacy by secret retrieval of appointment offers from the blockchain architecture. The integration of the ECC-based ring signature was considered to ensure maximized user data privacy with transparency in the blockchain exchanging environment. The results of this proposed framework confirmed that the included ECC-based ring signature was superior in facilitating better user data privacy compared with the classical bilinear pairing-inspired ring signature techniques integrated with blockchain architecture. The results proved that the utilization of ECC-based ring signature in the proposed appointment scheme provided an additional benefit in protecting the signer identity and security. The experimental evaluations of this proposed framework explained an excellent degree of privacy preservation with reduced computation and communication overhead.

References

K. Garson, C. Adams, Security and privacy system architecture for an e-hospital environment, in *Proceedings of the 7th Symposium on Identity and Trust on the Internet* (2008), pp. 122–130

N. Griffiths. Task delegation using experience-based multi-dimensional trust, *in Proceedings of the Fourth International Joint Conference on Autonomous Agents and Multiagent Systems* (2005), pp. 489–496

M. Gronwald, Y. Zhao, Crypto currencies and currency competition, in *Has Hayek Been Too Optimistic?* (2018)

M.A. Almaiah, A. Al-Khasawneh, Investigating the main determinants of mobile cloud computing adoption in university campus. Educ. Inf. Technol. **25**(4), 3087–3107 (2020)

M. Adil, R. Khan, M.A. Almaiah, M. Al-Zahrani, M. Zakarya, M.S. Amjad, R. Ahmed, MAC-AODV based mutual authentication scheme for constraint oriented networks. IEEE Access **4**(8), 44459–44469 (2020)

M. Adil, R. Khan, M.A. Almaiah, M. Binsawad, J. Ali, A. Al Saaidah, Q.T.H. Ta, An efficient load balancing scheme of energy gauge nodes to maximize the lifespan of constraint oriented networks. IEEE Access **8**, 148510–148527 (2020)

M. Adil, M.A. Almaiah, A. Omar Alsayed, O. Almomani, An anonymous channel categorization scheme of edge nodes to detect jamming attacks in wireless sensor networks. Sensors. **20**(8), 2311 (2020)

M.N. Khan, H.U. Rahman, M.A. Almaiah, M.Z. Khan, A. Khan, M. Raza, M. Al-Zahrani, O. Almomani, R. Khan, Improving energy efficiency with content-based adaptive and dynamic scheduling in wireless sensor networks. IEEE Access **25**(8), 176495–176520 (2020)

M. Adil, R. Khan, J. Ali, B.H. Roh, Q.T. Ta, M.A. Almaiah, An energy proficient load balancing routing scheme for wireless sensor networks to maximize their lifespan in an operational environment. IEEE Access **31**(8), 163209–163224 (2020)

M.A. Almaiah, Z. Dawahdeh, O. Almomani, A. Alsaaidah, A. Al-khasawneh, S. Khawatreh, A new hybrid text encryption approach over mobile ad hoc network. Int. J. Electr. Comput. Eng. (IJECE) **10**(6), 6461–6471 (2020)

M.H. Qasem, N. Obeid, A. Hudaib, M.A. Almaiah, A. Al-Zahrani, A. Al-khasawneh, Multi-agent system combined with distributed data mining for mutual collaboration classification. IEEE Access (2021). Accessed 20 Apr 2021

M.A. Almaiah, A. Al-Zahrani, O. Almomani, A.K. Alhwaitat, Classification of cyber security threats on mobile devices and applications, in *Artificial Intelligence and Blockchain for Future Cybersecurity Applications* vol. 107

M.A. Almaiah, A new scheme for detecting malicious attacks in wireless sensor networks based on blockchain technology, in *Artificial Intelligence and Blockchain for Future Cybersecurity Applications*, vol. 217

M.A. Almaiah, M.M. Alamri, Proposing a new technical quality requirements for mobile learning applications. J. Theor. Appl. Inf. Technol. **96**(19) (2018)

H. Halpin, M. Piekarska, Introduction to security and privacy on the Blockchain, in *2017 IEEE European Symposium on Security and Privacy Workshops (EuroS&PW)* (2017), pp. 1–3

J. Hur, D.K. Noh, Attribute-based access control with efficient revocation in data outsourcing systems. IEEE Trans. Parallel Distrib. Syst. **22**(7), 1214–1221 (2010)

R. Jalan, R.W.L. Szeto, S. Wu, Systems and methods for network access control. Accessed 2 Oct 2018. US Patent 10,091,237

T. Pflanzner, A. Kertész, *A taxonomy and survey of IoT cloud applications*. EAI Endorsed Transactions on Internet of Things **3**(12) (2018), p. Terjedelem: 14 p.-Azonosító: e2. Li, X., et al., *Energy consumption optimization forself-poweredIoT networks withnon-orthogonal multiple access*. Int. J. Commun. Syst. **33**(1), e4174 (2020)

S.-L. Peng, S. Pal, L. Huang, *Principles of Internet of Things (IoT) Ecosystem: InsightParadigm* (Springer, 2020)

A. Ghani, et al., *Security and key management inIoT-basedwireless sensor networks: Anauthentication protocol using symmetric key*. Int. J. Commun. Syst. **32**(16), e4139 (2019)

P. Azad, et al., The role of structured and unstructured data managing mechanisms in theInternet of things. Clust. Comput. 1–14 (2019)

Z. Ghanbari et al., Resource allocation mechanisms and approaches on the Internet of Things. Clust. Comput. **22**(4), 1253–1282 (2019)

K.-K.R. Choo, A. Dehghantanha, R.M. Parizi, *Blockchain Cybersecurity, Trust and Privacy* (Springer, 2020)

Z. Lv, *Security of Internet of Things edge devices* (Software, Practice and Experience, 2020)

N. Dong, G. Bai, L.C. Huang, E.K.H. Lim, J.S. Dong, A blockchain-based decentralized booking system. Knowl. Eng. Rev. **35** (2019)

H. Saini, et al., *Innovations in Computer Science and Engineering: Proceedings of the Sixth ICICSE2018*, vol. 74 (Springer, 2019)

M. Niranjanamurthy, B. Nithya, S. Jagannatha, Analysis of Blockchain technology: pros, consand SWOT. Clust. Comput. **22**(6), 14743–14757 (2019)

N.M. Kumar, P.K. Mallick, Blockchain technology for security issues and challenges in IoT. Procedia Comput. Sci. **132**, 1815–1823 (2018)

A.E. Hassanien, N. Dey, S. Borra, *Medical big data and internet of medical things: Advances, challenges and applications* (CRC Press, 2018)

D. Pavithran, et al., *Towards building a blockchain framework for IoT*. Clust. Comput. 1–15 (2020)

A. Sultan, M.A. Mushtaq, M. Abubakar, *IOT security issues via blockchain: a review paper*, in *Proceedings of the 2019 International Conference on Blockchain Technology* (2019)

P. Karthikeyyan, S. Velliangiri, *Review of Blockchain based IoT application and its security issues*, in *2019 2nd International Conference on Intelligent Computing, Instrumentation and Control Technologies (ICICICT)* (IEEE, 2019)

B. Alotaibi, Utilizing blockchain to overcome cyber security concerns in the internet of things: a review. IEEE Sens. J. **19**(23), 10953–10971 (2019)

R. Wattenhofer, *The Science of the Blockchain* (Create Space Independent Publishing Platform, 2016)

K. Gopalakrishnan, *Security Vulnerabilities and Issues of Traditional Wireless Sensors Networks in IoT*, in *Principles of Internet of Things (IoT)*

K. Gopalakrishnan, *Security vulnerabilities and issues of traditional wireless sensors networks in IoT,* in Principles of Internet of Things (IoT) Ecosystem: Insight Paradigm. (Springer, Cham, 2020), pp. 519–549

A.K. Al Hwaitat, M.A. Almaiah, O. Almomani, M. Al-Zahrani, R.M. Al-Sayed, R.M. Asaifi, K.K. Adhim, A. Althunibat, A. Alsaaidah, Improved security particle swarm optimization (PSO) algorithm to detect radio jamming attacks in mobile networks. Quintana **11**(4), 614–624 (2020)

M.A. Almaiah, M. Al-Zahrani, Multilayer neural network based on MIMO and channel estimation for impulsive noise environment in mobile wireless networks. Int. J. Adv. Trends Comput. Sci. Eng. **9**(1), 315–321 (2020)

A. Jain, T. Singh, Security challenges and solutions of IoT ecosystem, in *Information and Communication Technology for Sustainable Development*. (Springer, 2020), pp. 259–270

M.A. Khan, K. Salah, IoT security: review, blockchain solutions, and open challenges. Futur. Gener. Comput. Syst. **82**, 395–411 (2018)

V. Manjula, R.T. Rajasekaran, Security vulnerabilities in traditional wireless sensor networks by an intern in IoT, blockchain technology for data sharing in IoT, in *Principles of Internet of Things (IoT) Ecosystem: Insight Paradigm*. (Springer, 2020), pp. 579–597

P. Pandey, S.C. Pandey, U. Kumar, Security issues of internet of things in health-care sector: an analytical approach, in *Advancement of Machine Intelligence in Interactive Medical Image Analysis*. (Springer, 2020), pp. 307–329

G. Viale Pereira et al., Increasing collaboration and participation in smar city governance: across-case analysis of smart city initiatives. Inf. Technol. Dev. **23**(3), 526–553 (2017)

J.H. Yu, J. Kang, S. Park, Information availability and return volatility in the bitcoin market: analyzing differences of user opinion and interest. Inf. Process. Manage. **56**(3), 721–732 (2019)

Classification of Malicious and Benign Binaries Using Visualization Technique and Machine Learning Algorithms

Ikram Ben Abdel Ouahab, Lotfi Elaachak, and Mohammed Bouhorma

Abstract Malware detection and classification are crucial with malware evolution these days. More and more malware attacks are targeting various environments like businesses and hospitals. Our main goal is to deliver an intelligent classifier as fast as efficient able to classifier binaries into malicious or benign classes. In this work, we consider a malware-benign classification using machine learning algorithms by converting binary executable files into grayscale images. The concept of the visualization gives powerful results in the malware family classification topics. In this context, we adopt visualization technique for both benign and malicious samples. So, a hybrid feature extraction method is implemented using local and global image features by combining DAISY and HOG features. Then a comparative study of machine learning algorithms leads us to a final efficient classifier that reaches an accuracy of 97,36% using Random Forest classification algorithm.

Keywords Cybersecurity · Image processing · Visualization · Machine learning · Malware classification

1 Introduction

Malware refers to any malicious software including virus, ransomware, downloader, spyware, adware, worm, trojan, botnet, rootkit, keylogger and many other types of malicious software. The top 3 hacker's motivations behind these attacks are money, secrets or fun. Most often, financial gain is the main goal of a hacker for example the trend of ransomware attacks in earlier 2021. Then comes hackers motivated by espionage by stealing confidential informations and secrets. Here, we talk about three types of informations: political, manufacturing and social media credentials. After

I. Ben Abdel Ouahab (✉) · L. Elaachak · M. Bouhorma
Laboratory of Computer Science, Systems and Telecommunications (LIST), Faculty of Sciences and Techniques (FSTT), University Abdelmalek Essaadi, Tangier, Morocco
e-mail: ibenabdelouahab@uae.ac.ma

M. Bouhorma
e-mail: mbouhorma@uae.ac.ma

© The Author(s), under exclusive license to Springer Nature Switzerland AG 2022   297
Y. Baddi et al. (eds.), *Big Data Intelligence for Smart Applications*,
Studies in Computational Intelligence 994,
https://doi.org/10.1007/978-3-030-87954-9_14

that, we notice beginners' learners of hacking doing for fun malicious attacks. Others are also having ideological background or because of grudge. Understanding, why a hacker may target you, is the first step for defense. It help knows weaknesses and vulnerabilities of any infrastructure.

Recently, cyberattacks are continuing to grow in sophistication and scale. Especially, in the period of the covid19 pandemic that has affected the surface of attacks where cybercriminals are leading to probably a cyber-pandemic. In December 2020, hackers broke into SolarWind's systems and added a malware into the company's software system. The targeted system is being used by 33,000 customers ('Inline XBRL Viewer' 2021). The malicious code created a backdoor to customer's information technology systems; moreover hackers installed more malware that helped then spy on the SolarWind's customers companies and organizations.

More recently, in February 2021, hackers are getting more inhuman by targeting French hospitals in a critical sanitary period. A wave of ransomware attacks hit two hospitals in less than week in France, prompting the transfer of some patients to other facilities. The ransomware used is a crypto-virus called RYUK. Firstly, the said ransomware attacks Dax hospital and then five days later hackers target the Villefranche-sur-Saône hospital complex in France's eastern Rhone. Meanwhile, France's cybersecurity agency has revealed that a weakness in software from the company Centre on led to the defenses of several French companies being breached over a period of three years.

New variants of malware continue to increase every month. In March 2021, AV-TEST institutions have recorded 19.20 million of new malware ('Malware Statistics Trends Report I AV-TEST' 2021). These variants are becoming more dangerous, and making data protection a hard work. For instance, fake update is the new trend of distributing a malware in a legitimate looking. The worst thing is that anti-malware solutions are insufficient and can't recognize new and zero-day malwares especially if it presented in an update of a well-known provider name. Also, there are news malware attacks. This type is frequently used with the covid19 pandemic where people are very interesting in any Covid related news like vaccination and so on. In addition, malware targets also IoT devices, since most of it does not have a strong security measures, hackers find it easier to manipulate and access data in IoT devices like monitoring systems and any smart devices.

In order to defend against malware attacks, we proposed a cybersecurity framework (Abdel ouahab Ikram et al. 2020) which is an intelligent solution dedicated to malware detection and analysis. The particularity of our framework lies in the use of the visualization technique; convert PE files into grayscale images. It is composed by 4 layers as shown in Fig. 1: Discovery, Identification, Reporting and Reaction. In a previous works we developed the Identification layer using visualization technique and machine learning classifiers to know the malware family and behaviors. In this paper, we develop the first layer to know either the input PE is malicious or benign. To do so, we use various machine learning algorithms and PE files in the form of grayscale images. Once the malicious PE file is discovered, then we can know the malware family using Identification Layer. After that, the framework is able to report

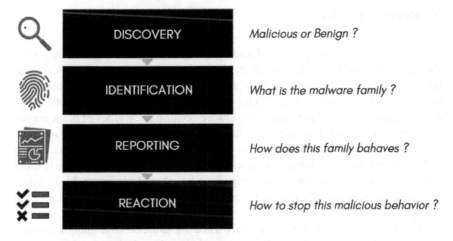

DISCOVERY	Malicious or Benign ?
IDENTIFICATION	What is the malware family ?
REPORTING	How does this family bahaves ?
REACTION	How to stop this malicious behavior ?

Fig. 1 The proposed cybersecurity framework architecture layers

the particularity and malicious behavior of the given malware to help react effectively against it.

More and more researchers are using Artificial Intelligence (AI) in different domains and applications. We found the use of machine learning algorithms in the healthcare domain and IoT. For instance, the diagnostic of breast cancer (Saoud et al. 2019), medical chatbots (Soufyane et al. 2021) and other smart applications dedicated to prevent the Covid19 pandemic by monitoring social distance and facemask detection (Abdel Ouahab et al. 2020). Also AI algorithms are also used in Agricultural applications (Bakkali Yedri et al. 2021) to classify plants easily. The power of AI is also been used in digital marketing (Ahsain and Kbir 2021), the detection of fake social media profile (Elyusufi et al. 2020) and the detection of illegitimate bitcoin transactions (Elbaghdadi et al. 2020). In cybersecurity both defenders and attackers are using powerful deep learning and machine learning algorithms to develop their tasks. We used machine learning algorithms to classify malwares into their corresponding families using various techniques in other works to ensure high quality defense framework (Ben Abdel Ouahab et al. 2019, 2020).

The remaining of this paper is organized as followed: Sect. 2 summarized some related works regarding highly to techniques used for malware-benign classification using various machine learning algorithms. Section 3 describes the proposed cybersecurity framework and the proposed method used to classify malware-benign samples. In Sect. 4, we gave an overview of the methodology used including machine learning algorithms, evaluation metrics, malware visualization technique and images descriptors taken as features. Later in Sect. 5 we present experimentations and discussed the results. Finally we conclude by giving our research perspectives.

2   Related Works

Malware detection is the process of finding a malware by scanning unknown files. This detection could be done using many traditional techniques as pattern matching, heuristic analysis, behavioral analysis and hash matching. These traditional techniques have many disadvantages. The pattern matching technique is not able to detect metamorphic and polymorphic malware because these types of malware can change their code and signatures (Vinod et al. 2012). Heuristic method use malware features like API calls, CFG, N-Gram or OpCode. It is a combination of data mining and machine learning techniques used to learn the behavior of unknown file in order to classify malware and benign ones. Signature-based methods are unable to detect unknown malware variants and also requires high amount of manpower, time, and costly.

In Tabish et al. (2009), authors proposed a non-signature based technique to detect unknown and zero-day malware. Their approach quantifies the byte level file content without the need of a priori information about the file type. To classify normal and potentially malicious block, they use data mining algorithms: decision tree and AdaBoosting, and get more than 90% detection accuracy. They tested the classifier on six different malware types.

In Anderson et al. (2011), we found the use of dynamic malware analysis to collect instruction traces of the executable target. Authors used graphs represented as Markov chains. A proposed processing graph based approach lead to a final similarity matrix used by SVM algorithms for the classification task. Using combined kernel features the classifier gives an accuracy of 96.41%. This approach shows improvement over signature-based and detection methods.

In Singh and Singh (2020), malware detection is based on dynamic analysis by extracting API calls of the malicious behavior. Using a huge dataset of 6434 benign and 8634 malicious samples, the authors train anti-malware classifier using ensemble algorithms as KNN, decision tree, naïve bayes, random forest, SVM and boosting. The average accuracy they got reaches a 99.1%.

In Bae et al. (2020), we found a focus on ransomware detection using a specific mechanism. To distinguish between ransomware, malware and benign samples authors used n-gram vectors based on system calls and CF-NCF method. A comparison of six machine learning algorithms was performed, includes: random forest, logistic regression, naïve bayes, stochastic gradient descent, k-nearest neighbors and support vector machine. The proposed ransomware detection system gives an accuracy of 98.65%.

Apart from that, researchers recently show more interest in malware visualization technique. The main concept is to convert a malware binary either to an image or to a signal. This idea was firstly introduced by Nataraj et al. (2011a) where authors demonstrate that binary texture analysis provides useful results for classification rather than dynamic and static features. However, their goal was the classification of malware into their corresponding families.

Table 1 Summary of related works: malware benign classification methods

Ref.	Features	Classifier	Accuracy (%)
Tabish et al. (2009)	Statistical analysis of byte-level file content	Decision tree and AdaBoosting	90
Anderson et al. (2011)	Graph based using dynamic analysis	SVM	96.41
Singh and Singh (2020)	API calls	Ensemble algorithms	99.1
Bae et al. (2020)	System calls and CF-NCF	Six machine learning algorithms	98.65
Azab and Khasawneh (2020)	Malware spectrogram analysis	CNN	96
	Grayscale images	CNN	95.5
Ours	Hybrid features (DAISY + HOG) from grayscale images	Random Forest	97.36

In Azab and Khasawneh (2020), the proposed framework is based on malware spectrogram image classification: MSIC. Using 2611 benign samples and 2626 malicious ones, authors perform a convolutional neural network (CNN) for the classification task. The average resulting accuracy is 96%. Based on their comparison the MSIC approach outperforms slightly the grayscale image based approach for malware classification.

Our proposed approach is based on grayscale images. We extract DAISY and HOG features, then we process and combine them to get the hybrid features. Next, we compare 5 machine learning algorithms using hybrid features, DAISY only and HOG only. Random forest based classifier gives the highest accuracy of 97.36% with hybrid features.

Altogether, we grouped all these related work in Table 1.

3   Methodology

Machine learning techniques identify malicious software based on features extracted from static analysis, dynamic analysis, vision-based analysis, or a hybrid solution that combine more than one method. We compare several machine learning algorithms to select the most appropriate one that give us the best performances. This process is widely used in machine learning applications. In our case, we have tested many algorithms, and then we present here the top 5. We have a binary classification, so we need algorithms that can handle it. We use: Gaussian Naïve Bayes, Decision Tree based on CART algorithm, K-Nearest Neighbor with $k = 4$, Random Forest, and Logistic Regression algorithm.

Algorithm	Definition
Gaussian Naïve Bayes (GNB)	The Naive Bayes classifier works on the principle of conditional probability, as given by the Bayes theorem. It is an easy to use algorithm that only estimates the mean and the standard deviation from training data. Naive Bayes is better suited for categorical input variables than numerical variables
Decision Tree (DT)	Decision tree creates the classifier model by building a decision tree. It is a non-parametric method which doesn't depend upon probability distribution assumptions. Decision trees can handle high dimensional data with a good accuracy. DT requires less data processing from the user. It can be used for classification and also for feature engineering like prediction of missing values and variable selection
K Nearest Neighbor (KNN)	KNN is a supervised machine learning algorithm used for classification and regression problems. The principle of KNN is that similar things exist in close proximity. Another important thing is choosing the appropriate K value. So, to have the right K value, we need to run the algorithm several times and reduces the errors
Random Forest (RF)	Random Forest is a supervised learning algorithm. It is a very used algorithm because of it flexibility, versatility and usage. It can be solve classification and regression problems and gives good results using default hyper-parameters. The main limitation of random forest is that a large number of trees can make the algorithm too slow and ineffective for real-time predictions
Logistic Regression (LR)	Logistic regression is a statistical based machine learning technique. LR is a linear model for classification. Using LR probabilities describing the possible outcomes of a single trial are modeled using a Logistic function

To evaluate these machine learning algorithms we use: accuracy, f1-score, recall, precision and ROC-AUC. The accuracy is one of the metrics to evaluate classification models, it represent the number of correct predictions by the total number of predictions. The precision is intended to know the proportion of positive predictions that was actually well classified. However, the recall leads us to the proportion of actual positives that was identified correctly. And by combining precision and recall we obtain the f-measure or what we call f1-score. ROC-AUC stands for the Area Under the Receiver Operating Characteristic Curve. We compute also this metric from prediction scores using test features.

Metric	Definition
Accuracy	The accuracy is a statistical measure of classifier. It is the proportion of correct predictions, true positives and true negatives, among all data. Accuracy = (TP + TN)/(TP + TN + FP + FN)
F1-score	F1-score is a combination of recall and precision using the harmonic mean
Recall	Recall refers to the true positive rate or sensitivity Recall = TP / (TP + FN)

(continued)

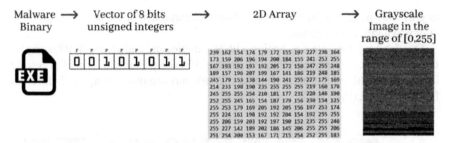

Fig. 2 From binary to image conversion process

(continued)

Metric	Definition
Precision	Precision = TP /(TP + FP)
ROC-AUC	Receiver Operator Characteristic (ROC) curve is an evaluation metric for binary classification problems

3.1 PE as Grayscale Image

Malwares from the Malimg database are already converted from PE files to grayscale images. However, the benign samples need to be converted using the same technique by converting the binary data (bytes sequence) to an 8 bits unsigned integers vector then to a 2D array in the range of [0, 255], so that we can visualize it as a grayscale image. Steps of the process are detailed in the Fig. 2. By converting a binary to images we can see visual similarities in samples belonging to the same family. And we found dis-similarities in variants belonging to different families. The advantage of the visualization technique is that new malware variants appear similar to variants of same family. Contrary to signature based detection, that can't detect new variants of malwares because of the new hashes in each variant.

3.2 Database

Benign samples are collected from official sources; we chose various types of executable files such as text editor, navigator, and others. After we download the PE files, we did an online analysis to ensure that the version is benign. To do so we used the VirusTotal online tool ('VirusTotal'. 2021). So, all the 50 samples results were 100% benign. Malicious samples are taken from the Malimg database (Nataraj et al. 2011b) where we found 25 families. We took 4 samples per family and a total of 100 malware samples belonging to all families. The malicious samples are provided as grayscale images using the visualization technique that allow us to convert a PE binary to a grayscale images.

Our database has 2 classes: benign with 50 samples and malware with 100 samples from 25 families. The size of final database is 150 samples in the form of grayscale images using the visualization technique. Malicious samples belong to 25 different families named (Fig. 3):

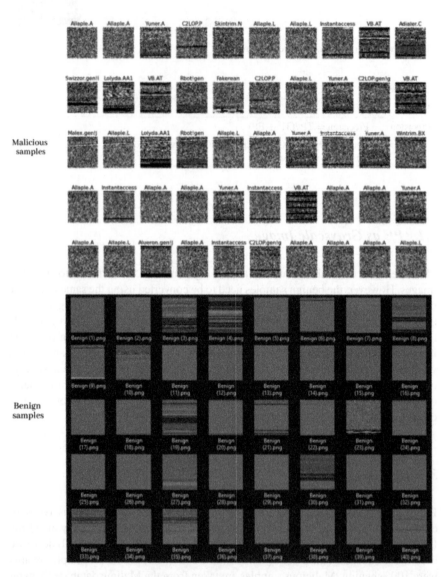

Fig. 3 An extract from the benign-malware database

- Allaple.L,
- Allaple.A,
- Yuner.A,
- Lolyda.AA 1,
- Lolyda.AA 2,
- Lolyda.AA 3,
- C2Lop.P,
- C2Lop.gen!g,
- Instantaccess,
- Swizzot.gen!I,
- Swizzor.gen!E,
- VB.AT,
- Fakerean,
- Alueron.gen!J,
- Malex.gen!J,
- Lolyda.AT,
- Adialer.C,
- Wintrim.BX,
- Dialplatform.B,
- Dontovo.A,
- Obfuscator.AD,
- Agent.FYI,
- Autorun.K,
- Rbot!gen and
- Skintrim.N.

Afterwards, we perform binary classifiers using many machine learning algorithms to classify grayscale images into two classes labeled as follow: *(0) Benign* and *(1) Malicious*. To split data into train and test parts we do cross-validation which split randomly the samples. We use 25% for test and the remaining 75% for training each classifier. Data distributes randomly while splitting data into train and test. Talking with numbers, we train our models using 74 malicious sample and 38 benign. Then for testing our model we use 26 malicious and 12 benign samples as shown in figure (Fig. 4).

3.3   Local and Global Image Descriptors

The DAISY local image descriptor is based on gradient orientation histograms similar to the SIFT descriptor. It is formulated in a way that allows for fast dense extraction which is useful for bag-of-features image representations. The DAISY descriptor is widely used in many applications of object recognition using grayscale images. It

Data Distribution after split

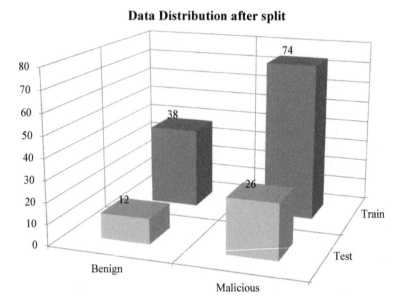

Fig. 4 Split database into train and test portions

shows rapid and efficiency so we are going to implement it in the malware Visio-
based approach. As indicated by his name, DAISY descriptor is similar to daisy,
which is constructed by some central-symmetrical circles, as shown in figure below.

The Histogram of Oriented Gradient (HOG) feature descriptor is popular and
gives excellent results especially in pedestrian detection applications. However, this
feature is never been applied to malware images. The idea of HOG features is that
local object appearance and shape can often be characterized rather well by the
distribution of the local intensity gradients or edge directions, even without precise
knowledge of the corresponding gradient or edge positions (Shu et al. 2011). The Hog
descriptor is characterized by its simplicity and rapidity in computation. In Fig. 5, we
show a sample of HOG and DAISY representation separately using 2 input images.

In our approach, we make use of global feature descriptors as well as local feature
descriptors simultaneously to represent each image, to improve recognition accuracy.
Inspired from this paper (Wilson and Arif 1702), that use this features combination
in object detection task. On fifteen scene categories dataset, the average accuracy of
their model was 76.4%.

First, we extract DAISY features (Tola et al. 2008) from the grayscale images
using the Scikit-image library.[1] We use the K-means algorithm to quantize DAISY
features into K clusters to form visual words based on the standard "bag-of-visual-
words" concept. Then, we extract a standard HOG descriptor. Second, we make
two levels pooling scheme. Starting from the DAISY features, a sum pooling is
performed by processing frequency of each visual word in every single image. To

---

[1] https://scikit-image.org/.

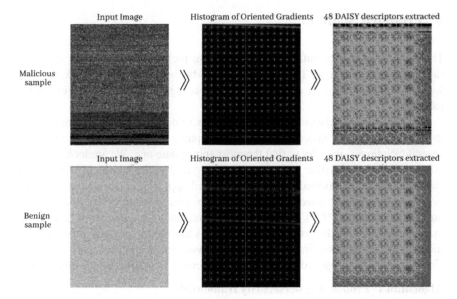

Fig. 5 Representation of HOG and DAISY features

form the DAISY histogram feature we do the L2 normalization. After that, the HOG global descriptor of each image is normalized as the previous one. The final feature is done by a concatenation of the normalized DAISY and HOG histogram features (Fig. 6).

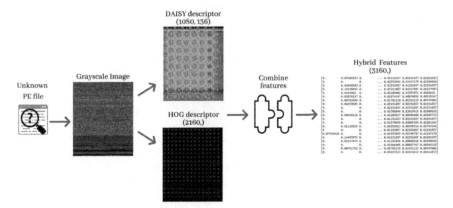

Fig. 6 The process of computing hybrid features

4 Proposed Solution

This section is divided into two parts: first we present our proposed cybersecurity framework where we'll need the malware-benign classification task. Second, we present the process of computing features using the hybrid feature extraction technique.

The proposed cybersecurity framework aims to defend against malware attacks using AI techniques. Recently, we are in the phase of benchmarking and testing several techniques, in order to find the most appropriate one. We are looking for an efficient malware classifier and we are focusing also in the processing time of the classifier. As we mentioned above, the proposed approach is an intelligent system dedicated for malware attacks. That we have initialized in a previous work (Abdel ouahab Ikram et al. 2020). It is a layered architecture composed of 4 layers. The main particularity of the proposed framework is the adoption of visualization technique in all the process, first to detect the malware and second for malware family classification. The proposed cybersecurity framework is summarized in algorithm 1.

Algorithm 1: Proposed cybersecurity framework

```
Input: Unknown binary file
Output: Reaction
Begin
    1.  X = unknown binary file
    #CONVERT BINARY TO GRAYSCALE IMAGE
    2.  X_img = Grayscale binary image
    #DISCOVERY LAYER
    3.  Malicious/Benign classification process
    4.  if the binary is malicious
            #IDENTIFICATION LAYER
        a.  malware families classification process
        b.  X_family = The resulting malware family of X
        c.  if the X_family probability is low AND/OR many families appear
                "ALERT: It could be a new family!"
        d.  else if the X_family probability is high
                "Good prediction"
                #REPORTING LAYER
                Return: X_family behavioral report
                #REACTION LAYER
                Return: Reaction step by step actions
End
```

In this paper, we focused on the first layer of the intelligent cybersecurity framework. Our goal is to classify binaries into malicious or benign class using machine learning and hybrid features. The first step is to convert the unknown binary into grayscale image, and then we extracted HOG and DAISY features and combined them. The resulting features are used as input to various machine learning algorithms to do the classification task. Finally, we evaluated these classifiers to obtain a final efficient one. In addition to the hybrid features, we also tested HOG and DAISY features separately to analyses the performance with each classifier. As describe in algorithm 2.

Algorithm 2: Malicious-Benign classification based on hybrid features

```
Input: unknown binary file
Output: benign or malicious
Begin
    1.  X = unknown binary file
    2.  classifier() function of a trained machine learning classifier
    3.  F = Combination of DAISY and HOG features

    4.  convert X to grayscale image
    5.  Extract DAISY features
    6.  Clustering DAISY features using KMeans algorithm
    7.  Extract HOG features
    8.  2-level Pooling
    9.  F = combine features
    10. classifier(F)
    11. if classifier return "1":
            "This is a malicious file"
    12. else if classifier return "0":
            "This is a benign file"
End
```

5 Results and Discussion

By implementing the proposed solution, we got interesting results in many sides. We use malware extracted features with 5 machine learning classifiers: Gaussian Naïve Bayes, Decision Tree, K-Nearest Neighbor, Random Forest and Logistic Regression. For each used algorithm, we calculated 6 evaluation metrics that are: training score, test accuracy, F1-score, recall, precision and ROC-AUC values. Firstly, we use only the HOG images descriptor. Secondly, we tried to improve classifiers performances by using the DAISY images descriptor. Thirdly, we use our proposed hybrid features composed by combining both HOG and DAISY features as described in previous section.

The Table 2 presents the results we obtained using only HOG features for all 5

Table 2 Using HOG features only for malware-benign classification

Model	Train score	Test accuracy	F1-score	Recall	Precision	ROC-AUC
Gaussian NB	0,5803	0,3421	0,4186	0,75	0,2903	0,4519
Decision tree	1	0,5789	0,3333	0,3333	0,3333	0,5128
K-Nearest neighbor	0,8214	0,8157	0,6666	0,5833	0,7777	0,7532
Random forest	1	0,8421	0,75	0,75	0,75	0,8173
Logistic regression	0,6607	0,6842	0	0	0	0,5

classifiers. In general, the results are very bad. The two top classifiers using only HOG features are KNN and RF with an accuracy of 81.57% and 84.21% respectively. In term of malware benign classification this could not be an acceptable accuracy. So, we conclude from performing a machine learning based classifier using the visualization technique that HOG features only are not suitable for malware images classification and are not recommended to be used in such case.

The Table 3 represents results using only DAISY features for all the 5 classifiers and by calculating all evaluation metrics. The performance of all classifier has been improves comparing to the HOG features. The top classifier is Logistic Regression based. The accuracy of this algorithm has moved from 68.42% (with HOG features) to 97.36% using the DAISY features. Same thing with other classifiers. We can see clearly that all evaluation metrics are improved including recall, precision, f1-score and roc-auc values. So using only DAISY features gave satisfactory performances. That we are going to improve even more using the proposed hybrid features.

The Table 4 present results of classifiers using hybrid features (HOG and DAISY). In general, we can see that performances have been improved. In particular, using decision tree algorithm the accuracy goes from 86.84 to 92.10% using the proposed hybrid features. Also, the others evaluation metrics have clearly increased. In addition, random forest classifier shows also an important increase. For instance, the accuracy goes from 94.73% using DAISY features to 97.36% using hybrid features.

Table 3 Using DAISY features only for malware-benign classification

Model	Train score	Test accuracy	F1-score	Recall	Precision	ROC-AUC
Gaussian NB	1	0,921	0,8695	0,8333	0,909	0,8974
Decision tree	1	0,8684	0,7826	0,75	0,8181	0,8365
K-Nearest neighbor	0,9553	0,9473	0,909	0,8333	1	0,9166
Random forest	1	0,9473	0,909	0,8333	1	0,9166
Logistic regression	0,9553	0,9736	0,9565	0,9166	1	0,9583

Table 4 Using hybrid features (DAISY and HOG) for malware-benign classification

Model	Train score	Test accuracy	F1-score	Recall	Precision	ROC-AUC
Gaussian NB	1	0,921	0,8695	0,8333	0,909	0,8974
Decision tree	1	0,921	0,8695	0,8333	0,909	0,8974
K-Nearest neighbor	0,9553	0,9473	0,909	0,8333	1	0,9166
Random forest	1	0,9736	0,9565	0,9166	1	0,9583
Logistic regression	0,9553	0,9736	0,9565	0,9166	1	0,9583

So the proposed combination of DAISY and HOG features give improvements to most machine learning algorithms.

Moreover, we notice that DAISY only and hybrid features give very close results. For example, we are going to analyze more the decision tree classifier results in the three cases. As we can see in the graph Fig. 7, hybrid features curve and DAISY curve are very close for all evaluation metrics. Another example we can analyze is Random Forest classifier results. We can see the graph in Fig. 8, curves of DAISY

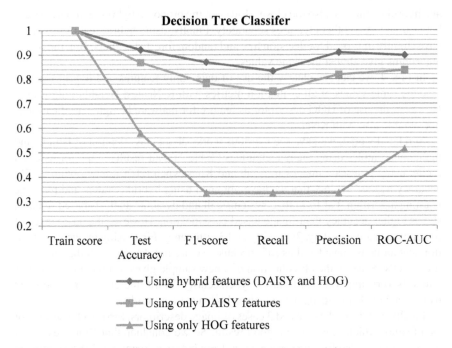

Fig. 7 Decision tree classifier using various features and evaluation metrics

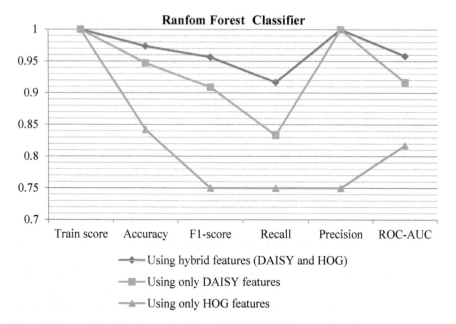

Fig. 8 Random forest classifier using various features and evaluation metrics

and hybrid feature are also too close. However, the proposed hybrid features perform better than to two others features separately. Finally, using hybrid features, metrics have slightly improved for decision tree and random forest classifiers. But, the others algorithms give almost the same values.

In Fig. 9, we can see clearly the use of daisy only features and hybrid features give the same values of all evaluation metrics with K-Nearest Neighbor algorithm. However, the use of HOG descriptor as feature for malware benign classification is not suitable also with KNN.

In related works, we have seen various works about classification of benign and malicious samples using many techniques (Table 1). Our results outperforms CNN model using grayscale images in Azab and Khasawneh (2020), they obtain 95.5% accuracy against 97.36% that we have obtained using hybrid features and machine learning algorithm. Far from that, systems call and dynamic analysis based features shows higher accuracy than Image based classification. However the advantages of our approach are: 1/ safety, 2/ rapidity and 3/ efficiency. First, using image visualization technique is considered as safe because we are not dealing with the malicious file directly. Second, the classification of a new sample gives results in few seconds which is very important in term of cybersecurity. Third, the classifier we trained using machine learning algorithms gives high accuracy results.

Finally, in this work we tested 3 cases of image descriptors. First, HOG descriptor as a feature which didn't give a good result. So we conclude that HOG descriptor is not recommended for images converted from binaries. Second, DAISY descriptor

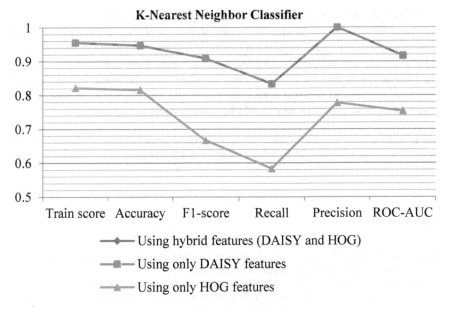

Fig. 9 KNN classifier using various features and evaluation metrics

is also tested with many machine learning classifiers. And third, a hybrid feature is been used. We got this hybrid feature by combining DAISY and HOG descriptors as described in section above (Algorithm 2).

6 Conclusion

To summarize, we perform a malware-benign classifier using classical machine learning algorithms. As input features we adopt the HOG descriptor, DAISY descriptor and we proposed a hybrid features. It is the combination of both HOG and DAISY features. Then we have demonstrated the effectiveness of the hybrid features using various machine learning algorithms. In order to know a PE file is malicious or benign, we perform five classifiers based on machine learning algorithms. We make a comparison of five classifiers: Logistic Regression, Decision Tree, Random Forest, Gaussian NB and K-Nearest Neighbor. To evaluate them we adopt these evaluation metrics: accuracy, f1score, recall, precision and ROC_AUC. Then we compare the classifiers using only HOG features, only DAISY features and the hybrid features approach. When using only HOG features, the classifier gives the worst results where the accuracy is less than 84%. However, using only daisy features gives interesting results with some algorithms. Finally, the hybrid DAISY and HOG features combination shows a global improvement in almost all the used algorithms.

We have described the algorithm of an intelligent cybersecurity framework based on AI and visualization technique. In the next step, we will adopt the malware-benign classifier based on Logistic regression or Random Forest with hybrid features which gave as an accuracy of 97.36%. As a future work, we are looking forward to improve classifier performances by adopting deep learning methods. Since we are dealing with images classification task represented from malware variants, we believe that Convolutional Neural Networks should be more suitable in this case.

Acknowledgements We acknowledge financial support for this research from the "Centre National pour la Recherche Scientifique et Technique", CNRST, Rabat, Morocco.

References

'Inline XBRL Viewer'. https://www.sec.gov/ix?doc=/Archives/edgar/data/1739942/000162828 020017451/swi-20201214.htm. Accessed 04 Mar 2021

'Malware Statistics & Trends Report I AV-TEST'. /en/statistics/malware/. Accessed 14 Apr 2021

B. abdel ouahab Ikram, B. Mohammed, E.A. Lotfi, A.B. Anouar, 'Towards a new cyberdefense generation: proposition of an intelligent cybersecurity framework for malware attacks', in *Recent Advances in Computer Science and Communications*, vol. 13 (2020), pp. 1–19. https://doi.org/10.2174/2666255813999201117093512

H. Saoud, A. Ghadi, M. Ghailani, Proposed approach for breast cancer diagnosis using machine learning, in *Proceedings of the 4th International Conference on Smart City Applications* (New York, NY, USA, 2019), pp. 1–5. https://doi.org/10.1145/3368756.3369089

A. Soufyane, B.A. Abdelhakim, M.B. Ahmed, An intelligent chatbot using NLP and TF-IDF algorithm for text understanding applied to the medical field, in *Emerging Trends in ICT for Sustainable Development* (Cham, 2021), pp. 3–10. https://doi.org/10.1007/978-3-030-53440-0_1

I.B. abdel Ouahab, L. Elaachak, F. Elouaai, M. Bouhorma, A smart surveillance prototype ensures the respect of social distance during COVID19, in *Innovations in Smart Cities Applications Vol. 4* (2020), pp. 1197–1209. https://doi.org/10.1007/978-3-030-66840-2_91

O. Bakkali Yedri, M. Ben Ahmed, M. Bouhorma, L. El Achaak, A smart agricultural system to classify agricultural plants and fungus diseases using deep learning, in *Emerging Trends in ICT for Sustainable Development* (Cham, 2021), pp. 229–239. https://doi.org/10.1007/978-3-030-53440-0_25

S. Ahsain, M.A. Kbir, Data mining and machine learning techniques applied to digital marketing domain needs, in *Innovations in Smart Cities Applications Vol. 4* (Cham, 2021), pp. 730–740. https://doi.org/10.1007/978-3-030-66840-2_55

Y. Elyusufi, Z. Elyusufi, M.A. Kbir, Social networks fake profiles detection using machine learning algorithms, in *Innovations in Smart Cities Applications Edition 3* (Cham, 2020), pp. 30–40. https://doi.org/10.1007/978-3-030-37629-1_3

A. Elbaghdadi, S. Mezroui, A.E. Oualkadi, SVM: an approach to detect illicit transaction in the bitcoin network, in *Innovations in Smart Cities Applications Vol. 4* (2020), pp. 1130–1141. https://doi.org/10.1007/978-3-030-66840-2_86

I. Ben Abdel Ouahab, L. El Aachak, B.A. Abdelhakim, M. Bouhorma, Speedy and efficient malwares images classifier using reduced GIST features for a new defense guide, Marrakech, Morocco (2020). https://doi.org/10.1145/3386723.3387839

I. Ben abdel ouahab, M. Bouhorma, B.A. Abdelhakim, L. El Aachak, B. Zafar, Machine learning application for malwares classification using visualization technique, in *Proceedings of the 4th*

International Conference on Smart City Applications (Casablanca MA, 2019), pp. 110:1–110:6. https://doi.org/10.1145/3368756.3369098

P. Vinod, V. Laxmi, M.S. Gaur, G. Chauhan, MOMENTUM: MetamOrphic malware exploration techniques using MSA signatures, in *2012 International Conference on Innovations in Information Technology (IIT)* (2012), pp. 232–237. https://doi.org/10.1109/INNOVATIONS.2012.620 7739

S.M. Tabish, M.Z. Shafiq, M. Farooq, Malware detection using statistical analysis of byte-level file content, in *Proceedings of the ACM SIGKDD Workshop on CyberSecurity and Intelligence Informatics* (New York, NY, USA, 2009), pp. 23–31. https://doi.org/10.1145/1599272.1599278

B. Anderson, D. Quist, J. Neil, C. Storlie, T. Lane, Graph-based malware detection using dynamic analysis. J Comput Virol **7**(4), 247–258 (2011). https://doi.org/10.1007/s11416-011-0152-x

J. Singh, J. Singh, Assessment of supervised machine learning algorithms using dynamic API calls for malware detection. Int. J. Comput. Appl. , 1–8 (2020). https://doi.org/10.1080/1206212X. 2020.1732641

S.I. Bae, G.B. Lee, E.G. Im, Ransomware detection using machine learning algorithms. Concurr. Comput.: Pract. Exp. **32**(18)(2020). https://doi.org/10.1002/cpe.5422

L. Nataraj, V. Yegneswaran, P. Porras, J. Zhang, A comparative assessment of malware classification using binary texture analysis and dynamic analysis, in *Proceedings of the 4th ACM Workshop on Security and Artificial Intelligence—AISec '11* (Chicago, Illinois, USA, 2011), p. 21. https://doi. org/10.1145/2046684.2046689

A. Azab, M. Khasawneh, MSIC: malware spectrogram image classification. IEEE Access **8**, 102007–102021 (2020). https://doi.org/10.1109/ACCESS.2020.2999320

'VirusTotal'. https://www.virustotal.com/gui/. Accessed 04 Mar 2021

L. Nataraj, S. Karthikeyan, G. Jacob, B.S. Manjunath, Malware images: visualization and automatic classification, in *Proceedings of the 8th International Symposium on Visualization for Cyber Security - VizSec '11* (Pittsburgh, Pennsylvania, 2011), pp. 1–7. https://doi.org/10.1145/2016904. 2016908

C. Shu, X. Ding, C. Fang, Histogram of the oriented gradient for face recognition. Tsinghua Sci. Technol. **16**(2), 216–224 (2011). https://doi.org/10.1016/S1007-0214(11)70032-3

J. Wilson, M. Arif, Scene recognition by combining local and global image descriptors', arXiv: 1702.06850 *[cs]*, Feb 2017, Accessed 10 Feb 2021. http://arxiv.org/abs/1702.06850

E. Tola, V. Lepetit, P. Fua, A fast local descriptor for dense matching, in *2008 IEEE Conference on Computer Vision and Pattern Recognition* (2008), pp. 1–8. https://doi.org/10.1109/CVPR.2008. 4587673

FakeTouch: Machine Learning Based Framework for Detecting Fake News

Abu Bakkar Siddikk, Rifat Jahan Lia, Md. Fahim Muntasir, Sheikh Shah Mohammad Motiur Rahman, Md. Shohel Arman, and Mahmuda Rawnak Jahan

Abstract Fake news is any content or information that is false and often generated to mislead its readers in believing something which is not true. Fake news has become one of major threats that can harm someone's reputation. It often circulates wrong or made up information about various products, events, people or entity. The deliberate making of such news is escalating drastically these days. Fake news deceives us in taking wrong decisions. Therefore, Fake News Detection has attained immense deal of interest from researchers all over the world. In this chapter, a machine learning approach has been proposed named FakeTouch starting with Natural Language Processing based concept by applying text processing, cleaning and extraction techniques. This approach aim to arrange the information to be "obeyed" into each classification model for training and tuning parameters for every model to bring out the optimized and best prediction to find out the Fake news. To evaluate the proposed framework, three use cases with three different datasets has been developed during this study. The proposed framework will also help to understand what amount of data is responsible for detecting fake news, trying to stage the linguistic differences

A. B. Siddikk (✉) · R. J. Lia · Md. F. Muntasir · S. S. M. M. Rahman · Md. S. Arman ·
M. R. Jahan
Department of Software Engineering, Daffodil International University, Dhaka, Bangladesh
e-mail: abu35-1994@diu.edu.bd

R. J. Lia
e-mail: rifat35-1845@diu.edu.bd

Md. F. Muntasir
e-mail: fahim35-1900@diu.edu.bd

S. S. M. M. Rahman
e-mail: motiur.swe@diu.edu.bd

Md. S. Arman
e-mail: arman.swe@diu.edu.bd

M. R. Jahan
e-mail: jahan.swe@diu.edu.bd

A. B. Siddikk · R. J. Lia · Md. F. Muntasir · S. S. M. M. Rahman
nFuture Research Lab, Dhaka, Bangladesh

© The Author(s), under exclusive license to Springer Nature Switzerland AG 2022 317
Y. Baddi et al. (eds.), *Big Data Intelligence for Smart Applications*,
Studies in Computational Intelligence 994,
https://doi.org/10.1007/978-3-030-87954-9_15

between fake and true articles providing a visualization of the results using different visualization tools. This chapter also presents a comprehensive performance evaluation to compare different well known machine learning classifiers like Support Vector Machine, Naïve Bayes Method, Decision Tree Classifier, Random Forest, Logistic Regression as well as to develop an ensemble method (Bagging & Boosting) like XGBClassifier, Bagging Classifier of different combinations of classification models to identify which will give the best optimal results for three part of datasets. As a result, it has been found that with an appropriate set of features extracted from the texts and the headlines, XGB classifier can effectively classify fake news with very high detection rate. This framework also provides a strong baseline of an intelligent anti-fake news detector.

Keywords Fake news detection · Scraping · Social media · Text classification · Comparison of algorithms · Machine learning · Natural language processing

1 Introduction

The term 'fake news' has become an alarming issue particularly the spreading of factually incorrect and misleading articles published mostly for the aim of creating money through page views. Researchers [8] at Indiana University found these two sorts of information often go viral because "information overload and finite span of users limit the capacity of social media to discriminate information on the idea of quality." Shlok et al. Mugdha et al. (2020) conducted a study considering the fake news within American political Speech which was the topic of considerable attention, particularly following the election of President Trump. In another study Granik and Mesyura (2017) the authors developed a model based on naive Bayes classifier to examine how this particular method works for this particular problem given. Because of the availability of cheaper hardware and publicly available datasets, machine learning algorithms have started to work much better for making intelligent system in different areas. Shabani et al. Shabani and Sokhn (2018) have applied Hybrid Machine-Crowd Approach as an advantageous solution to tackle the fake news problem in general. Author focuses on distinguishing satire or parody and fabricated content using the Fake vs Satire public dataset, this approach provides higher accuracy at an acceptable cost and latency. Tanvir et al. Tanvir et al. (2019) experimented several machine learning and deep learning approaches to detect fake news. Different types of classification models including Support Vector Machine (SVM), Naïve Bayes, Logistic Regression, Long short-term memory, Recurrent Neural Network are implemented for that task. A combination of these classification models is also tested to enhance further the accuracy of prediction. Those models are implemented to prepare from the training dataset using k-fold (k = 2) cross-validation, and then predict using the data set. Ahmad et al. Ahmad et al. (2020) proposed their ability to require a choice relies totally on the sort of data consume. On the basis of information digest, their worldview was shaped. There is an increasing evidence that buyers have reacted absurdly to the news that later proved to be fake stated by the authors. One recent case is the spread of the novel coronavirus where fake reports cover the web about the origin, nature, and behavior of the virus. Things worsened as more people examine the fake content online. Identifying such news online may be a daunting task. An investigation based on the purpose of to let the developers and researchers can understand further which model and technique is the best also will help to implement the tools or models in future has been performed during this study. In this chapter, several machine learning techniques are applied to evaluate and improve the accuracy of the model performance. Thus, the major contribution of this work is to identify the best outcome-based model using the desired framework. Along with the machine learning algorithms, ensemble methods are also considered during the experiments including Support Vector Machine, Naïve Bayes Method, Decision Tree Classifier, Random Forest, Logistic Regression, XGBClassifier. On one standard dataset divided in three way using a novel set of features and statistically validated the results by calculating accuracy and F1 scores. Then, those machine learning algorithms which are out-performers are implemented for this task. A combination of these classification models is also examined and evaluated to check the enhancement of the accuracy to classify or identify the fake news. The techniques are implemented to prepare from the training dataset using TF-IDF technique, and then predict using the data set whether it is fake or real with the help of sci-kit learn python package Tanvir et al. (2019). The rest of the chapter is organized as follows. The implemented framework for fake news detection has been presented in Sect. 2. Evaluation parameters of this approach are briefly explained in Sect. 3. The experimental information including the result and analysis has been broadly discussed and debated in Sect. 4. Finally, Sect. 5 concludes this chapter.

2 Methodology

The proposed and implemented methodology will be described in this section from dataset collection to detection in details.

2.1 Collect Data

From background study, it has been found that most of the fake news threatens not only for our society but also all over the world. Thus, this data includes the information about the political news (most uses), social news as well as world news. At the very beginning fake news dataset have been collected from Kaggle Bali et al. (2019) where this dataset is related to the fake news spread during the time of the 2016 U.S. Presidential Election, IEEE Dataport and some of True news have been collected using Web-scraping from relevant and trusted source such as CNN, BBC, and The New York Times etc. Finally, fake and true data has been concatenated which

For selected 100% document the calculation is:

$$CS\text{-}1 = \frac{100}{100} * \textbf{Total Number of Document}$$

For selected 75% document the calculation is:

$$CS\text{-}2 = \frac{75}{100} * \textbf{Total Number of Document}$$

For selected 50% document the calculation is:

$$CS\text{-}3 = \frac{50}{100} * \textbf{Total Number of Document}$$

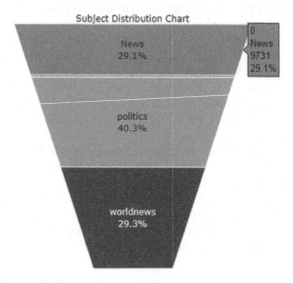

Fig. 1 Distribution of Different class for the Subject level feature

are already collected from different sources. There is a total 44909 (https://doi.org/ 10.6084/m9.figshare.13325198.v1) data in the dataset. Total dataset has been divided into three case studies to identify whether the proposed approach is dependent on the size of data or not to detect more efficiently. Total 44,909 of document has been assign as CS-1, 33,681 documents have been assigned as CS-2 and 22,454 documents have been assign as CS-3.

This dataset includes the Label class as F and T where 'F' stands for Fake news also indicating Fake and 'T' stands for True news indicating Real or Fact which set as a label. The dataset also includes one column for statements where all the data including the links and mentions. There are 4 features namely Title, Text, Subject and Date. Then have been labelled the data as a Fake news and a True news (Fig. 1).

2.2 Data Pre-processing

In Machine Learning field, after the assortment of the information, it is of most extreme significance that the information will be arranged and brought to a structure that is effectively deciphered and examined by machines. At the beginning of data preprocessing have checked where data are inconsistent or not then have converted all data into lowercase as if the data not misconduct with the machine. After that, the unwanted features which are not necessary (for example date) have been removed. In addition, the unnecessary signs, punctuation's and stop words have been removed because these can create a noise on the dataset. After preprocessing the processed data is ready for the next steps.

2.3 Features Extraction

Features extraction strategies should be executed when managing enormous measure of information, as its greater part can be repetitive and unimportant making calculations be tedious and bring deceiving arrangement results. Additionally, feature extraction is a general term for a method for developing the blends of the factors to get around those issues while as yet depicting the information with adequate exactness. Many machine learning professionals and practitioners believe and accept that appropriately enhanced component extraction is the way to successful model development. Term Frequency-Inverse Document Frequency (TF-IDF) has been used in this part to extract and convert the features as numerical representation.

In the Sect. 2.3.1 have been clearly described how one of the best feature extraction technique TF-IDF applied on our experimental text dataset.

2.3.1 Term Frequency-Inverse Document Frequency (TF-IDF)

TF-IDF Tanvir et al. (2019) is a mathematical measurement that speaks to the significance of a term in a report among others in a bunch of records. It consolidates two measurements, Term Frequency and Inverse Document Frequency. The initial measurement considers the occasions that a term shows up in a record and gives higher position. TF-IDF is widely used in Natural Language Processing Rahman et al. (2018, 2020, 2020) for example sentiment analysis, classification and so on.

TF-IDF (implemented from sci-kit learn Shukla et al. (2021)) In Python, TF-IDF is applied on text-data using sklearn. TfidfVectorizer from sklearn.feature extraction at very first have imported for perform the work. After that, it have been setup the vectorizer and then run the program fit and transform over it to compute the TF-IDF score for the text Dataset using the TF-IDF algorithm. Finally, the sklearn fit transform function have been applied the following fit and transform functions, yielding the result of TF IDF.

Following the Two Mathematical Formula have been compute the TF-IDF score as follows (Fig. 2)

$$TF(t) = \frac{Number of Times appares in a document}{Total number of term in the document}.$$

$$IDF(t) = log\frac{Total number of document}{Total number of document with term t in it}.$$

- Words Level TF-IDF: TF-IDF value of each term represented in a matrix format.
- N-gram Level TF-IDF: Combination of N termism spoke to by N-gram level. This matrix portrays TF-IDF N-gram scores
- Character Level TF-IDF: TF-IDF values of the n- grams character level in corpus represented in a matrix.

2.4 Model Generation

For this proposed framework 8 different types of machine learning algorithms have been used. Decision Tree (DT), Support Vector Machine (SVM), Random Forest (RF), K-Neighbours (KNN), Naïve Bayes (NB), Passive Aggressive Classifier (PAC), Logistic Regression (LR) and Extrembaselinee Gradient Boosting (XGB) Bali et al. (2019) have been implemented to fit and construct the model.Essentially, this model was created based on our workflow, in which we performed many techniques using the Python programming language and many useful Python packages to demonstrate our structure. This method was created by taking a step-by-step approach from this work. And this study found that this workflow produces the best results for detecting false news ever. Using this method, researchers will continue to identify false news in order to obtain the best estimation outcome.

3 Evaluation Metrics

3.1 Performance Parameters

Performance measures are based on data, and represent a clear story about whether a representation or activity is achieving its objectives and if progress is being made toward attaining policy or organizational goals. In order to assess the performance of the proposed model, as evaluation matrices precision, recall, F1-score, true negative rate, false positive rate and accuracy Siddikk et al. (2021) are used.

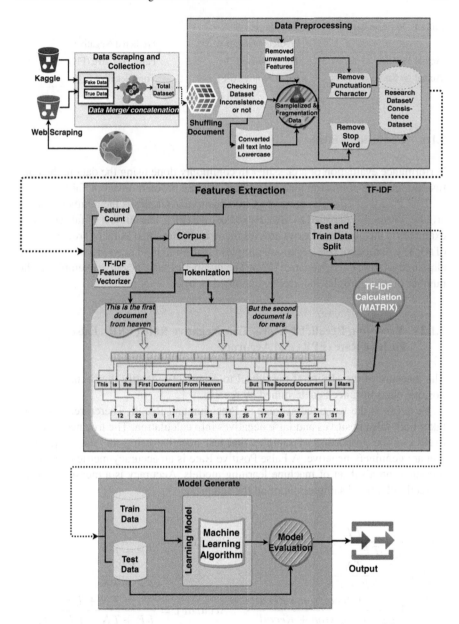

Fig. 2 FakeTouch Framework for training algorithms and classification of news articles

Table 1 Representation of confusion matrix

	Predicted Negative	Predicted Negative
Actual Positive	True Positive (TP)	False Negative (FN)
Actual Negative	False Positive (FP)	True Negative (TN)

3.1.1 Confusion Matrix:

Performance evaluation of a classifier in commonly done using the data in the confusion matrix. A confusion matrix for the two-class problem given in Table 1.

True Positive (TP): The number of true positive examples is the number of news articles, correctly classified as fake; False Positive (FP): The number of false positive examples is the number of news articles incorrectly classified as fake; True Negative (TN): The number of true negative examples is the number of news articles, correctly classified as true; False Negative (FN): The number of false negative examples is the number of news articles incorrectly classified as true.

3.1.2 Precision, Recall, F1-Score, True Negative Rate (TNR), False Positive Rate (FPR) and Accuracy

The measure of the ability of the model to accurately identify the occurrence of a positive class instance is determined by recall.

F1 Score is the weighted average of Precision and Recall. Therefore, this score takes both false positives and false negatives into calculation. The true negative rate (TNR), is the proportion of samples that test negative using the test in question that are genuinely negative. A False Positive Rate is an accuracy metric that can be measured on a subset of machine learning models. Accuracy is a measure of total correctly identified samples out of all the samples.

$$Recall = \frac{TP}{TP + FN} \qquad Precision = \frac{TP}{TP + FP}$$

$$TNR = \frac{TN}{FP + TN} \qquad FPR = \frac{FP}{FP + TN}$$

$$F1 = \frac{2 * (Precision * Recall)}{Precision + Recall} \qquad Accuracy = \frac{TN + TN}{TP + TN + FP + FN}$$

4 Results and Discussion

4.1 Environments and Tools

The implementation of proposed Framework was performed on the configuration of Processor: Intel(R) Core(TM) i3-7100U CPU @ 2.40GHz, 2401 MHz, Installed RAM 4.00 (3.67 GB usable), system type 64-bit operating system, X64-based processor, System model HP ProBook 450 G4, running under Windows 10 Pro operating system. The algorithm was implemented on Python based Jupyter Notebook 2020. Matplotlib software, Pandas tools, Sci-kit learning tools and so on are also used in addition to that.

4.2 Result Analysis

There are three type of case studies which have been experimented to evaluate the assessment in this section. Here three case study will be described and evaluated based on the multifarious machine learning techniques.

4.2.1 Case Study #1

For the case study #1 have been used feature extraction technique TF-IDF. After that, Confusion Matrix have been found using the selected Classification Model as follows:

Tables 2, 3 and 4 strongly represents the confusion matrix for every model according to Table 1 format.For example Table 2 represents data followed by the Table 1. from top left represent True Positive(TP), top right represent False Negative(FN), bottom left represents False Positive(FP) and bottom right represents True Negative(TN) for every confusion matrix.

Table 2 represents the confusion matrix for the assigned machine learning classifiers where using the value of confusion matrix can find the evaluation metrics from dataset one. From the mathematics, calculation, analysis and investigation, have been found that eXtreme Gradient Boosting as classifier have achieved highest an accuracy among all classifier. For the precision, recall and f1-score Decision tree given the highest score among all classifier. The accuracy, precession, recall and f1-score have been presented Tables 6 and 7. In Fig. 3 represent the true negative rate and false positive rate where have been found that in level 1, maximum 99.74% of TNR attained from XGB individually with minimum 0.25 FPR. In case TNR, FPR given the lowest rate for the KNN whereas XGB given a better performance individually (Table 5).

Table 2 Representation of Confusion Matrix; DT: Decision Tree; SVM: Support Vector Machine; RF: Random Forest; KNN: k-Nearest Neighbors; MNB: Multinomial Naive Bayes; PAC: Passive Aggressive Classifier; LR: Logistic Regression; XGB: eXtreme Gradient Boostingn Matrix

4361 34	4332 63
25 7304	28 7301
Confusion Matrix for DT	ConfusionMatrix for SVC
4244 151	4257 34
28 7301	53 7276
Confusion Matrix for RF	Confusion Matrix for LR
4374 21	3122 1273
19 7310	23 7306
Confusion Matrix for XGBC	Confusion Matrix for MNB
4339 56	2824 1571
30 729	226 7103
Confusion Matrix for PAC	Confusion Matrix for KNN

Table 3 Representation of Confusion Matrix; DT: Decision Tree; SVM: Support Vector Machine; RF: Random Forest; KNN: k-Nearest Neighbors; MNB: Multinomial Naive Bayes; PAC: Passive Aggressive Classifier; LR: Logistic Regression; XGB: eXtreme Gradient Boosting

3320 23	3282 61
18 5432	22 5428
Confusion Matrix for DT	ConfusionMatrix for SVC
3204 139	3209 134
17 5433	31 5419
Confusion Matrix for RF	Confusion Matrix for LR
3334 9	2256 1087
7 5443	13 5437
Confusion Matrix for XGBC	Confusion Matrix for MNB
3277 66	2087 125
16 5434	6158 5292
Confusion Matrix for PAC	Confusion Matrix for KNN

4.2.2 Case Study #2

In this section case study #2 conduct with the feature extraction technique TF-IDF. After trained the model Confusion Matrix have been found using the selected Classification Model as follows:

Confusion matrix based on selected classifier have been assigned in Table 3 where using the value of confusion matrix can find out the accuracy, precession, recall, f1-score etc. on dataset two. The accuracy, TNR and FPR have been visualized in Table 3 and Fig. 4. It has been found that eXtreme Gradient Boosting as classifier have

Table 4 Representation of Confusion Matrix; DT: Decision Tree; SVM: Support Vector Machine; RF: Random Forest; KNN: k-Nearest Neighbours; MNB: Multinomial Naive Bayes; PAC: Passive Aggressive Classifier; LR: Logistic Regression; XGB: eXtreme Gradient Boosting

2195 11	2153 53
10 3647	22 3635
Confusion Matrix for DT	ConfusionMatrix for SVC
2135 71	2108 98
13 3644	28 3629
Confusion Matrix for RF	Confusion Matrix for LR
2199 7	1449 757
7 3650	7 3650
Confusion Matrix for XGBC	Confusion Matrix for MNB
2165 41	1336 870
26 3631	94 3563
Confusion Matrix for PAC	Confusion Matrix for KNN

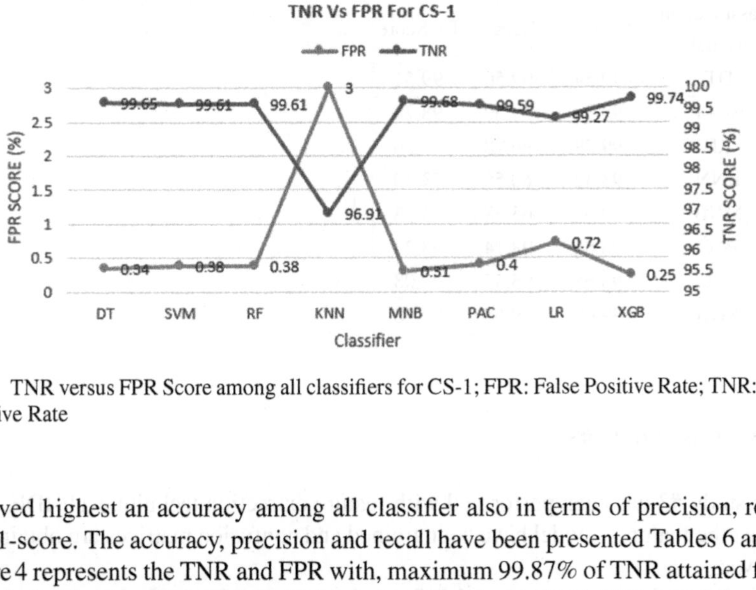

Fig. 3 TNR versus FPR Score among all classifiers for CS-1; FPR: False Positive Rate; TNR: True Negative Rate

achieved highest an accuracy among all classifier also in terms of precision, recall and f1-score. The accuracy, precision and recall have been presented Tables 6 and 7. Figure 4 represents the TNR and FPR with, maximum 99.87% of TNR attained from XGB individually which is little bit higher and minimum 0.12 FPR which is given little bit high performance than dataset one. In case study 2 given little bit better performance than the case study 1

Table 5 Precision, Recall and F1-Score for Machine Learning algorithm; CS: Case Study; DT: Decision Tree; SVM: Support Vector Machine; RF: Random Forest; KNN: k-Nearest Neighbours; MNB: Multinomial Naive Bayes; PAC: Passive Aggressive Classifier; LR: Logistic Regression; XGB: eXtreme Gradient Boosting

Classification Model	CS-1			CS-2		
	Precision	Recall	F1-Score	Precision	Recall	F1-Score
DT	99.43	99.22	99.53	99.46	99.31	99.38
SVM	99.35	98.56	98.95	99.33	98.17	98.74
RF	99.34	96.56	97.93	99.47	95.84	97.62
KNN	92.59	64.25	75.85	92.96	62.42	74.68
MNB	99.26	71.03	82.80	99.42	67.48	80.19
PAC	99.31	98.72	99.01	99.51	98.02	98.75
LR	98.77	96.86	97.90	99.04	95.99	97.49
XGB	98.63	98.49	98.55	99.79	99.73	99.75

Classification Model	CS-3		
	Precision	Recall	F1-Score
DT	99.54	99.50	99.51
SVM	98.98	97.59	98.28
RF	99.39	96.78	98.06
KNN	93.42	60.56	73.48
MNB	99.51	65.68	79.13
PAC	98.81	98.14	98.23
LR	98.68	95.55	97.08
XGB	99.68	99.68	99.68

4.2.3 Case Study #3:

Case study #3 has been performed with feature extraction technique TF-IDF. After extract the features, model have been trained and confusion matrices are obtained as follows:

Similarly, the case study 1 and 2, Table 4 represents the confusion matrix for the assign machine learning classifier where using the value of confusion matrix can find out the accuracy, precession, recall, f1-score etc. on dataset three. It has been found that eXtreme Gradient Boosting as classifier have achieved highest accuracy among all classifier which are similar to the case study 1 and 2. The accuracy, precession, recall and f1-score have been presented Tables 6 and 7. Figure 5 represents the true negative rate and false positive rate have been found that in level 1 with maximum 99.80% of TNR attained from XGB individually which is a little bit higher from the dataset one and also significantly higher from the dataset two and minimum 0.19 FPR

Table 6 Accuracy of word embedding model (TF-IDF) for machine learning algorithm

		CS-1	CS-2	CS-3
Word embedding model	Classification model	Accuracy	Accuracy	Accuracy
Term Frequency–Inverse Document Frequency (TF-IDF)	DT	99.49	99.53	99.64
	SVM	99.22	99.09	98.72
	RF	98.47	98.23	98.57
	KNN	84.67	83.92	83.56
	MNB	88.95	87.49	86.97
	PAC	99.27	99.09	98.86
	LR	98.37	98.12	97.85
	XGB	99.66	99.82	99.76

Table 7 Comparison-based Classification result using Kaggle Fake News Dataset

Authors	ML techniques	Environment	Accuracy(%)
Arvinder et al. Bali et al. (2019)	XGB	Intelprocessor, core i7, DDR4 8GB, 2400 MHz, Ubuntu X64 OS.	88.0%
Dimitrios et al. Katsaros et al. (2019)	CNN	TeslaK20x, GPU 2.30 GHz, 128 GB main memory	88.0%
Shaban et al. Shabani and Sokhn (2018)	HybridMachine Crowd approach	NA	84.0%
Rohit et al. Kaliyar et al. (2020)	FNDNet	NA	98.36%
Abdullah et al. Tanvir et al. (2019)	SVM	NA	89.34%
Proposed Model (FakeTouch)	XGB	HP ProBook 450 G4, Intel processor, core i3,DDR4 4GB, 2400 MHz, Windows X64 OS.	99.66%

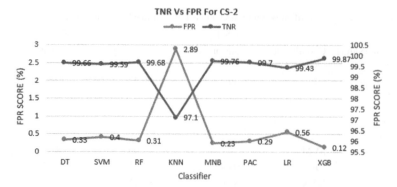

Fig. 4 TNR vs FPR Score among all classifiers for CS-2; FPR: False Positive Rate; TNR: True
Negative Rate

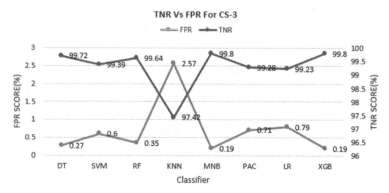

Fig. 5 TNR versus FPR score among all classifiers for CS-3; FPR: False Positive Rate; TNR: True
Negative Rate

which has given a little bit high performance than dataset one and less for the dataset
two. In case study 2 performs better performance than the case study 1. According
to the investigation, for the three case studies with machine learning-based models
have found that performance decrease as the scale of data increases and after a certain
period of data level performance increase gradually. In this case, the amount of data
extremely related for how accurately detect fake news. On the other hand, eXtreme
Gradient Boosting outperforms with higher accuracy which proves it to be the best
to detect the fake news accurately in every cases.

4.3 Comparative Discussion

The Classification results for all machine learning models which have been obtained
for CS-1, CS-2, and CS-3 (Sect. 2.1) as Illustrates Fig. 6.

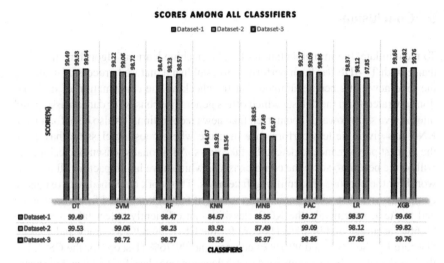

SCORES AMONG ALL CLASSIFIERS

	DT	SVM	RF	KNN	MNB	PAC	LR	XGB
Dataset-1	99.49	99.22	98.47	84.67	88.95	99.27	98.37	99.66
Dataset-2	99.53	99.06	98.23	83.92	87.49	99.09	98.12	99.82
Dataset-3	99.64	98.72	98.57	83.56	86.97	98.86	97.85	99.76

CLASSIFIERS

Fig. 6 Different machine learning accuracy results; CS: Case study

It has been noticed that classification models which investigated have given highest performance for XGB classifier and provided less performance for KNN models for the given three part of dataset (Sect. 2.1). But the important point is that the DT classification have been increased when this propose document gradually decrease. On the other hand, SVM, KNN, MNB, PAC and LR classifier has been decrease their performance. Using the word embedding model TF-IDF in this experiment have achieved the score of 99.66, 99.82 and 99.76% (Table 7) highest accuracy respectively using three amounts of dataset (Sect. 2.1) for XGB classifier among all classification models. Apart from that, lowest accuracy has found for KNN models. It's been clearly said that if someone implement the proposed model by applying on framework for detect fake news, Boosting Technique will be help to find out the best estimate output only of accurately detect Fake News. In addition, for the DT classifier will give the highest performance if dataset gradually decrease. For the DT classifier, number of documents depend on the estimate result. Model performance for detecting fake news will be decrease while dataset gradually decrease for the classifier named SVM, RF, KNN, MNB, PAC and LR. XGB classifier always provides the highest performance of accuracy for any number of documents. Table 8 shows the comparison among different proposed models with FakeTouch. Following is the configuration of the computing environment for carrying out the experiment. The proposed Framework shows comparatively best results and effectively (training accuracy, testing accuracy also light-weight model for training). From the proposed Framework, this investigation achieved maximum accuracy of 99.66%. It's also been claimed that the proposed approach achieved better result with the real-world text based fake news dataset as compared to existing works. Finally, the proposed approach strongly encourages all researchers that is most significant framework to detect fake news.

5   Conclusion

To recapitulate, the implementation of eight machine learning algorithms in accordance with TF-IDF has been performed to establish a matured process of detecting the fake news. It's been examined that for checking the confirmation of information separated from the dataset which offers general responses for data amassing and interpretive show towards counterfeit fake news recognition. It's also been found that KNN has given the less performance measure where boosting classifier has given the highest performance to detect the fake news. Additionally, ensemble techniques will be the best fit to solve the complex issues which have fall impact for all over the world for the purpose of identifying fake news. This work explained with extensive comprehensive that with a hot and various set of features extracted from the heading and the text, specifically the XGB classifier can efficiently detect fake news with 99.66% accuracy and 98.55% f1-score for the CS-1, 99.82% accuracy and 99.75% f1-score for the CS-2 and 99.76% accuracy and 99.76% f1-score for the CS-3. For the less data it can be recommended to use decision tree classifier to detect fake news.

References

A.P.S. Bali, M. Fernandes, S. Choubey, M. Goel, Comparative performance of machine learning algorithms for fake news detection, in *International Conference on Advances in Computing and Data Sciences* (Springer, Singapore, 2019), pp. 420–430

T. Antipova, (ed.), *Integrated Science in Digital Age 2020* (Springer, 2020)

M. Granik, V. Mesyura, Fake news detection using naive Bayes classifier, in *2017 IEEE First Ukraine Conference on Electrical and Computer Engineering (UKRCON)* (IEEE, 2017), pp. 900–903

W.Y. Wang, liar, liar pants on fire: a new benchmark dataset for fake news detection (2017). arXiv:1705.00648

S.B.S. Mugdha, S.M. Ferdous, A. Fahmin, Evaluating machine learning algorithms for bengali fake news detection, in *2020 23rd International Conference on Computer and Information Technology (ICCIT)* (IEEE, 2020), pp. 1–6

D. Katsaros, G. Stavropoulos, D. Papakostas, Which machine learning paradigm for fake news detection?, in *2019 IEEE/WIC/ACM International Conference on Web Intelligence (WI)* (IEEE, 2019), pp. 383–387

S. Shabani, M. Sokhn, Hybrid machine-crowd approach for fake news detection, in *2018 IEEE 4th International Conference on Collaboration and Internet Computing (CIC)* (IEEE, 2018), pp. 299–306

R.K. Kaliyar, A. Goswami, P. Narang, S. Sinha, FNDNet-a deep convolutional neural network for fake news detection. Cogn. Syst. Res. **61**, 32–44 (2020)

A. Jain, A. Kasbe, Fake news detection, in *2018 IEEE International Students' Conference on Electrical, Electronics and Computer Science, SCEECS 2018*, pp. 1–5 (2018)

I. Ahmad, M. Yousaf, S. Yousaf, M.O. Ahmad, Fake news detection using machine learning ensemble methods. Complexity **2020** (2020)

T. Thomas, P.S. Nair, Analysis of various machine learning models for detecting fake news. Int. Res. J. Eng. Technol. (IRJET) **07**(07) (2020). ISSN: 2395-0072

A. Tanvir, E. Mahir, S. Akhter, M.R. Huq, Detecting fake news using machine learning and deep learning algorithms. Outlook India **19** (2019)

Y. Shukla, N. Yadav, A. Hari, A unique approach for detection of fake news using machine learning. Int. J. Res. Appl. Sci. Eng. Technol. (IJRASET). ISSN, 2321-9653

D.M. Lazer, M.A. Baum, Y. Benkler, A.J. Berinsky, K.M. Greenhill, F. Menczer, J.L. Zittrain, The science of fake news. Science **359**(6380), 1094–1096 (2018)

V. Pérez-Rosas, B. Kleinberg, A. Lefevre, R. Mihalcea, Automatic detection of fake news (2017). arXiv:1708.07104

K. Shu, S. Wang, H. Liu, Beyond news contents: The role of social context for fake news detection, in *Proceedings of the Twelfth ACM International Conference on Web Search and Data Mining*, pp. 312–320 (2019)

X. Zhou, R. Zafarani, Fake news: a survey of research, detection methods, and opportunities (2018). arXiv:1812.00315

T. Quandt, L. Frischlich, S. Boberg, T. Schatto-Eckrodt, Fake news. Int. Encycl. Journal. Stud. 1–6 (2019)

N. Ruchansky, S. Seo, Y. Liu, Csi: a hybrid deep model for fake news detection, in *Proceedings of the 2017 ACM on Conference on Information and Knowledge Management*, pp. 797–806 (2017)

M. Potthast, J. Kiesel, K. Reinartz, J. Bevendorff, B. Stein, A stylometric inquiry into hyperpartisan and fake news (2017). arXiv:1702.05638

X. Zhou, R. Zafarani, K. Shu, H. & Liu, Fake news: fundamental theories, detection strategies and challenges, in *Proceedings of the Twelfth ACM International Conference on Web Search and Data Mining*, pp. 836–837 (2019)

J. Albright, Welcome to the era of fake news. Media Commun. **5**(2), 87–89 (2017)

E.C. Tandoc Jr., Z.W. Lim, R. Ling, Defining, "fake news" a typology of scholarly definitions. Digit. Journal. **6**(2), 137–153 (2018)

H. Karimi, P. Roy, S. Saba-Sadiya, J., J. Tang, Multi-source multi-class fake news detection, in *Proceedings of the 27th international conference on computational linguistics*, pp. 1546–1557 (2018)

X. Zhou, R. Zafarani, A survey of fake news: fundamental theories, detection methods, and opportunities. ACM Comput. Surv. (CSUR) **53**(5), 1–40 (2020)

K. Shu, X. Zhou, S. Wang, R. Zafarani, H. Liu, The role of user profiles for fake news detection, in *Proceedings of the 2019 IEEE/ACM International Conference on Advances in Social Networks Analysis and Mining*, pp. 436–439 (2019)

R. Oshikawa, J. Qian, W.Y. Wang, A survey on natural language processing for fake news detection (2018). arXiv:1811.00770

X. Zhang, A.A. Ghorbani, An overview of online fake news: characterization, detection, and discussion. Inf. Proc. Manag. **57**(2), 102025 (2020)

K. Shu, L. Cui, S. Wang, D. Lee, H. & Liu, Defend: explainable fake news detection, in *Proceedings of the 25th ACM SIGKDD International Conference on Knowledge Discovery & Data Mining*, pp. 395–405 (2019)

R. Zellers, A. Holtzman, H. Rashkin, Y. Bisk, A. Farhadi, F. Roesner, Y. Choi, Defending against neural fake news (2019). arXiv:1905.12616

K. Shu, S. Wang, H. Liu, Understanding user profiles on social media for fake news detection, in *2018 IEEE Conference on Multimedia Information Processing and Retrieval (MIPR)*, pp. 430–435 (IEEE, 2018)

H. Karimi, J. Tang, Learning hierarchical discourse-level structure for fake news detection (2019). arXiv:1903.07389

R.K. Nielsen, L. Graves, "News you don't believe": audience perspectives on fake news (2017)

A. Kucharski, Study epidemiology of fake news. Nature **540**(7634), 525–525 (2016)

S.S.M.M. Rahman, M.H. Rahman, K. Sarker, M.S. Rahman, N. Ahsan, M.M. Sarker, Supervised ensemble machine learning aided performance evaluation of sentiment classification. J. Phys.: Conf. Ser. **1060**(1), 012036) (2018). IOP Publishing

S.S.M.M. Rahman, K.B.M.B. Biplob, M.H. Rahman, K. Sarker, T. Islam, An investigation and evaluation of N-Gram, TF-IDF and ensemble methods in sentiment classification, in *International Conference on Cyber Security and Computer Science* (Springer, Cham, 2020), pp. 391–402

M.M. Rahman, S.S.M.M. Rahman, S.M. Allayear, M.F.K. Patwary, M.T.A. Munna, A sentiment analysis based approach for understanding the user satisfaction on android application, in *Data Engineering and Communication Technology* (Springer, Singapore, 2020)

A.B. Siddikk, M.F. Muntasir, R.J. Lia, S.S.M.M. Rahman, T. Islam, M. Alazab, Revisiting the approaches, datasets and evaluation parameters to detect android malware: a comparative study from state-of-art. Artif. Intell. Blockchain Future Cybersecur. Appl. **125** (2021)

Printed in the United States
by Baker & Taylor Publisher Services